CAMBRIDGE LIBRARY COLLECTION

Books of enduring scholarly value

Life Sciences

Until the nineteenth century, the various subjects now known as the life sciences were regarded either as arcane studies which had little impact on ordinary daily life, or as a genteel hobby for the leisured classes. The increasing academic rigour and systematisation brought to the study of botany, zoology and other disciplines, and their adoption in university curricula, are reflected in the books reissued in this series.

Linnaeus

This biography of Swedish botanist Carl Linnaeus (1707–78) was originally published in 1903 by another Swedish botanist, Theodor Magnus Fries (1832–1913). This English version, published in 1923, was translated and edited by British botanist Benjamin Daydon Jackson (1846–1927). Jackson is best known for founding (with J. D. Hooker) the *Index Kewensis*, the register of all botanical names of seed plants; he also acted as Secretary of the Linnean Society for twenty-two years, becoming its General Secretary in 1902. This biography covers the early life of Linnaeus (including some family history), his student years at Uppsala, his travels, his years as a teacher, notes on his personality and his relationship with the wider scientific community. Jackson provides further information in the appendices, enhancing Fries' work with Linnaeus' genealogy, a list of his pupils, his private notes on 'Divine Punishment' and a brief history of Sweden during Linnaeus' lifetime.

Cambridge University Press has long been a pioneer in the reissuing of out-of-print titles from its own backlist, producing digital reprints of books that are still sought after by scholars and students but could not be reprinted economically using traditional technology. The Cambridge Library Collection extends this activity to a wider range of books which are still of importance to researchers and professionals, either for the source material they contain, or as landmarks in the history of their academic discipline.

Drawing from the world-renowned collections in the Cambridge University Library, and guided by the advice of experts in each subject area, Cambridge University Press is using state-of-the-art scanning machines in its own Printing House to capture the content of each book selected for inclusion. The files are processed to give a consistently clear, crisp image, and the books finished to the high quality standard for which the Press is recognised around the world. The latest print-on-demand technology ensures that the books will remain available indefinitely, and that orders for single or multiple copies can quickly be supplied.

The Cambridge Library Collection will bring back to life books of enduring scholarly value (including out-of-copyright works originally issued by other publishers) across a wide range of disciplines in the humanities and social sciences and in science and technology.

Linnaeus

Theodor Magnus Fries
Edited and Translated by
Benjamin Daydon Jackson

CAMBRIDGE UNIVERSITY PRESS

Cambridge, New York, Melbourne, Madrid, Cape Town,
Singapore, São Paolo, Delhi, Tokyo, Mexico City

Published in the United States of America by Cambridge University Press, New York

www.cambridge.org
Information on this title: www.cambridge.org/9781108037235

This edition first published 1923
This digitally printed version 2011

ISBN 978-1-108-03723-5 Paperback

LINNÆUS

CARL VON LINNÉ
(Portrait after Per Krafft the Elder, 1774).

LINNÆUS

(AFTERWARDS CARL VON LINNE)

THE STORY OF HIS LIFE, ADAPTED FROM THE SWEDISH OF THEODOR MAGNUS FRIES, EMERITUS PROFESSOR OF BOTANY IN THE UNIVERSITY OF UPPSALA, AND BROUGHT DOWN TO THE PRESENT TIME IN THE LIGHT OF RECENT RESEARCH

BY

BENJAMIN DAYDON JACKSON

Knight of the Swedish Order of the Polar Star, Hon. Ph.D. (Upsal.), General Secretary of the Linnean Society, London; Author of "Index Kewensis," "Glossary of Botanic Terms," "Index to the Linnean Herbarium," etc.

LONDON

H. F. & G. WITHERBY

326 HIGH HOLBORN, W.C.

1923

MANUFACTURED IN GREAT BRITAIN

INTRODUCTION

THE following pages have been written to make known to English readers the monumental " Life of Carl von Linné " by the late Professor T. M. Fries (1832-1913) by far the most detailed and accurate account of the great Swedish naturalist ever published.

The author was admirably equipped for his task. His father, Professor E. M. Fries (1794-1878), was born at Femsjö in the same province, Småland, as Linné; he spent his scientific career at Lund and Uppsala and in his household cherished the Linnean traditions. His eldest son was born at the father's birthplace, but, at the early age of four was taken to Uppsala, where his father was then a professor; thanks, however, to the Linnean atmosphere maintained in his home, he became imbued with the phrases and dialect of his eminent predecessor. In due time, after many years as Docent, he became professor in 1877 in the subjects of botany and practical economy, with control of both botanic gardens, the old one having Linné's house in it; later, when by his exertion the residence of Linné at Hammerby became the property of the state, he was appointed the first administrator. For six years he filled the distinguished office of Rector magnificus, during which time he gave eight rectorial addresses on the first half of Linné's life, and three years after vacating his chair, he published the life which is the basis of the account here presented, the result of more than thirty years of constant research. Thus he was not only nurtured in the Linnean tradition, but he had access to the

University records from which he gathered much; he likewise gleaned from Swedish and other sources every allusion to his great countryman's career. He visited London several times, as in 1871, 1896 and 1904 when he devoted his time to the examination of the Linnean MSS. and letters, preserved in the Linnean Society's collections. Further, he was able to make use of the life-work of Dr. J. E. E. Ährling, and between the two men, hardly anything escaped notice. In due course these researches were incorporated in his *magnum opus*. After 1903 he was still busily engaged in gathering fresh material; he translated and printed unpublished MSS. and by 1907 when the bicentenary celebration of the birth of Carl von Linné was celebrated in Sweden with great enthusiasm, he was undertaking the editorship of the "Bref och skrifvelser" (Letters and communications of and to Linné) of which he lived to complete six volumes with illuminative notes of the contemporaries of Linné, which work is still in progress.

Shortly after the issue of the "Life" an offer was made, I do not know by whom, to an eminent London publisher to bring out an English translation, and I was asked by Mr. George Murray, at that time Keeper of Botany at the British Museum, to interview one of the partners of the firm in question. This I did, though I could learn nothing of the firm's intentions, but could only impress upon them my readiness to further the project to the utmost of my power. I heard nothing more of the proposition and presumed it was declined on account of the extent of the work. Since then I have been repeatedly urged to draw up an adaptation of Professor Fries's work for English readers, but pressure of official work has hitherto hindered my compliance. Now, however, the time seems opportune to lay before the scientific and reading public an adequate account of the great naturalist. I have found myself obliged to adapt the work, for a complete translation would not only be very long, but many of the details of persons, places and things, comprehensible to the Swedish reader, would need explanation to those of any other nationality. Consequently even when I have closely followed the original, I have compressed the translation, giving the sense, I hope, with accuracy, but omitting those portions which could not be verified by anyone in this country, or which seemed superfluous, such as certain details, quotations from minutes, letters, etc. Again, the original has numerous and copious notes, which I have incorporated in the text so far as they are essential to the meaning. Further, the author was very careful to give references to many inaccessible sources; I have avoided copying these by adhering to the form of the author's volumes, so that any seeker for references will not find it difficult to note them in the original. Most of the illustrations have been omitted, though portraits of Linné, views of his houses, etc., could not well be passed over. Professor Fries's method of using the birth-name "Linnæus" during the early half of his life, adopting the Swedish form "Linné" from the time of his settling as professor in Uppsala, has been followed. In a letter to P Wargentin, dated 10th February, 1764, Linné says, "Linnæus or Linné are the same to me; one is Latin, the other Swedish."

I have added a glossary of Swedish titles, a short history of Sweden during the lifetime of Linné, a select bibliography and an index. I am confident that in consequence of the fullness and accuracy of Professor Fries's work, this volume will give a better idea of the life and aims of the Father of Modern Biology than any previous publication in the English language.

In conclusion I must express my hearty thanks to Professor Robert Fries, the author's son, not only for his kind and prompt response to my suggested adaptation of his father's admirable biography of Linné, but for constant help and advice during the progress of the work, by which the volume has so greatly benefited.

I have now the pleasing task of recording my gratitude, in the first case, to my wife, who read the whole of the manuscript, and, in the second case, to Mr. John Ramsbottom, Secretary for Botany, Linnean Society of London, who read it in proof; from their helpful emendations the present volume has gained immeasurably.

B. D. J.

PREFACE

By the late Professor T. M. Fries, in his "Linné," 1903.

Amongst the Swedes there is hardly to be found anyone whose life and activity at home or abroad has been so often described as that of Carl von Linné. It may therefore be thought a wasted effort to put forward another biography, as much of it must be a repetition of well-known facts.

It needed no long investigation to find with astonishment how much remained for elucidation and arrangement, and how imperfect and misleading are all Linnean biographies hitherto published. This was frankly admitted by Linné's pupil, Dr. J. G. Acrel, in his address on relinquishing the presidency of the Academy of Science in August, 1796, and the lapse of more than a century has repeatedly emphasized the fact, that a new, comprehensive and accurate representation is needed, based upon investigation of our great countryman's life and work, as a legitimate object and a dutiful testimony to his memory.

The reason why no such account has hitherto been attempted, is due to the difficulties inseparably bound up with it; the chief difficulty being that the materials had to be gathered from a very wide field, and not from Sweden only. It is lamentable that Linné's extensive and important correspondence, both home and foreign, and many of his manuscripts are now in England, with the Linnean library and collections, to the delight of their purchaser and the shame of Sweden. In addition to these, letters and documents required for a complete biography are dispersed over the whole learned world, in public and private libraries. Nor is this all; the numerous writings, great and small, which flowed from his pen; the more prominent of his biographies; the letters from his contemporaries; his pupils' notes from his lectures, and conversations with them; all had to be gone through, so as to glean from them everything which could throw light upon material brought together from other quarters.

Fully conscious of these difficulties, I fear I may have over-estimated my own powers, when taking upon myself the burden of producing a new and detailed "Linné-biography." My excuse is, that, animated by my father's admiration and love for the "Flower-king," I have for more than thirty years devoted my time, when other compulsory objects have permitted, assiduously collecting material for such a work. I am quite aware that this material, the result of persistent search, may still be added to, but increasing age warns me no longer to delay the drawing up of a narrative, which I wish to put forward as a small tribute of respect and gratitude to the Master's memory.

How far I have succeeded, must be left to the judgment of others. A few remarks as to the principles I have kept before me during the collecting and writing of these pages, may be permitted.

Warned by experience of the untrustworthiness of certain current statements, hitherto considered as almost infallible dogmas, I have made it a rule, critically to test everything, even to the smallest detail. Hence it has been needful to go back to original sources, particularly official records and other documents to be found in London, Stockholm, Uppsala, Lund, Växjö and elsewhere. I am especially indebted to the collections belonging to the Swedish Academy of Science which were made by the late Dr. Ährling, who, by his short biography of Linné in the "Nordisk Familjebok," and his accurate and invaluable notes in his "Carl von Linné's Swedish works" 1878-80, has more than any other person thrown light upon previously dark or erroneous statements.

During the elaboration of these materials it soon became clear that I ought not to restrict the account to those items only which have a direct bearing upon the events of his life. Any true idea of him, his great breadth of view, his winning personality and his powers of work, can hardly be understood without ascertaining the conditions under which he lived, and the difficulties of every kind against which he had to contend. Seen against such a background, his image appears distinct and striking.

In conclusion it must not be omitted to state that the period of Linné's lifetime which is depicted in the first part of this narration, formed the substance of eight rectorial programmes, which were published in Uppsala from 1893 to 1898 under the title "Bidrag till en lefnadsteckning öfver Carl von Linné" [Contribution to an account of the life of Carl von Linné]. For the later portion relating to Linné's professorship there has been no such preliminary attempt.

TH. M. FRIES.

UPPSALA,
September, 1903.

CONTENTS

APPENDICES

LIST OF ILLUSTRATIONS

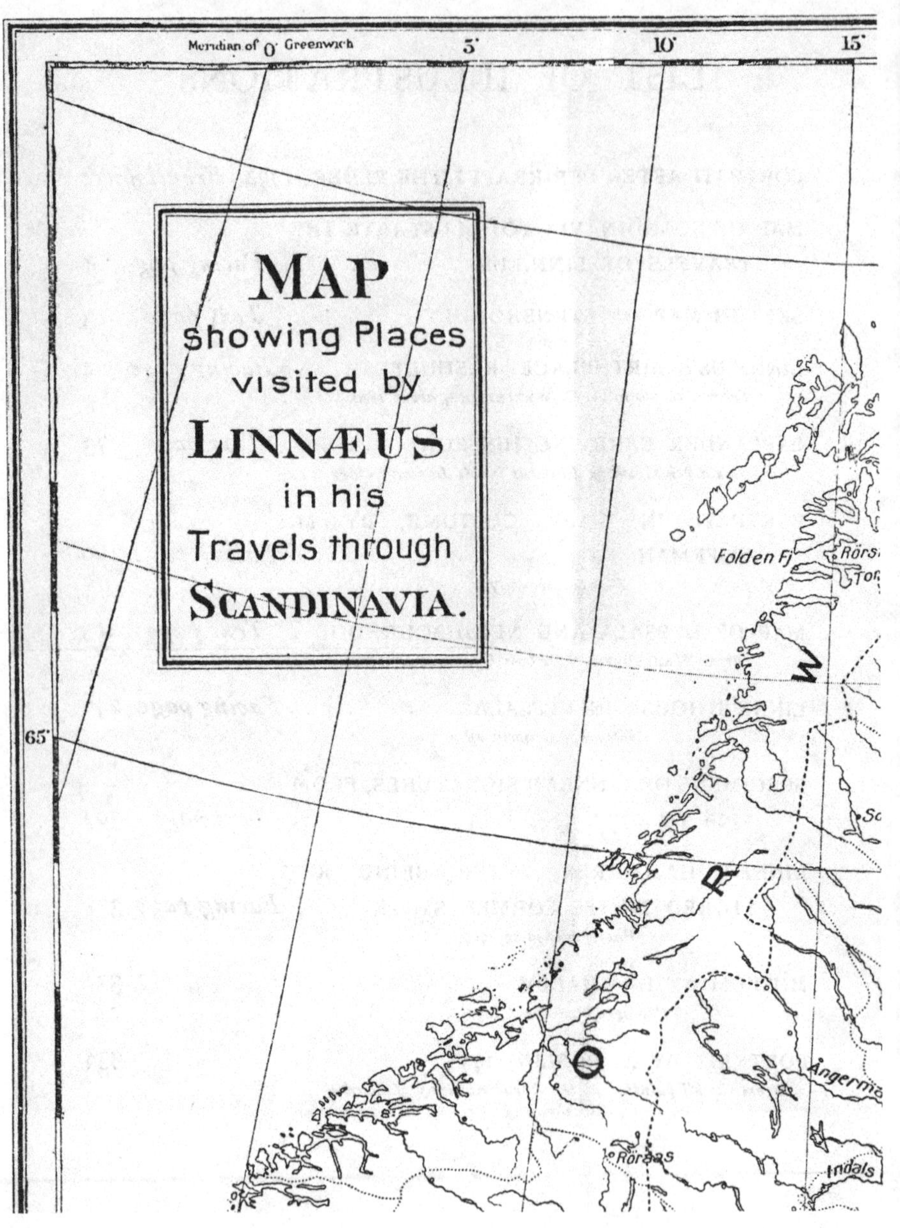
Meridian of 0° Greenwich
5°
10°
15°
65°
MAP
showing Places
visited by
LINNÆUS
in his
Travels through
SCANDINAVIA.
Folden Fj
W
R
O
Rôraas
Angerma
Indals

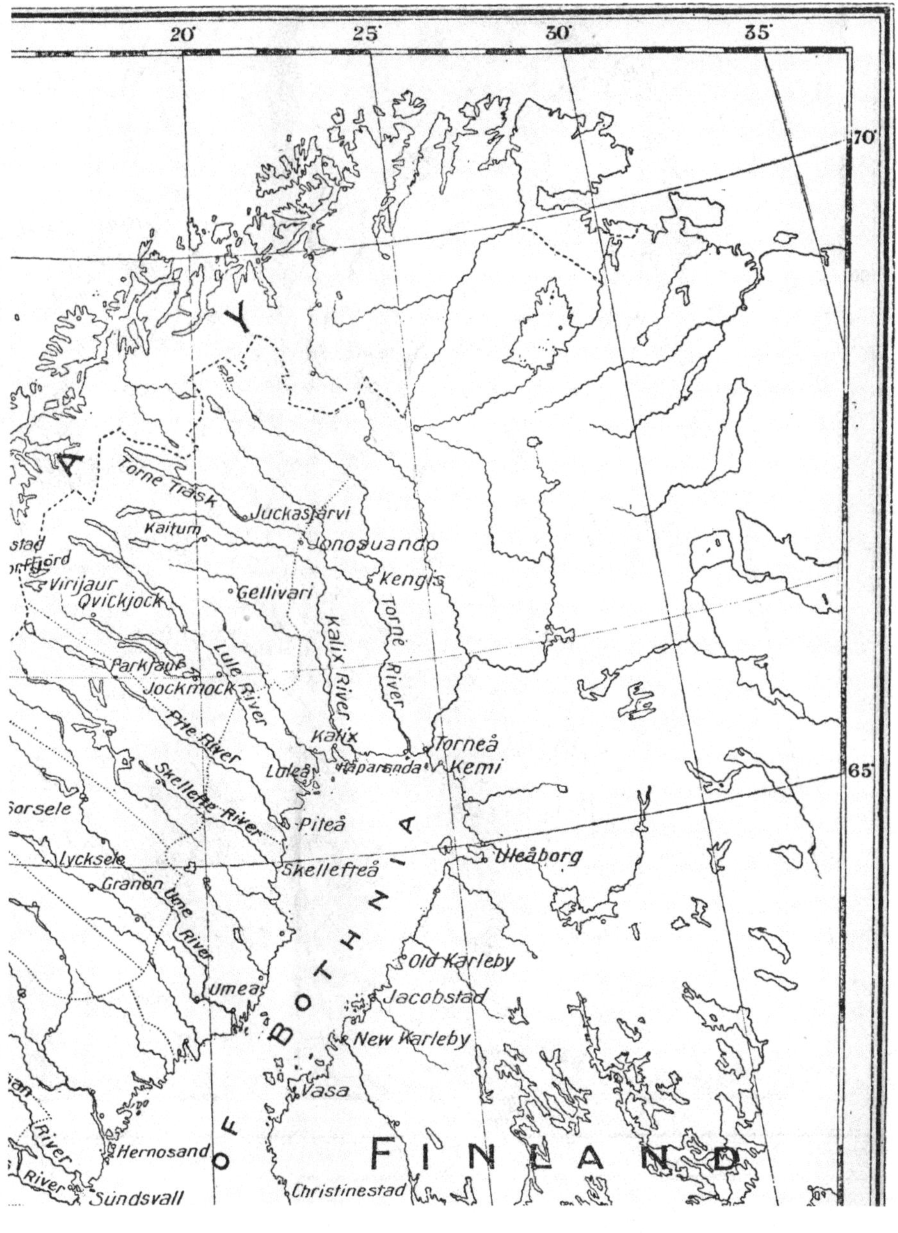

20°
25°
30°
35°
70°
65°
Torne Träsk
Juckasjärvi
Kaitum
Jonosuando
Kengis
Virijaur
Qvickjock
Gellivari
Kalix River
Torne River
Lule River
Parkjaur
Jockmock
Pite River
Kalix
Torneå
Haparanda
Kemi
Luleå
Skellefte River
Sorsele
Piteå
Lycksele
Uleåborg
Skellefteå
Granön
Ume River
Old Karleby
Jacobstad
Umea
New Karleby
Vasa
Hernosand
Christinestad
Sundsvall
River
FINLAND
OF BOTHNIA

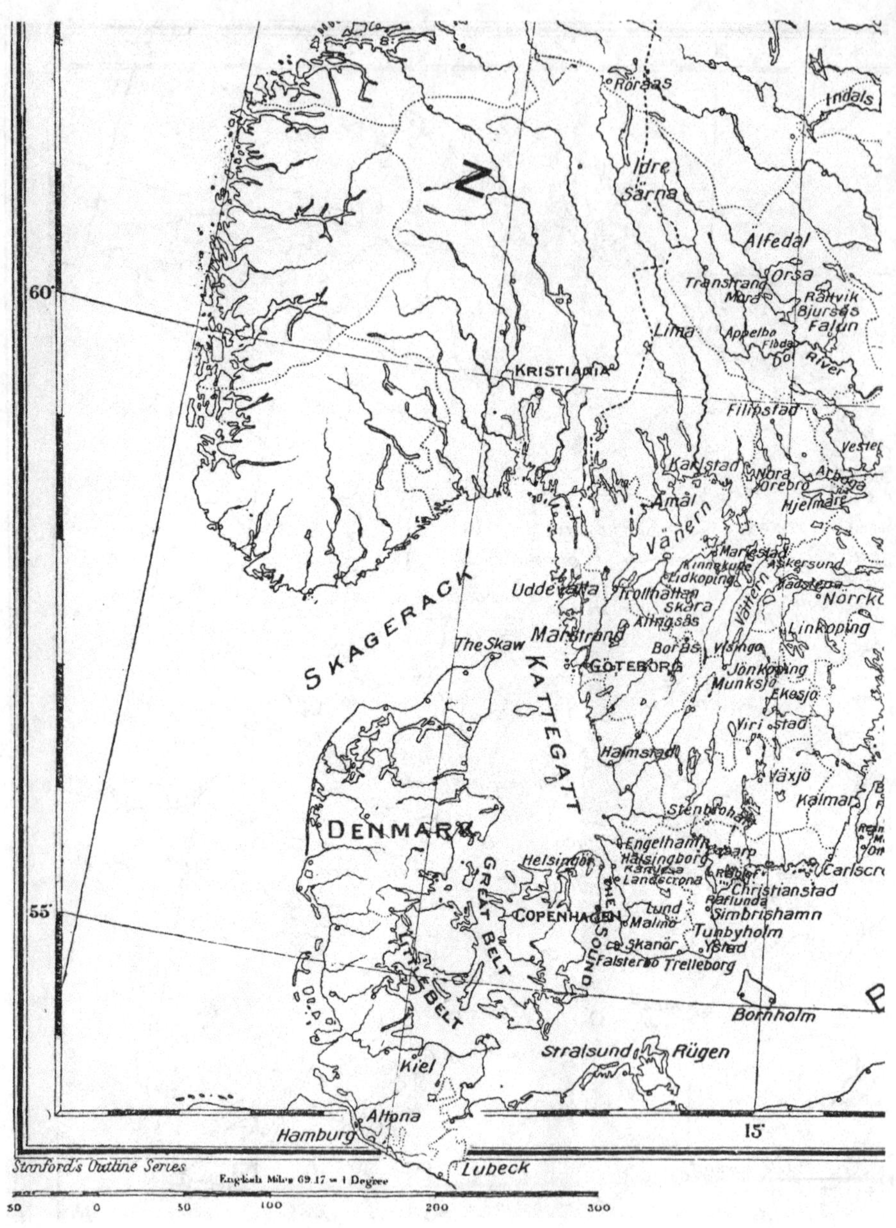
Röraas
Indals
Idre
Särna
Alfedal
Orsa
Transtrand
Mora
Rättvik
Bjursås
Falun
Lima
Appelbo
River
KRISTIANIA
Filipstad
Karlstad
Nora
Örebro
Arboga
Åmål
Hjelmare
Vänern
Mariestad
Kinnekulle
Askersund
Lidköping
Vadstena
Uddevalla
Trollhättan
Skara
Vättern
Alingsås
Linköping
SKAGERACK
The Skaw
Marstrand
Borås
Visingsö
GÖTEBORG
KATTEGATT
Jönköping
Munksjö
Eksjö
Halmstad
Växjö
Kalmar
Stenbrohult
DENMARK
Engelholm
Hälsingborg
Landscrona
Christianstad
Helsingör
GREAT BELT
THE SOUND
COPENHAGEN
Lund
Malmö
Simbrishamn
Tunbyholm
Skanör
Ystad
Falsterbo
Trelleborg
LITTLE BELT
Bornholm
Stralsund
Rügen
Kiel
Altona
Hamburg
Lubeck
60°
55°
15°
Stanford's Outline Series
English Miles 69.17 = 1 Degree
50
0
50
100
200
300

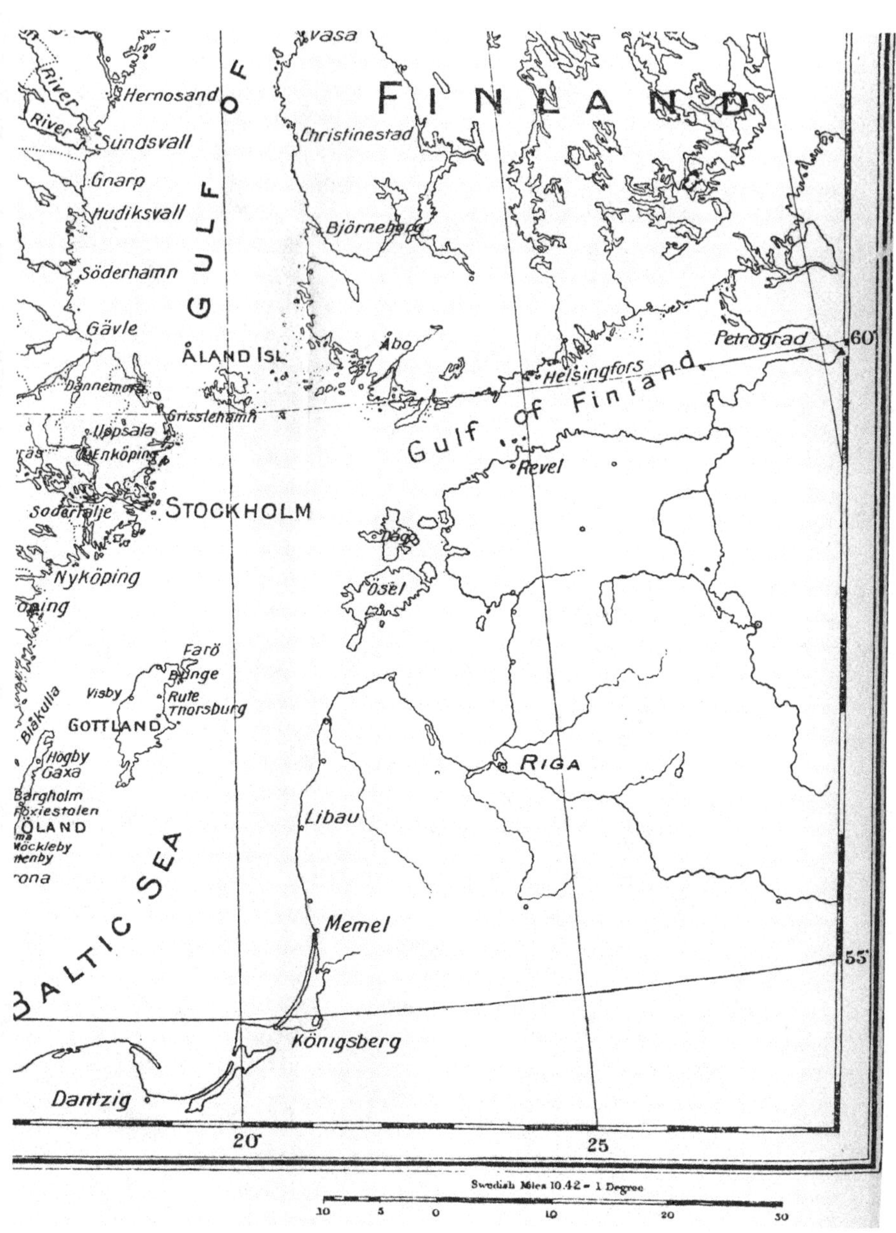
Vasa
FINLAND
GULF OF
Hernosand
River
River
Sundsvall
Gnarp
Hudiksvall
Söderhamn
Gävle
Christinestad
Björneborg
Åbo
ÅLAND ISL.
Dannemora
Grisslehamn
Upsala
Enköping
Södertälje
STOCKHOLM
Nyköping
Petrograd
Helsingfors
Gulf of Finland
Revel
Dagö
Ösel
Farö
Bunge
Visby
Rute
Thorsburg
GOTTLAND
Bläkulla
Högby
Gaxa
Borgholm
Föxiestolen
ÖLAND
Möckleby
RIGA
Libau
BALTIC SEA
Memel
Königsberg
Dantzig
60°
55°
20°
25
Swedish Miles 10.42 = 1 Degree
10
5
0
10
20
30

CHAPTER I

BIRTH, PARENTAGE AND CHILDHOOD—RESIDENCE AT VÄXJÖ SCHOOL AND LUND UNIVERSITY (1707-1727)

THE ancestors of Carl von Linné were, as he said, peasants and priests, plain and simple farmers, who by dint of thrift managed to procure education for a son, or even two, to fit them for the church. Thus, Ingemar Bengtsson (1633-1693), the grandfather of Linnæus, had a son, Nils Ingemarsson, who took a surname from a famous lime-tree—Linnæus—when he entered upon his school and university career. This tree had served the same object when his cousins derived their surname, Tiliander, *Tilia* (the Latin for lime-tree), and the suffix "ander," from the Greek, ἀνήρ, ἀνδρὸς, a man, familiar to us in the names of the two Swedes who were successively librarians to Sir Joseph Banks, Bart.—Daniel Solander and Jonas Dryander. A third branch of the family assumed the name Lindelius, from lind, the Swedish name of the same tree, but possibly it may have been taken from their farm, Linnegården. The special tree, popularly supposed to supply these three surnames, had acquired a sanctity amongst the neighbours, who firmly believed that ill-fortune surely befell those who took even a twig from the grand and stately tree. A further and widespread superstition was, that if and when one of the three main branches died, the corresponding family would die out. Samuel Linnæus, in 1778, shortly after his famous brother's death, wrote that the tree stood between Jonsboda and Hvittaryds parish, close to the southern boundary of Småland. The twigs which fell from it, considered dangerous to remove, were heaped on the roots, which they nourished and kept fresh. By 1823 it had perished, but its relics overspread a great heap of stones in the cultivated ground.

Nils Linnæus (1674-1748), after being educated at home, proceeded with his cousins Tiliander to the provincial school at Växjö, finally journeying to the University of Lund. Possessing only one daler eight öre in silver coinage (about two shillings in value), his poverty soon forced him to seek a tutor's place in Denmark, but he afterwards returned to a similar position in the province of Skåne (Scania). At midsummer in 1703 he came home, hoping to obtain ordination; this he did not obtain, but instead, received a licence to preach, and in October of that year he was sent by Bishop Olof Cavallius to assist Samuel Brodersonius, Rector of Stenbrohult parish, in the county of Kronoberg, province of Småland, succeeding in 1704 to ordination. A few weeks later he was licensed to become Comminister or perpetual curate in the parish. In little more than twelve months he married his Rector's eldest daughter, Christina Brodersonia, on the 6th March, 1706, and eleven days later, the young married couple removed to the official residence at South Råshult, where, on the 13th May, 1707, Old Style [23rd New Style], their eldest son was born, and christened Carl, on the 19th of the same month.

The date of the birth has been disputed owing to the peculiar state of the Swedish calendar in the early part of the eighteenth century. In 1696 King Carl XI., wishing to bring the calendar into accord with most European countries, ordered the omission of the 29th February in every leap-year until 1744; 1700 being a common year in the Gregorian reckoning, no difference was made, but it brought Sweden by one day nearer New Style, and one day different from Russia. This gradual change was then abandoned, but the *one* day's difference was maintained until 1712, when the Old Style was resumed, the final correction taking place in February, 1753. Sir J E. Smith made the not uncommon mistake of

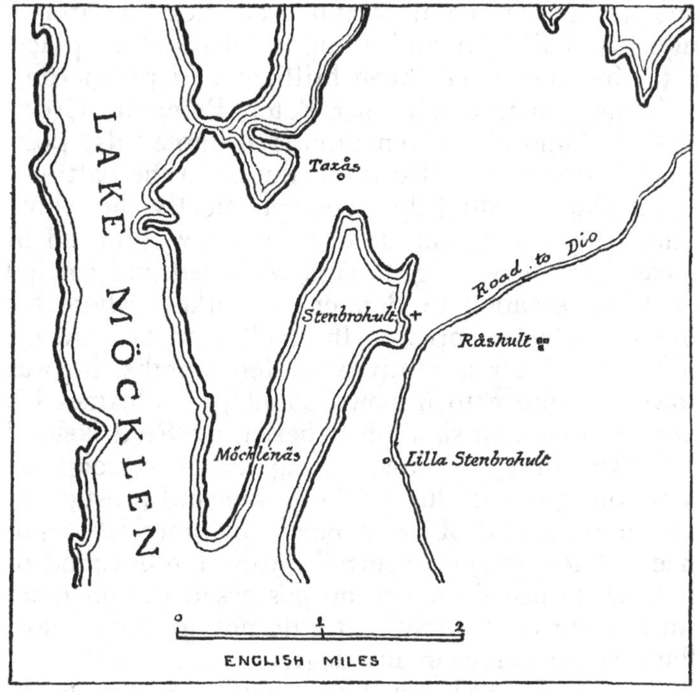

SKETCH MAP OF STENBROHULT

believing that Sweden and Russia employed the same calendar, and that the 13th May in Sweden corresponded to the 24th May, New Style; he therefore enjoined the Anniversary of the Linnean Society, which he founded, to take place on the 24th May, really one day later than the true equivalent.

Carl's birth took place in the tenth regnal year of that intrepid but self-confident King Carl XII., who succeeded to the throne of Sweden at the early age of fifteen. His tender years encouraged Russia, Denmark and Poland, to unite against him, but his enemies found him equal to the occasion. Denmark being defeated, he turned his arms against Russia, and in the famous battle of Narva in 1700, he is said to have slain thirty thousand of his opponents, and made twenty thousand prisoners, though his own force was under ten thousand. He next dethroned Augustus of Poland, and set up Stanislas in his place. So far his career had been brilliant and prosperous, but in striving to crush Tsar Peter, Peter the Great, he allowed himself to be manœuvred into a false position, and sustained a disastrous defeat in the battle of Pultowa on the 8th July, 1709, practically the whole of his troops being captured, save a few hundred of his cavalry. The king, though wounded and carried in a litter, escaped to Bender in Turkey, where his violent conduct compelled the Sultan to besiege his residence. After a captivity of ten months, he was allowed to return to his own country. He met his death by a cannon-shot when besieging Fredrikshald in Norway in 1718, thus, during his short reign of twenty-one years, reducing the power and prestige of Sweden from one of great power to practical impotence. These events occurred during the boyhood of Linnæus, to use the name he possessed during more than two-thirds of his life, and do not appear to have influenced his career in any degree.

The family did not long remain in the lowly cottage where he first saw the light, for on the last day of the year 1707, his grandfather Brodersonius died, and the chaplain of Växjö, Petrus Comstadius, was appointed to succeed him, but he too died before taking up the pastorate. A powerful patron, State Secretary Josias Cederhielm persuaded Carl XII., who was then in Poland, to issue a licence for Comminister Nils Linnæus to become Rector of Stenbrohult, on the 12th August. Through this

LINNÉ'S BIRTHPLACE, RÅSHULT

(From a Sketch by A. C. Wetterling about 1820).

arose the favourable circumstance that the pastorate descended from father to son, or from father-in-law to son-in-law, for a series of years, Brodersonius being successor to his father, and being followed by his son-in-law, and in turn by the latter's son, Samuel, altogether making five successions in the family.

On the 31st June, 1709, the removal took place from South Råshult to the rectory at Stenbrohult, distant about an English mile by a beautiful woodland path. In this new abode the parents of Carl Linnæus had four other children—one son, Samuel, and three daughters. All accounts show that the modest home was the abode of complete affection, simple habits, and sincere piety. The father, on the testimony of his younger son, was very honest and trustworthy, knew nothing of the world's deceits, distrusted its fashion and vanity, always friendly, merry and glad, and very jocose, and by no means vindictive. If he saw anyone suffering from whatever cause, he was so tender-hearted that he could not refrain from tears. His spiritual duties he discharged faithfully and honestly. At the same time he applied himself to putting the church and rectory into better order, both as to the buildings and their surroundings; he was a good householder and thrifty. (The rectory was reduced to ashes by a disastrous fire in the night of the 20th April, 1748.) Assuredly he was not rich in earthly treasures, but by wise forethought, he had money to lend, or to buy small pieces of land in the parish. Besides his rectorial duties, he had, until 1720, to combine them with the curate's functions.

The mother's disposition was, according to her elder son's testimony, quick and active. She left behind her the reputation of living with her husband in true love, displaying great common sense, for twenty-seven years and three months, until her death, and her five children were reared in praiseworthy fashion. She comported herself so well with high and low, rich and poor, in the parish and outside it, that no one could do other than praise her. She was pious, and kept her house excellently, being economical and energetic, mild and earnest, and endowed with high intelligence.

Such was the home from which Linnæus came, and therefore his gratitude to his parents and his remembrance of his happy childhood in that dear home never waned. In after years, in his printed writings, and intimate letters to his brother, sisters and kinsfolk he gave free utterance to his feelings. He referred with emotion to " our parents' tears for their children passed up above the clouds, and stayed not till they came into God's presence, who cared for their welfare." With a touch of melancholy he recalled how " it commonly happens that the young ones, hatched in the same nest, fly away as soon as fledged, each in its own direction, which they seldom leave together from the same tree," and that " fate had been gracious to his brother and sisters, that they were vouchsafed to dwell together on their father's land, whilst I [Linnæus] was driven forth far from my kinsfolk, to live alone as a stranger." More than once in after life, he refers to Stenbrohult in his frequent blending of Swedish and Latin, as his " ljuva natale "—sweet birthplace. Assuredly it deserved that love, for no fairer spot for the training of a naturalist could be found than in this broken country of hill and dale, mixed woodland and a delightful lake, where deciduous trees grew with pine and fir, and scarce plants throve round the paths.

During Linnæus's tenderest years he was, through his parents, awakened to pleasure in nature, which lasted during the whole of his life. Both cherished a strong love for flowers, especially his father, in whose relations it seems to have been characteristic. His uncle and benefactor, Sven Tiliander, whom he often visited, had travelled in Germany, and devoted himself to gardening, and at Bremen had laid out a garden in the style of the period. This interest continued undiminished after his return home, and caused him to lay out a garden at Pjetteryd rectory, to which his friends in Germany contributed by sending rare plants not previously cultivated in Sweden. He inoculated the young Nils Linnæus with a like devotion, which ended only with his life. During his university career he learned—an uncommon attainment then—the Latin names of certain plants, and "had himself laid in with his own hand, fifty plants in a 'herbarium vivum.'" He had hardly entered into his official residence at Råshult, before he began to employ his leisure hours in laying out a garden "more for the sake of the plants themselves, than for any advantage to himself, and his young, newly wedded wife, who had till then hardly ever seen a garden, was delighted with its charm." In accordance with the taste of the time, he had, with his own hands, raised an eminence and a surrounding border round the field, with plants or shrubs to represent guests, and flowers to adorn the table.

When he removed to the rectory he was able to develop his ideas on a larger scale, making "a fine garden where formerly there was not a twig, so that it surpassed all gardens in the province; for in it were several hundred different foreign plants." Here the parents spent their leisure hours, and flowers became Carl's first and choicest playthings. It is also related that "the father took the little year-old son out with him sometimes into the garden, putting the child on the ground in the grass and leaving a little flower in his hand with which to amuse himself," also that "when the boy was unreasonable and by nothing else could be pacified, he became silent at once, so soon as one put a flower into his hand." When somewhat older, the child laid out a little garden of his own, which was always being enlarged, and there he had in a small plot a sample of all that was found in the large garden. Still later when he sat as flower king in the rich Uppsala garden, he recalled his father's garden at Stenbrohult, as he had there "with his mother's milk excited his imagination with a never-extinguished love of flowers."

But it was not only in the garden that the father excited the love of plants in his son. "Carl," as he related himself, "was barely four years old when he accompanied his father to a picnic at Möklanäs, the promontory which jutted out into Lake Möklen, forming a bay in front of the church," in the most beautiful summer time, and when the guests towards evening rested in a green meadow, the Pastor told the company how each flower had its name with specially remarkable and marvellous characters, describing the roots of *Succisa*, *Tormentilla* and *Orchis*, with many others. The boy received these descriptions with delight, the subject being one so sympathetic with his temperament. From this time his father had no peace from the lad, whose demands for the names of plants came faster than could be answered. He often forgot the names certainly, whereupon he was admonished by his father, who threatened that he would never give him the name of another plant, if he forgot the last; consequently the boy's whole care afterwards was to remember the names, lest he should be deprived of his most cherished delight.

With increasing age Carl naturally enlarged his field of observation outside the limits of his garden, and thus he attained a knowledge which, in after years, gave such splendid returns. Wonder and love for his birthplace were early awakened in the child's bosom, and the feeling was so strengthened during his boyhood, manhood and old age, that he could never think upon it without emotion. "Stenbrohult is a church," he says in one of his autobiographies, "furnished with the delightful plants which Sweden displays, for it lies near the lake Möklen, which here extends in a quarter of a mile [English mile and three-quarters] long bay, and almost reaches the foundations of the church. The level farmlands surround the church on all sides except the west, where Möklen displays its limpid waters. A little way off, the fine beechwoods show themselves towards the south, with Taxås high hill to the north and Möklanäs beyond the lake to the west. To the east the fields are sheltered from the north by coniferous woods, and from east and south are pleasant fields and leafy trees." In another place he says "the meadows resemble more the most splendid groves and richest flower gardens, than their actual selves, so that one may sit in summer and hear the cuckoo with other different birds' songs, insects piping and humming, and at the same time view the glowing and splendidly coloured flowers. One cannot but turn giddy at the Creator's magnificent arrangement." Lastly in yet another place he declares "Stenbrohult parish is like a queen amongst sisters, she has predominance of rare and scarce plants, which in other localities in the country seldom or never show themselves. Yes, the Rector's surroundings seemed as if they had been adorned by Flora herself. . I doubt if there is a spot in the whole world set out in more pleasant fashion, so that it is not surprising if I had cause to complain 'Nescio qua natale solum dulcedine cunctos ducit et immemores non sinit esse sui' (I know not how the natal soil draws all with delight, and permits us never to forget it.)" To this loved abode he returned from his journeys as often as was possible, and with great gladness he sought, and again saw, the rarest plants, which grew wild in that spot.

With floral playthings soon were blended earnest things. Book knowledge was early sown, and the 15th February, 1714, was, in its outlook, an important day for little Carl, as then he received his first tutor, Johan Telander, of the Gymnasium (1694-1763). "Quick intelligence was not wanting, though his fancy turned mostly to the garden. The mother forbade this, but the father, tolerant, took his only boy always under his protection and defence. That his studies should be less disturbed than at home, and that its enticements should be withdrawn, the youngster in September, 1714, accompanied his teacher to Växjö, there to benefit by his private tuition. On his entrance he was inscribed in the school matriculation as a pupil in the lowest class, and he spent seven years in the normal studies, being one year more than the usual time.

" It is well known that the instruction of the time was not clement, but on the contrary, very hard. Thus it was the case with Linnæus, for in later days he was accustomed to speak of his first teacher as a severe tutor, who taught with strokes and not with enticement, and was little adapted to bringing up children. He in after years passed the hard judgment on the Lower School at Växjö, that coarse teachers and coarse methods were in vogue to give children a taste for science, such as might raise the hair on their heads. Some amelioration, however, took place in 1717, when he gained a new tutor, Gabriel Höök, also from the Gymnasium, who treated the boy with conspicuous gentleness, though unable to implant in him a liking for study, for which the lad already showed an aversion. The result was that so long as he remained under private tuition, he was equal in general knowledge to his schoolfellows, though in his leisure hours, he delighted to gather flowers in the fields and to teach his comrades about them, thus gaining for himself, when barely eight years old, the nickname of 'the little botanist.' An essential alteration took place in 1721 by his being removed to the Rector's class, when, according to established custom, he became free from the tutor's superintendence, the result being that he enjoyed his liberty, and employed his newly acquired freedom by neglecting his books and rambling about in search of plants. It may have been that his neglect in great measure prevailed only at that time of year when the flowers tempted him to an excursion at large, or that his quick powers of comprehension atoned for the wanting industry, or that his comrades in diligence and attention were not specially better than himself; suffice it to say, nothing occurred to prevent his transference to the Gymnasium at the normal time. Since the annual examination took place on the 8th-10th July, 1723, Carl Linnæus with fifteen comrades on the day following, himself the eleventh in order, passed into the Gymnasium."

During the period when the young Linnæus passed through the lower school, he made the acquaintance of the man who, without doubt, exercised no small influence in his development as a naturalist. This was Daniel Lannerus (1679-1761) who in 1719, was appointed Rector of Växjö school. He was a great lover of plants, and as he was also an intimate friend of Nils Linnæus he gave permission to the little Carl to go often into the garden and enjoy himself by eating berries. The Rector asked Carl about certain plants and whether he knew their names. Linnæus answered readily and in turn put questions to the Rector, thus not only receiving instruction but also having the opportunity to ascertain the names of many plants. The Rector's favour, which increased more and more, made Carl's stay at Växjö a very pleasant time, especially as he was introduced by him to Lector and Provincial-Medicus Rothman, who, taking an uncommon liking to the lad, gave him permission to visit his garden, where many kinds of plants were found. Through this latter acquaintanceship were awakened thus early thoughts of a medical career. During the holidays at home, Carl amused himself with his brother and sisters, made a lancet of wood, as though he would bleed them, tested their slightest symptoms by their pulse, and sometimes sought for plants by which to cure their ailments.

At the Gymnasium the pupils' freedom was in a certain degree curtailed, although the object; with regard to which the studies were directed, was the same, being regulated by an ordinance of King Carl XI. Theology was foremost, and the Greek and mathematical teachers usually ended their days as pastors, with the exception of those professors of Logic and Physics, whose subjects were regarded as leading to a provincial doctor's career. Everything pointed to theology dominating studies, so as to prepare the pupils for priestly functions; therefore it was hardly an exaggeration in Linnæus's words, "that no other science was practicable, than that which made priests."

It was the cherished wish and expectation of his parents, especially his mother, that their first born should become a priest, to which he was destined from the cradle. Though deep and warm religious feeling was by no means wanting, this did not agree with his inclination, for he had no call to become a priest, and when the then gymnasiast enjoyed a further extension of liberty, he employed it by an increased application to his botanical studies. Within the town of Växjö itself he sought what there was to be found, not neglecting to herborise the many flowers and mosses on the roofs. Excursions were made to the Solberg and other places in the neighbourhood, and the journeys to and from home served also to widen his knowledge, as he always turned his eyes to the roadsides to discover flowers, being able to locate any plant in those five miles (nearly thirty English miles). It is remarkable that he neither in youth nor in mature age attained more than an inconsiderable acquaintance with the flora of his native province. He said, afterwards, that "I was a lynx abroad, but a mole at home, and knew more about Virginia in North America, the Cape, Ceylon and the East Indies, than of my own province, which I left before I was properly awake or able to chase sleep from my eyes. I had hardly seen more than Stenbrohult, my birthplace, and Växjö, my first school, leaving both before I was grown up, and since then have only seen them as a migratory bird, as I only visited them a few times, when passing to Öland and Skåne, and then without a long stay." Besides this he busied himself to gain knowledge from certain old botanic books, such as Månson's "Örtabook," Til-landz's "Catalogus Plantarum prope Aboam inventarum" and Palmberg's "Serta Florea Suecana," which nevertheless were found "wretched guides," with Bromelius's "Chloris Gothica" and the elder Olof Rudbeck's "Hortus Upsaliensis," although the latter were yet too learned for him. Still, whatever these books were, they tempted him more than schoolbooks, so that he read them day and night, till he had them at his fingers' ends, with extracts from Pliny and Colerus, chiefly botanical and medical.

His occupation with a "useless science," as they called it, drew the attention of his comrades and teachers. His exertions were approved, however, by two of his professors, Lannerus and Rothman, and even his father, a warm friend of flowers, encouraged him in these occupations. Partly in order to avoid hindering him in these pursuits during his summer holidays, and partly because he knew Carl's too hasty disposition, he freed the Gymnasium scholar from supervising his eleven years old brother Samuel. The mother also seems to have thought that his time could not be better employed, and was glad that he occupied himself with diligence on virtuous tasks, and not on vicious ones.

During this period it became evident to all members of the family that their earnest hopes regarding the ultimate priesthood for Carl must be abandoned. Linnæus himself realized that he was amongst the worst of his schoolfellows in the subjects of eloquence, metaphysics, morality, Greek, Hebrew and theology; but on the other hand was always one of the best as regards mathematics and particularly physics. Notwithstanding this his knowledge in at least some of the above named subjects was not below the general standard, this being proved by many circumstances. Especially was he good in Latin, for he had, during his school and Gymnasium period, applied himself particularly to this subject. The results appear from the ease with which afterwards he expressed himself in that language, both in speech and writing, and this in spite of his own testimony that in learning languages easily he was never an adept. On the other hand it may be taken as fairly certain that his knowledge of theology, Hebrew and the like, left much to be desired, even though it was not below the average. He seems to have been regarded by his teachers throughout his schooldays as belonging to the medium class. In the prescribed time he was moved from one division to another, as is shown by the place he occupied as the eleventh in order of fifteen which he took among the lower division of the Gymnasium, when in May, 1727, he was sent up to the University.

Before Linnæus reached this point, in September 1726, an occurrence happened which determined his whole career. His father then came to Växjö to hear about his dear son and to consult Dr. Rothman about a disorder which had troubled him for several weeks. In the first case, he received information which came upon him like a thunderclap, and that the expectation which he had till then cherished, that his son would become a priest, was instantly destroyed. The professor of whom he inquired, declared emphatically that his son in the indispensable subjects for an intended priest was utterly deficient, and the words seem to have been used, that he was far better fitted for a workman, a joiner or tailor. Linnæus himself at first attached but little importance to this statement, but in his later autobiographies, when his memory was failing, he seems to have attached too much weight to it. Deeply cast down, he afterwards came to Dr. Rothman, to whom he confided his trouble in both aspects; and he was not unhelped. Specially regarding Carl, Rothman declared that "the professor was right in this, that he could never become a priest, but on the other hand he was assured that Carl would become a famous *Doctor*, by which he could gain a living equal to that of any priest." He went so far as to assure the father sacredly that amongst all the studying deacons in Växjö, there was not one who gave such hope for his future as his Carl; and he offered to take him into his house, and give him private lectures in physiology, etc., while he would love him as his own child. By this interview the father was not a little comforted, and gave his willing consent to the arrangement. The mother, however, received the news passionately, and in her distress blamed the garden and its flowers, wherefore she forbade her younger son, Samuel, on any account to dare to concern himself with this calamitous syren.

The quick result of this occurrence was, as related, that the young Gymnasium student received the advantage of Rothman's personal instruction. Gradually there arose between them the most intimate relations, resembling those of father and son, which continued unclouded until death severed the bond. The youth's previous desire to devote his future to the practice of medicine, of which botany at that time was looked upon as an important and essential part, was strengthened by the teacher more and more. He chose particularly the "Lectures on Boerhaave's 'Institutiones Medicæ'" (employed in the early part of the eighteenth century as a textbook in almost the whole of Europe), explaining with the greatest learning to his pupil, who after each lecture was examined and found to answer readily on every point that had been set out. He also became Carl's first teacher in scientific botany, and showed him that the knowledge of plants he had till then acquired, did not correspond to the time's demands as to scientific research. "To know a fluent Latin word or name for a plant was nothing, but for proper naming of plants, in describing and classifying, the greatest weight must be laid upon the structure of the flower such as was set forth by Tournefort." His valuable work "Institutiones Rei Herbariæ" (Paris, 1700) seems certainly not available, but in its place the young investigator was lent Valentin's "Historia Plantarum" (probably "Tournefortius contractus," Francof. 1715), so that he could copy the figures in their "classes plantarum." Henceforth his whole effort was to know and refer each and every plant to its class after Tournefort's method. Many of the plants that he had already met with in his home, gave him trouble enough, as he was not sufficiently a botanist to disentangle or to know them.

At last the day dawned when Carl Linnæus should quit the scholastic dust of Växjö which he had trodden for twelve whole years. His friends bade good-bye to him on the 1st May by holding a feast at the house of Munthe, one of the most respected citizens, on the night of that day. The following morning he betook himself to Stenbrohult, taking with him his testimonial, that is, a Latin notice to the Rector of the University where his studies should be continued, which Nils Krök, the then Rector of the Gymnasium, had drawn up for him.

This certificate has played a notable part in the description of Linnæus's youth, and has caused a belief in some people on the ground of its supposed contents, to pass a harsh judgment upon those who gave it to a youth proceeding to the University as of an unjust and disgraceful character. Later on, the document has been presented in its original and true form—and not in a very free and highly coloured version, which Linnæus in his later days, left as a representation of its contents. All such accusations must be silenced, and in place thereof, it must be realized that Rector Krök was a professor very well disposed to Linnæus, who did what he could to prepare for him a good reception at the University. "As nature," said he, "in the vegetable kingdom offers a delightful spectacle, when by removal of plants from one place to another leads to their happy and early growth, so the Muses by a specially graceful way of partiality, invite youths with uncommon gifts, sometimes to change their place of learning, thereby the more quickly to transpose their studies' sweet nectar into blood and sap. With this view the High School Muses call from our Gymnasium Carl Linnæus, a specially distinguished youth of a worthy family, that he may be the more welcome and at his first entrance may be able to settle himself under their favourable protection, he seeks to exhibit a testimony of his studies and his behaviour. To this end I certify that with regard to morals, he has displayed a godly, good and honourable disposition; in his studies has applied unwearied diligence, and has kept himself from all evil intercourse. Therefore I recommend to you, Rector Magnificus, and to your patronage, your favour and good will this well-behaved youth, and obligingly and obediently beg you, on the ground of your friendly graciousness to all who are noble, discreet and virtuous, praying you to take him under your guardianship and comfort him with your favour. Your benevolent and paternal tenderness he will never forget; remembrance of it will ever be retained, so I, as long as I live, will offer up pious wishes for your welfare."

After returning to the paternal home, Carl's time was divided between collecting plants and reading for examination for his entrance to the University. Then was decided the definite settlement of his future life's object by repeated discussions between him and his parents, who even yet had not given up expectations of his entering upon a clergyman's career. He himself determined to become a medical man and botanist and nothing else; his mother was more disturbed at that, than if her lad changed his religion, and his father took up a mediating position, though he, from an economic reason, seemed disposed to share his wife's view. An important cause of this stubborn resistance on the part of his parents undoubtedly was partly that it was most unusual that a youth who had passed through the Gymnasium should choose any other life career than a priest's or school teacher's, and partly the very poor prospect which a student of medicine enjoyed at that time, and the scanty income, which, in most cases, awaited such, after the completion of the relatively long and expensive student's course. Whilst these discussions were taking place, as Samuel Linnæus relates, "some good friends came to Stenbrohult. His father took them into the garden, where they seated themselves round a little table and talked, while sundry glasses of beer were drunk. During this conversation his father said, 'Yes, it always happens that what a man has delight in, always succeeds.' Carl, who was present, took this speech to heart. When the company had departed and his father had come back to the table, Carl went to him asking what was that which he had said when the friends were there? The father, who was always of happy temper and jocular, asked, 'What was it that I said?' But Carl, insisting on a positive reply, was answered. 'So far as the liking is for that which is good, I stand by it.' Thereupon Carl said, 'Yes, father, but do not urge me to be a priest, for I have no inclination that way.' Then the following objection was raised, 'Thou knowest thy parents' poor condition, and the study thou wishest to choose is very costly.' But Carl caught up his father's words and said, 'If that is correct, God will certainly provide the sacrifice. Should I succeed as I wish, the way will be made for me.' His father replied with tears in his eyes and with a troubled mind, 'May God grant you success; I will not compel you to follow that for which you have no liking.' Thereupon it was settled that the course should be to Lund, because a distant friend, stepson to Carl's great-uncle (father's mother's brother), the well-to-do Canon Bonde Humerus, it was hoped might help the poor student. On the 14th August he began the journey from Stenbrohult, and on the 17th he arrived at Lund."

There his first problem was to seek his former tutor, Gabriel Höök, now Master of Philosophy, to obtain from him counsel and enlightenment. First he must be registered at the University, for which end he must be examined by the Dean of the Faculty of Philosophy, which was the custom down to the year 1831. The Dean was the Professor of Rhetoric, Carl Papke, afterwards Bishop of Lund. After he had been passed on the 19th August, the same day he inscribed his name in the matriculation list of the faculty of philosophy Accompanied by Magister Höök, he proceeded to the Rector, and Divinity Professor, Martin Hegardt, who, after he had read the testimonial from Växjö schools, received the oath of the student, whose name was then inscribed in the University Matriculation List. He ought strictly after this, to have inscribed himself in the Småland's Nation, but he neglected to do so, which was—without reason—interpreted as his desire to live the life of a recluse. The reason for this was nothing else than his unwillingness to subject himself to the then "penalism" or fagging system prevalent at Lund. This had been entirely forbidden by a royal decree of the 25th November, 1691, and Professor Döbeln had, as Rector, in 1717, acted strongly to outroot this flourishing penalism, which persisted in spite of the king's letter styling it "invincible and diabolical"; but at the time of Linnæus's arrival at Lund, the coarsest nuisance and the grossest bullying which were inflicted by the senior members of the Nations on the newly arrived country youths were nominally suppressed, but they still continued. Therefore it was not uncommon for students newly arrived at Lund to neglect to inscribe themselves in any Nation. Linnæus found a special reason for this, namely, that he had already, on his arrival, come to the determination not to stay more than a year, so that he held it unnecessary to "go penal," and to devote himself to any Nation, as he was impartial. Probably he calculated that when he inscribed himself in Småland's Nation at Uppsala, he would be regarded as a senior student, and thus escape having to "go penal," an idea which proved correct.

Besides this it was soon seen that the young student had miscalculated. Shortly before his arrival at Lund, Dean Humerus died, and thereby his not groundless hopes of help from his relation were frustrated. Samuel Linnæus relates that "when Carl came to the town gate of Lund, all the bells were tolling. He asked someone the cause of this and was answered 'For Professor Humerus.'" Not less was the hope dashed—which he had hitherto cherished—that he would find competent and zealous professors in medicine and botany. In the latter subject he found then no academic professor, and the whole medical faculty was carried on by a single man, the previously named Johan Jacob von Döbeln, who was both learned and experienced, but who, as he himself declared, "could not procure the new things which he required, because there were no means provided for the support of the study of medicine, nor for Anatomy, Botany or Chemistry." He probably too, as the result of age, when Linnæus was residing in Lund, had already lost somewhat of his former strength, whereby he seems to have given too little regard to private practice and academic objects. Naturally, Linnæus attended his lectures, in the autumn term of 1727, on miscellaneous topics, also in the spring term of 1728, the subject being then the "Physics" of Budaeus.

Fortunately there was at that time in Lund a man whose great services to medicine and natural history in Sweden are both generally known and recognised. This man was Dr. Kilian Stobæus (1690-1742). To avail himself of his instruction was now Linnæus's earnest desire; a preliminary step was taken when he, through Höök's care, obtained lodging in the house of Stobæus, to which he removed on the 21st August, and where he continued to reside during the whole of his stay in Lund.

This Stobæus—assuredly the most eminent of the five distinguished professors of the same name, who were at Lund—is described as a "sickly man, one-eyed, lame in one foot, constantly troubled with sick-headache, hypochondria and backache, but nevertheless of unsurpassed genius." At his house the young student was enabled to see an excellent museum of all kinds of natural objects; stones, shells, birds and herbarium of collected and glued-down plants, such as he had never seen before. At first he did not attain his wish to obtain private tuition, for Stobæus "saw the youth, but found neither in appearance, dress nor habits, anything to recommend him, more than an ordinary stranger who wished to devote himself to the study of medicine." For this reason the only means Linnæus had to gain his favour, was to attend Stobæus's lectures on shells, which he did in the company of Benzelius (afterwards Secretary of State, who died in 1791) and Nils Retzius, later Provincial Doctor in Skåne, who died in 1757 A good opportunity for a nearer approach between teacher and pupil was lost as he himself narrates. Stobæus "had become medical man for the higher ranks in Skåne, and as his consultations gave him no rest, on one occasion he called Linnæus to help him by writing a letter and attending a patient, but Linnæus's unpractised hand in writing caused him to be rejected."

How the relations between Stobæus and Linnæus, through a happy occurrence which immediately led to the most intimate and affectionate relations, may best be sketched in the latter's own words. "I had no books nor money to buy any, but I became acquainted with the amanuensis of Dr. Stobæus, a German student, David Samuel Koulas (died 1743), whom I induced to lend me each evening books from Stobæus's library, which I returned each morning before the doctor rose. This went on for three months. But Dr. Stobæus's old mother, whose bedroom was next to mine, noticed that a light was burning each night, and thinking that going to sleep with the light burning made it dangerous on account of fire, she told her son, who, to avoid so great a danger, came up at half-past one in the morning, expecting to find me sleeping with the candle still alight, but to his surprise found a pile of his own books on the table by the bedside and myself awake." Our youngster related how he had succeeded in getting them, and returning them in the morning after using them the whole night. Stobæus's angry look at once was changed to a generous sympathy; he told him to go to bed immediately, saying that he would himself lend him his books the next day. From this time onward he had liberty to take out of Stobæus's library any book he wanted. Stobæus noted from this night the youth's diligence, for he allowed him to use all his lectures gratis, admitted him to meals at his table without payment, sent him to visit patients, and practically treated him as his son. It was noticed that the young fellow began to distinguish himself in the University; he learned from Stobæus about fossils and shells, received the ground-work of certain special parts of medicine, while Stobæus let him see that if he continued his diligence as he had begun, he was disposed to make him his heir, so highly did he esteem him. All this kindness which he received, Linnæus reciprocated with the warmest gratitude and the highest regard. This is shown by the letters which, after he had left Lund, he wrote to his generous benefactor; shown also in his autobiographies, where he warmly mentions his protector Stobæus, "to whom I am indebted so long as I live, for the love he bore me, and that he loved me, not as a pupil, but as if I were his son."

The influence of this especially distinguished instruction, with the use of the beloved museum of Stobæus soon showed itself. Before everything was Linnæus's endeavour to provide himself with an arranged herbarium of dried plants secured on white paper, such as seemed at that time, and far into that century, the proper thing. For this he visited not only the little academic garden, but such as Dr. Hegardt's in Lund, whence he procured sundry plants for his "Herbarium vivum," among them being in November, 1727, flowering specimens of Jerusalem artichoke, *Helianthus tuberosus*. Besides this, he undertook, whenever the season permitted, flying excursions into the neighbourhood, for here one found entirely different plants from those occurring in Småland. He and his companions—for he had by 1728 initiated certain students in botanic matters—directed their course to Malmö and Lomma, where one also was able to get fossils from the sands by the seashore; or to Fogelsång, where nature had its theatre; here was a high hill of pyrites, and also a glen through which a stream ran. On both sides above the brook were thickets where the rarest plants were to be found. The floral treasures here gathered were investigated by help of Johrenius's "Hodegus Botanicus," which Linnæus had bought as soon as he became a student, Tournefort's method not being practicable.

These excursions, however, came to a sudden end, for on a hot day, 26th May, 1728, at Fogelsång his whole arm swelled up like a log, and Linnæus was obliged to go to bed. His condition grew worse, although Stobæus employed all his skill trying to cure the evil. The latter was soon obliged to journey to Ramlösa to drink the waters, and he parted from the sick youth with scant hope of his life, but entrusted the care of him to Veterinary and University Surgeon Carl Christopher Schnell. The latter "made a great incision from the elbow to the armpit," after which his recovery was so rapid, that by the 28th June, Linnæus was able to travel homeward to Stenbrohult, which he reached the day following. That he did not, after his health was restored and strength regained, continue his excursions in Skåne, he gave as the reason, that he received a letter from his benefactor Rothman, who urgently insisted that he should exchange Lund for Uppsala. In Linnæus's earlier autobiographies this illness was ascribed to a virulent abscess in the right arm, or a severe inflammation; afterwards it was attributed to the attack of a small hair-like worm which found a place in the "Fauna suecica," Ed. II., 503, as *Furia infernalis*, by many regarded as a myth. The latest investigator, Sir Arthur E. Shipley, F.R.S., thinks what probably stung Linnæus, was a virulent insect, which might very well have conveyed some pathogenic germs to his system, unknown in the time of the great naturalist.

The summer of 1728 was spent at home busy on his usual employments, but with this change, that now he did not occupy himself only with plants, but also animals and minerals, which became his objects for later research. He hastened to write to Stobæus about his discoveries and sent specimens. His father, who regarded his son's career as settled, did not remark on this, but his mother did; with an almost pathetic obstinacy, she still clung to the expectation that his stay at Lund would have resulted in a change in his plan of life, but when she now saw that Carl did nothing but glue plants to paper, she became at last convinced that her desires for her dear son's future career were now hopeless.

During the course of the summer the rectory of Stenbrohult received a short visit from Dr. Rothman.

He strongly stated how much more advantageous it would be for a prospective physician to study at Uppsala. The young student embraced the proposition with delight, his parents gave their sanction, and the Rector of Lund was applied to for an academic testimonial. This was issued on the 6th September by Arvid Moller, Professor of the Laws of Nature and of Morals, "to the well-practised and richly gifted Carl Linnæus, Smålander," with a certificate that "he had well conducted himself at the High School, and that he had with no less industry made himself beloved by those to whom he was known." At the same time his proclivity for the study of medicine is mentioned "in which he had enjoyed the fortunate guidance of Dr. Kilian Stobæus" (Döbeln's name not being included), henceforth intending public instruction, wherefore the Rector of the University "prays that Almighty God may favour his departure and future in all his undertakings," also "recommends his praiseworthy endeavours, as strongly as we can, to the patrons and guardians of Science." This testimonial he only received after his arrival at Uppsala. After he had obtained from his parents "100 dalers in silver [£7 10s.] once for all, as they could not afterwards further assist him," he set out on the 23rd August from Stenbrohult to Uppsala, which—such were the tardy means of communication—he reached on the 5th September, a journey of nearly four hundred English miles.

With this began a new and important stage in the life of Linnæus.

CHAPTER II

EARLY STUDENT YEARS AT UPPSALA (SEPTEMBER, 1728—DECEMBER, 1731)

UPON his arrival at Uppsala, Linnæus still lacked his Academic testimonial, and he was unable before he obtained it, to become inscribed in the University, whose pupil he now desired to become. How he spent the weeks during which he awaited the said certificate, nothing is recorded; probably he did as other new arrivals, spent at least the first few days in making himself at home in a strange place, where he had the intention of remaining for a number of years. At last, the Lund certificate came, and he applied to the then Rector, Professor John Hermansson, and wrote his name in the register in which newcomers to the University noted their contributions to the University library; thus his signature appears: "Carolus Linnæus, Smolandus." He paid as prescribed by law, as "others than the nobility," six dalers in copper [three shillings]. His name was also inscribed on the same day by the Rector himself in the University Matriculation List.

With this, Linnæus had completely exchanged Lund for Uppsala. The motives which his old teacher Rothman alleged caused him to take this step, were principally that Lund University was not serviceable for his aim in studying medicine, whilst at Uppsala, there were Professors of Medicine, the learned Roberg in medicine itself, and the great Rudbeck in botany; there was a splendid library and a handsome University garden, with many scholarships, royal or patronal, by which a clever but needy student of medicine could progress." From what follows, we shall find that Rothman by this advice made himself responsible for the belief, that he regarded the then existing University's circumstances as similar to those during his own life as student. During the twenty years which had passed since he left Uppsala, they had greatly changed, and certainly not for the better.

The professors in the medical faculty were two, Olof Rudbeck the younger, and Lars Roberg, both without doubt very distinguished, learned and experienced men. Since being appointed, they had between themselves so divided the duties which were then looked upon as belonging to that faculty, that the former undertook anatomy, botany, zoology and pharmacology, while the latter took up theoretic and practical medicine, surgery, physiology and chemistry. Rudbeck was the senior, both in age and service, and was then sixty-eight years old. In his strength, he had been both a zealous and distinguished teacher. During his travels in and outside his fatherland—especially in his journey to Lule Lapland undertaken in 1695—he had amassed extremely valuable botanic and zoological collections, with accurate reports. In collaboration with his father, Olof Rudbeck the elder, on the great botanic work "Campus Elysii," he had been both a zealous and skilful partner, and his father's intention was that he should, after his death, continue its publication. By this he would have obtained, without doubt, a very distinguished place for all time amongst the pre-Linnean botanists. Unfortunately in 1702 there occurred the great and destructive fire in Uppsala, which destroyed the greater part of his collections and notes, also most of the copies of the two volumes of "Campus Elysii," then printed, with the many thousand woodcuts prepared for its continuation. It is no wonder that through these disasters his manifest interest in natural history, especially botany, cooled considerably. On the other hand, his liking for another science blazed up, to which he had already felt himself drawn, namely philology.

When he began to issue his colossal report on the Lapland journey, he only included the small portion as far as the Dal-Elf river, in the first and only volume which appeared, since the mention of the ferryman at that place led him to the most daring and unrestrained philological and geographical diversions concerning "that cruel and fierce Karen [Charon]." After the Uppsala fire he devoted himself to his "Thesaurus linguarum Asiæ et Europæ harmonicus," a work "surpassing the 'Atlantica' in extent, genius and boldness," to quote the words of the elder Fries. In order to work upon it undisturbed, in 1721 he requested to have a period of release from lecturing. The Consistory urged amongst other things, that "as Dr. Rudbeck's late father, during his lifetime, had published a learned and laudable work to the entire nation's lustre and honour, of which the four volumes were lost in the fire, and now after his death were appreciated in foreign lands, especially in Denmark, some have already begun to censure and refute the same. Both father and son, in the Swedish transactions, possessed profound science, and as it may be possible to replace in part what in the forementioned work was lost, and partly to vindicate it, therefore his wish should be supported." To this the king replied that Professor Rudbeck, "in recognition of his long professorial career of thirty-one years, also in order to complete the work in hand, may dispense with his public lectures for a given period."

This was respect as well as solicitude for the elder Rudbeck's "Atlantica," but the continued research in his usual style which he practised in the medical instruction in the University was thrust on one side. The execution of Rudbeck's remitted lectures devolved on his son-in-law, Dr. Petrus Martin, who, however, died on the 27th June, 1727, wherefore Rudbeck during the succeeding years gave a few lectures, after which he was accorded an extension of his release from duty, that he might "apply all the rest of his time to completing that philological work, upon which he had been labouring many years, and that Medical Adjunct Nils Rosén, should receive the commission to enter upon the forenamed subject as deputy." The latter, at the time when Linnæus came to Uppsala, was travelling abroad, and, in his stead, Elias Preutz officiated, acting as deputy during a part of Linnæus's early studentship. Preutz said of himself, that in Rosén's duties he fulfilled his functions with all diligence to the satisfaction of the medical professors, but neither Rudbeck nor the medical students shared that view. How far this influenced Linnæus's career will be set out in the following narrative.

One of the two ordinary professors in the medical faculty was, as previously stated, Lars Roberg, a more than usually gifted man, but who, at the time when Linnæus arrived, was almost sixty-five years of age, consequently no longer possessing the strength and perseverance necessary for the discharge of his weighty and extensive duties in a satisfactory manner. Besides his peculiar temperament, he had a fiery genius, spoke with special politeness, was an entertaining companion, and full of quaint ideas, but with these brilliant powers he combined a curious method of living. In his old age he stretched still further his contempt for any other than a rich competency, which he loved more to possess than to enjoy. With his uncommon powers he seemed moreover to have greatly withdrawn himself from the prosecution of his duties, in that he gave private—and less valuable—lectures, from which he could expect economic advantage.

Neither of the medical professors can be acquitted from the charge of waning energy in teaching but on the other hand, it must be admitted that in a great measure extenuating circumstances may be pleaded on their behalf. One is—the advanced age of both; for one, scientific activity in other directions, and, for the other, a too volatile temperament; also, the wretched condition in which the institutions were. It must be conceded that few men would not become wearied, if (in spite of repeated complaints, reminders and petitions to remedy the worst evils, and without which suggested improvements professorial activities were paralysed) matters continued as before. How did conditions stand in these respects at Uppsala at this time?

In answering this question, we must first concern ourselves with the University hospital. The means assigned for its maintenance were so insufficient, that Roberg was obliged to let a room in it as a public house or beer-shop, but on account of the great disturbance and scandal caused thereby, the arrangement was forbidden by the Consistory after much discussion, without any substitute being voted for this economic advantage. Professor Roberg still lamented so great discredit to the hospital, and the Consistory readily agreed to help him, so far as it could; but until the finances of the hospital revived, the Consistory could not find a way. Shortly afterwards, Roberg handed in a document in which he stated "that inspection of the fireplaces and chimneys showed the outhouse to be so unusable, together with the chimney stack and cooking stoves, that the servants refused to stay." As the worthy professor was then at a loss as to what ought to be done, he solicited counsel and was advised that the Consistory should remove these difficulties as far as possible. Later on, he renewed his complaints, but nothing was done, except that it was acknowledged to be dangerous to retain people near such conditions that might also set on fire the largest and finest houses in the town, "which misfortune may God graciously avert."

This is not the place to set out all the reports given for many years in the Minutes of the Consistory; enough to show that no clinical teaching was available for medical aspirants; neither was it promised, being entirely excluded from the syllabus of 1728, and not reappearing during Roberg's remaining professorial career.

The want of the requisites might have been less felt if the medical students had opportunities under the professor's guidance of visiting patients in their own homes, a method of teaching which Roberg should in some measure have employed. There was nothing of this during Linnæus's student life, either because Roberg tired of it, or the patients were tired of him, by reason of his increasing covetousness, or his summary orders.

It was no better as regards the botanic garden, which, wrecked in the fire of 1702, had never since been even in a decent state. Certainly Rudbeck and his colleague Roberg did what they could; both of them possessing knowledge of, and interest in, botany, but attempts to improve matters ended unsuccessfully. Thereupon ensued Rudbeck's practically complete transition from botany to philology, as previously mentioned. When it concerned the gardener "that he need not have skill in dressing the garden as is usual, and soberly not to neglect his duty," it is not surprising that Linnæus soon after his arrival at Uppsala, lamented at the state of the garden, "which declines daily, so that now hardly 200 species are to be found in the whole place, and not more than 100 rarities." Soon after, Professor Roberg begged that the Consistory would think about the botanic garden, which was then in ruins; they admitted as usual that the business was urgent, but there it ended.

The conditions as regards anatomy were still more unsatisfactory, and the requirements of the time for a hall of anatomy had to be met by the younger Olof

Rudbeck allotting to it an outbuilding in the Gustavianum. In spite of this during the first ten years of the eighteenth century anatomical teaching had sunk to such insignificance as at the present day is inconceivable. A complaint in 1715 to the Consistory, resulted in a promise by Professor Roberg that an anatomic demonstration should be held. Three years later he issued, for students' use, his well-known text-book "Lijkrevnings-tavlor" [Plates for dissections]. That this was followed by autopsies, is not reported, but it is evident that just before Linnæus's arrival anatomic teaching under the Adjunct Martin's guidance had been prosecuted with no little ardour. It advanced so that the Professor of Law, Reftelius, lamented in the Consistory concerning anatomy, that it was prosecuted on the days and at the hours when public and private lectures were given, and that youths were thereby kept from their other exercises. The Consistory therefore decided that the anatomical demonstrations should be held only on certain days. Work in the anatomy school was carried on more diligently since Nils Rosén's return from his travels abroad. According to Dr. Wallin's account, Dr. Rosén, during the anatomy lectures, used lights every day in the school, so it was resolved that Rosén should be informed at once that lights should not be used, for fear of fire in the library. It must be taken as a special piece of bad luck that Linnæus came up to Uppsala immediately after Martin's death, and before Rosén came home from abroad, thus at a time when there was no instruction in anatomy nor in chemistry.

As regards the latter subject, it may be enough to state that the University did not possess a chemical laboratory. Chemical lectures were seldom given, but when they were, the students assembled at the University apothecary's, where a few simple chemical experiments were shown. It was still worse as regards zoology; not a trace of the collections belonging thereto were available, with the exception of Professor Roberg's small collection of rarities, amongst which were a speckled snake a quarter-ell [6 inches] in length, with two heads, also a dragon; whilst in the University library a few zoological objects were kept, but never utilized for instruction.

To sum up; it may be said with reason, that a worse provision for medical teaching could hardly exist. Linnæus said that he worked at medicine during the greatest barbarism at Uppsala. But in considering this, it can only awaken surprise and wonder, that almost without guidance, he developed under such conditions in a few years into a great man and pioneer, not only in natural history, but also in the domain of pure medicine. The copiousness with which the foregoing has been narrated regarding the disgraceful state at Uppsala, should find its explanation and excuse in the desire to set out the contrasts in this aspect. Without knowing how to obtain the slightest help from teachers such as he formerly encountered, Linnæus displays most plainly his uncommon endowments and energetic mind in their clearness and greatness against this dark background.

Naturally Linnæus did not neglect to make use of even the crumbs of instruction, which were available in the medical faculty. During his first year of study, these were restricted, so far as regards Rudbeck, who, however, in the autumn term of 1728 gave a few public lectures on Swedish birds, during which he reached no further than the domestic fowl and some of the smaller waders, put forward in a very unpretentious way. To Linnæus and probably to the other hearers, these lectures were of great interest, for, in the course of them, Rudbeck showed his drawings of birds drawn from life in their proper colours. In the following term, Professor Rudbeck gave three lectures before Easter on the raptorial birds, and these were attended by Linnæus, but when after Whitsuntide, Rudbeck gave two lectures in the University garden, Linnæus had left the town. These lectures were the last delivered by Rudbeck, and thus it happened that during the whole of his student-life, Linnæus never had the chance of hearing any botanic discourse, either public or private.

The instruction afforded by Professor Roberg during the same period was even more insignificant. Linnæus, in the autumn term of 1728, was among the auditors, and seems to have been dissatisfied when Roberg confined himself to allowing his audience to explain in Swedish, Langius's "Theses physiologicæ," himself making a few annotations, though privately, for during this term he gave no public lectures. The spring term following, a change was made, for before midsummer he delivered four public lectures on certain questions extracted from the "Problemata" of Aristotle, according to the principles of Des Cartes. Linnæus tried to get to five lectures on practical medicine, but was disappointed as before, so concluded that it would be better to buy the book. So it happened in later terms, for, according to careful notes by Linnæus, he failed during the rest of his University career to obtain better instruction, except a small amount he received from Adjunct Rosén in 1731. The reason seems that it was partly due to the absence of proper teaching, and partly that he was much engaged in more important work.

Linnæus's share of Roberg's private lectures brought him into closer connection with the professor, and it is apparent that he felt himself strongly drawn to him on the ground of Roberg's extensive reading, great ardour, and special methods. On his side, Roberg displayed to the young naturalist no small favour, permitted him access to his own library, and imparted counsel and exposition. With Rudbeck a little later, he came into closer relations. Both to him and also to Roberg he showed in the spring term in 1729, a catalogue of the rarer plants he had met with in Småland and Skåne.

Of even greater importance for Linnæus's early years at Uppsala, as regards his scientific development, was the acquaintance he made with a medical student, Petrus Artedi. Like Linnæus, he had been destined from his cradle to become a priest *avita premere vestigia* [to follow the ancestral traces], but even when at school, his taste for natural history was kindled, and also for alchemy, to which he devoted all his spare time. He left Hernösand's Gymnasium *summa cum laude* in 1724, betook himself to Uppsala so that he might study divinity as his relations wished, but soon turned to natural history. In spite of his father's exhortations to fly from the tempting sirens, he entered the medical faculty, and it was soon said of him, that he was the only medical student who then had a reputation for vivacity. It is therefore not surprising that Linnæus, after arrival at Uppsala, wanted to make his acquaintance, but Artedi had then gone home to Ångermanland, to bid farewell to his father, then seriously ill. After his father's death, Artedi came back to Uppsala, where he was soon sought out by Linnæus, who relates, "I found him pale, cast-down and tearful; the talk at once fell upon plants, minerals and animals. The ideas which he propounded were new to me, and the knowledge which he disclosed, astonished me." Though very different both in stature and temperament (Artedi being tall, deliberate and earnest, while Linnæus was small, active, hasty, quick-witted) they struck up a lasting friendship, which not even death could sever. It became a necessity for them to meet every day to share their common beloved objects, and to impart to each other what each had in the interval gathered or observed. An ardent disposition was the same in both; both desired to appropriate knowledge from the entire field of natural history, but each had with greater predilection devoted himself to certain branches. Artedi loved chemistry and particularly alchemy, as much as Linnæus loved plants. Artedi had some previous insight in botany, just as Linnæus had in chemistry, but as each recognized that he could not outstrip the other, he neglected the other's subjects. They both began at the same time on fishes and insects, but as Linnæus could not outvie Artedi, he left the subject entirely, just as Artedi left insects alone; Artedi studied amphibia, and Linnæus birds. There was between them a constant jealousy to keep secret what they discovered, but that gave way in about three days, before the temptation to boast to each other of their discoveries.

The young searchers gained essential help in the University library, which Linnæus soon found was excellent. Besides the array of books which they had at their disposition, there was a great botanic treasure preserved in the University, namely the learned Burser's precious plant-book, which, in a hundred and thirty large folio volumes had been bound by Chancellor Cojet and presented to the library. Linnæus did not neglect to solicit permission, nor had he to wait long, before he made use of so many books on botany, that he well-nigh surprised the staff of the library

An opportunity in another direction, in some measure completing what Uppsala University could deliver in instruction, divulged itself at this time and was embraced eagerly by Linnæus. Partly to see Stockholm, and partly to attend some members of the medical college, he travelled on the 14th January, 1729, to Stockholm, and there gained intelligence, that at the end of the same month and beginning of the following month, there would be an anatomical demonstration on the body of a woman who had been hanged. This was an opportunity for inquisitive and curious persons, also an event of no small importance, and that the greatest possible use should be made of it, the

Medical College was induced to have it done in the best way. Special meetings were held that they should agree how the anatomic event should be apportioned and who should undertake to demonstrate the various parts, and as it was known that judgment had been pronounced on the offender, a petition was sent up that the execution should take place after the New Year so that the dissection could happen conveniently. This was granted, and in a new meeting it was decided that as regards tickets, all Master-Veterinary Surgeons should get free tickets under the great seal of the College to the number of eighteen; that all Doctors should have free seats, but for all other spectators or hearers there should be an entrance fee each time of sixteen dalers in copper coinage [eight shillings].

This was naturally a great occasion for Linnæus. He certainly returned on the 19th January to Uppsala, but on the 29th he came back to Stockholm and attended the six lectures and demonstrations in the anatomic room in Södra Malms Townhall, by the chief surgeons. He noted how each acquitted himself, as " learnedly, elegantly, most learnedly, excellently," once only reporting " moderately." Encouraged by the increasing interest in anatomy, after he had returned on the 23rd February to Uppsala, he procured admission to the Anatomic Theatre. The building was well contrived with seven entrances, but it lacked a teacher.

Now began a period full of trouble for Linnæus. All the money he had from his parents on his departure for Uppsala, was spent on his journey from Småland, university fees and maintenance, two journeys to Stockholm, with more than a month's living there, entrance fees to the dissections, etc., etc. Certainly he had from the beginning cherished the hope of obtaining a scholarship, which was quickly realized inasmuch as on the 16th December, 1728, he received a Royal Medical Scholarship of the lowest class, but the help derived therefrom only amounted to ten dalers in silver [fifteen shillings] in each term. It is probable that the Stockholm visits exhausted his funds. The coming term therefore "was very wretched for him," and he began really to suffer want; he had to run into debt for food, and to go almost barefoot, as he could not sole his shoes, but had to substitute paper which he laid in his shoes. The prosperous, childless Roberg could easily have succoured him, but his friendship was no longer than his pupil's purse; it stopped as soon as Carl's money was gone. No employment by which poor youths used to push themselves with the academicians, could be entrusted to Linnæus as a medical student, for at this time it was no honour to study medicine. No wonder therefore, that he began to think regretfully how different it had been at Lund; the prodigal son would have gone willingly to his Stobæus again, but he had no money in his purse for so long a journey of nearly five hundred English miles. Moreover he feared that Dr. Stobæus would be thoroughly displeased when he again saw a youth for whom he had done so much, and who had left him so ungratefully.

From these economic troubles, which doubtless weighed more upon his conscience than on that of many of his comrades, because he was in a high degree always afraid of debt, he was freed before the end of the term through the acquaintance, which he had the good fortune to make, of the old and venerable Dr. Olof Celsius, an acquaintance which was advantageous in more than one respect for Linnæus, and without which, Sweden might never have reckoned him as amongst its great men. By means of the help he received from Celsius, and also the income which he derived from other sources, Linnæus's economic position hereafter so improved, that he cannot be said during his remaining student life, to have found himself pressed by difficulties as to subsistence, although occasionally finding himself in temporary pecuniary embarrassment. It may be truly said that Linnæus was a favoured son of fortune, in view of the fact that wherever he went, he found hearty friends and generous patrons, who cleared his path of the severest economic shocks. The tales related by some of his biographers that his penury at Uppsala lasted long, are inaccurate, for his distress lasted only for a few months. Dean Olof Celsius, D.D., the elder, who now became a paternal friend and benefactor, was at that time one of the most eminent and esteemed professors of the University. Besides a solid and extensive knowledge in theologic and philosophic sciences, he possessed a great acquaintance with natural history especially comprising a lively interest in botany. In his garden he cultivated many rare plants, which he had obtained, not only from other gardens in the country, but also from learned foreigners with whom he was in correspondence. He took a special interest in investigating the flora of the province of Uppland, in which Uppsala is situated, and in 1729 he informed the Royal Society of Science in that town of his botanic work, which the Society promised to publish in their Transactions. At the period in question he seems to have stayed in Stockholm as a member of the Ecclesiastic Deputation which in 1727, began its work in revising Church law, and bringing it in consonance with the government of the "Frihetstiden" [Era of Liberty] and partly too, with the new common law which was also in progress. This stay in Stockholm explains how it was that Linnæus until now, had failed to know him even by sight.

It was during a sojourn at Uppsala in the spring of 1729, more closely stated as 8th April, Old Style, that Celsius, led by his love of flowers, paid a visit to the botanic garden, which, although dilapidated, could still gratify his eyes and mind with some of the firstlings of spring. There he became aware of an unknown student sitting and describing certain plants.

Astonished at that unaccustomed sight, he entered into conversation with him, and asked Linnæus what he was writing, if he knew plants, where he came from, and how long he had been there; enquired the names of many plants, to which Linnæus replied with the Tournefortian nomenclature; he finally asked how many plants he had dried, the reply being that he had over six hundred native plants. From this, Celsius discovered that the young man possessed an insight into botany, which he had not suspected in any student, therefore he bade the young man to follow him home, and when he came to his house, he went in, by which Linnæus knew who his interlocutor was. Linnæus was dispatched to bring his herbarium, whereby Dr. Celsius became even more convinced of Linnæus's acquirements in botany. With this meeting the acquaintance was begun, which developed, in mutual use and satisfaction, so that Linnæus had reason "to thank God who had so graciously given him another Stobæus in Uppsala."

The first advantage which Linnæus gained by this acquaintanceship was, and in consideration of his small resources it was for him a special benefit—by his being treated almost as a son in Celsius's house. In writing to Stobæus he said: "I have plenty to do here, which has been doubled the last week, as Celsius has married off his daughter." Celsius could not fail to see the poverty of Linnæus, who remarks, "that Dr. Celsius was so good that he invited me (16th July, 1729) to take my meals with him for nothing, which happened from Midsummer Day till Michaelmas—when he journeyed to Stockholm—and that twice a day, gratis. Later he gave me a room in his house without rent," and this benefit he enjoyed till the beginning of the following year. Linnæus's economic position was improved too, by receiving on the 20th June, a Royal Scholarship in the medical faculty in the second class, namely 20 dalers in silver [30 shillings] each half-year. Besides this, in the autumn term, in place of the Adjunct Preutz, the fees came to Linnæus, who then lectured in botany, physiology and chemistry, so that he was able to buy shoes and to repay the debts which he previously incurred for sustenance. At the close of the academic year, 16th December, he was promoted to the first class of the Royal Scholarship with 30 dalers in silver [£2 5s.], thanks to Professor Rudbeck's urging, and in spite of Professor Roberg's "intrigues," though nothing of this appears in the Consistory's minutes. This kindness by Rudbeck for a student almost unknown to him may be ascribed to this circumstance, that amongst those who attended Carl's lectures, was a son of the Professor, who each evening related to his father, what he had learned of the acting professor (Docent).

During this year through Professor Roberg's support, an attempt was made to provide him with a fixed appointment. The minutes of the Consistory state: "Student Linnæus's application to become the gardener of the botanic garden in place of the deceased M. Winge was considered." In this application it was stated that former gardeners had little skill in reading or writing, and that through the troubles of cultivation and those of their households, the public garden was neglected, as is now visible to all. That no German garden labourer from Stockholm or cabbage-planter from abroad, should be appointed to this place, he offered himself to undertake the same and declared himself willing to find and engage a labourer under him, he himself helping with the day's work, until the garden should be re-arranged. Moreover he engaged to draw up a catalogue of the plants for printing, provided that he only should have the right to this, and the printer should not, without his knowledge, print and sell any copies.

This matter was deferred till Professor Rudbeck, who was then at Surbrunn taking the waters, should come back. The Professor meanwhile in the autumn after a stay in Stockholm, probably after his cure, engaged an excellent, virtuous and intelligent man, namely gardener Christopher Herman, who had approved himself during twenty-three years' service in good situations. Hereto may be added Rudbeck's remark, that he had a better opinion about Linnæus than that he should remain in the situation of a gardener, is certainly not improbable, but not to be found in the minutes of the Consistory.

But it was not only in an economic aspect that Linnæus profited by his intimacy with Celsius. It was an invaluable advantage that the latter offered free use of his library, which was rich in botanic books. Presumably he had also the opportunity of making use of Celsius's extensive botanic notes, which testify to a wide reading in the whole of the literature belonging to that subject. These notes form four volumes, preserved in the manuscript department of the University library.

It was natural that Linnæus should be specially eager to know the productions of the three kingdoms of nature which occurred in the new tract of country to which he had been transferred. In the castle garden laid out by Rudbeck, though now in decay, for the first time he saw some less common plants, and that to the said garden he paid not merely a hasty visit, is testified by the plan of it, which is to be found in his manuscript "Hortus Uplandicus." The gardens also belonging to Olof Celsius and the apothecary Lambert, yielded welcome contributions to his herbarium. Besides the excursions taken in the nearer neighbourhood of Uppsala, it is evident from his manuscript "Spolia Botanica," that more distant parts were visited.

The first long excursion was at Whitsuntide to the tract around Dannemora, twenty-eight miles north of Uppsala and famous for its iron, the best in Sweden, when he took with him his student comrade, Johannes

Humble, in order to teach him botany. He had good reason to be gratified with the result, as he discovered there many plants, which he had not before seen living, and many of which had not been known previously in Uppland. These discoveries gave great joy to Celsius. Of cultivated plants, the garden and orangery at Leufsta, which were also visited, showed a few rarities. In the animal world, a few things of interest were observed, particularly the shrew mice down in the Dannemora mines, which were as gentle and tame as dogs, coming and feeding out of people's hands. They were held as sacred, and no one harmed them. A harvest also was gathered of rare minerals. He wrote to Stobæus: "I was a hundred yards down in the mines, and searched for stones, of which I had so many, I could easily have brought away a portmanteau full." He noted also the iron workings and the great depth, with ores and instruments. Here, too, he saw as the greatest rarity, the fire-machine which Mårten Triewald had introduced, driven by water and air pressure, the only one at work in Sweden. A short distance farther he saw Ästerby ironworks with tilt hammers; the workmen being only in their shirts, with socks and slippers on their feet.

Excursions were, however, principally made in the company of Celsius. Such a trip took place on the 24th June to Börje parish, about seven English miles from Uppsala. Celsius betook himself thither for the single purpose of "showing 'Sceptrum Carolinum' [*Pedicularis*] to me, but it was not then in flower; we went over the stoniest place without doubt in Sweden." On the 7th September the same locality was again visited, and the desired plant then had fully ripe capsules.

Another, and in a botanic sense, particularly successful jaunt which the two made took place in June and July when they went to the islands off the coast, to find out what plants grew there, gathering twenty-six not previously recorded. That Celsius was particularly pleased with the result is shown by his suggesting to the Royal Society of Science the repayment to Linnæus of the expenses incurred, which was done, the payment being actually in excess of the expenditure.

It was after this specially pleasantly spent summer, when Carl's thoughts were devoted to plants, that he began the duties of the autumn term. These seem principally to be the determining and describing of the summer's harvest, collecting insects from among tree-mosses, with instructing to other students with good economic results. "In November," he wrote to Stobæus, "I gave lectures in botany and had many noblemen and barons besides others among my audience. I received generally a ducat from each" [nine shillings and twopence].

But added to this—and it was not the least important occupation—he studied diligently for his own advancement, with assiduous use of the books of the University, those of Celsius, and his own little store. It was not a thoughtless, uncritical storing in his memory of what he found in the old authors' ponderous folios, and in the insignificant, long-forgotten pamphlets, but on the contrary he began to distinguish for himself the different characteristics, to sort his collections critically, to notice results thus obtained. The description which he in old days applied to himself that "he wrote briefly and strongly all at which he laboured; regarding himself entirely as a born methodizer," may even be applied to his earliest youthful writings. It was during that period, when he first began to write many of the works which he afterwards elaborated and amplified, formed the skeleton of the volumes which a few years later were issued in quick succession, and even made for the previously unknown young man from the far north, a great name in the annals of natural history. Who, reading through these first early attempts, can refrain from wondering at the widely extended knowledge of literature, clear exposition, and able conviction of the correctness of his views, on the part of the twenty-two years old student?

Amongst these youthful writings there is one which deserves to be spoken of somewhat at length, especially as it had a considerable influence on the career and scientific development of Linnæus. To understand this rightly we must first remember that amongst the new principles which at that period began to be attacked, was that of the sexuality of plants. What Camerarius, Ray, Grew, Bradley and others had written upon the subject, was unknown to the young student, but nevertheless he had read in "Actis Lipsiensibus," a review of Vaillant's tract on the sexes of plants, which especially pleased him. This was the address with which Sebastien Vaillant on the 10th June, 1717, at the Jardin du Roy in Paris, began his public lectures, and it was printed in the following year both in French and Latin. Herein the sexuality of plants was set forth as an indisputable fact. Linnæus was warmly attracted by these new views and therefore he began to examine flowers for stamens and pistils, soon finding that they were not less different than the petals, and were the essential parts of the flower.

The adoption might have been delayed a little before presenting his views and observations on this, but at the end of 1729, an academic treatise came out with the title "Γάμος φυτῶν. sive nuptiæ arborum . præside Georgio Wallin . . submittit . . . Petrus Ugla." [The marriage of plants . . . under the presidency of G. Wallin, submitted by Petrus Ugla.]

Linnæus, who had been plodding away on the subject, and had no chance of opposing this thesis, felt himself called upon to enter the lists and therefore wrote some sheets on the true connection, according to botanic fashion. He delivered this little tract as a New Year's gift to his benefactor Celsius, and in his preface he says, " It is an old custom to awaken one's eminent patrons on New Year's Day with verses and good wishes, and I also find myself obliged to do so. I would gladly write in verse, but must bewail that it is true as the old proverb has it, 'Poets are born not made.' I was not born a poet, but a botanist instead, so I offer the fruit of the little harvest which God has vouchsafed me. In these few pages is handled the great analogy which is found between plants and animals, in their increase in like measure according to their kind, and what I have here simply written, I pray may be favourably received."

It is not without interest to note that though written superficially, these pages show the rapid development of the young student, his views which most people at that time would have regarded as bordering upon insanity being frankly put forward. Its full title is " Caroli Linnæi . . Præludia sponsaliorum plantarum in quibus Physiologia earum explicatur. Sexus demonstratur, modus generationis detegitur, nec non summa plantarum cum animalibus analogia concluditur. [Preliminaries on the marriage of plants in which the physiology of them is explained, sex shown, method of generation disclosed, and the true analogy of plants with animals, concluded.] After first setting forth the enlivening influence of the sun in spring, on all bodies which have been dormant during the cold winter, and that love animates plants themselves, the author shows how the old botanists seemed groping through thick darkness toward sex amongst plants, but for the most part so unsuccessfully, that one must shudder. Those who wish to see their points, may refer to the disputation, which is a compendium of all that the ancients said about it.

On the other hand, later botanists found many analogies between peculiarities of animal and plant life, how they suffer the same kinds of sicknesses, how plants, like many animals, are dormant in winter, but with returning warmer seasons again awaken to life; that both plants and animals are barren when young, most fertile in middle age, but when old, waste away; that as Malpighi and Grew showed, plants have vessels, fibres, and numberless other parts, just like animals. From this the conclusion is drawn, that in plants also, organs of generation are found, and this Vaillant set himself to work out. That these organs are to be looked for in the flower follows from this, that no fruit is produced without previous flowering. The parts of the flower such as calyx or petals play no such rôle, for many plants are destitute of them, but yet are fruitful. What the petals specially accomplish is that they contribute nothing to generation, but only serve as bridal beds, so splendidly devised by the great Creator, but furnished with such noble wrappings and perfumed with so many sweet odours, that the Bridegroom and his Bride may there celebrate their nuptials, with due ceremony. If one considers the stamens and pistils, one finds that most flowers possess them in the same flower, while certain plants have two distinct kinds on the same stalk; those being sterile which have only stamens, while those with pistils set fruit. Others again have stamens and pistils on different individuals. Tournefort noted certain kinds with separate sexes, and Linnæus now added a number of others, whose descriptions he found in various authors. After a short statement of Vaillant's views, and an attempt to show why most flowers are hermaphrodite, he sets forth Morland's erroneous opinion which was prevalent, though he could not determine how fertilization took place, but that it really occurs is evident in so many cases; in one group of plants with long styles and short stamens, the styles bend down when the stamens open and receive the pollen, after which they rise up again to their former position; that fertilization fails, or is partial only, if rain washes away the pollen, as with rye or fruit-trees; that in plants with both sexes on the same individual, such as maize or *Typha*, the male flowers are uppermost so that the pollen may fall upon the pistils; that if the male flowers are removed, no fructification takes place; that if, in hermaphrodite flowers, the anthers are removed, true seed cannot develop, but they are infertile and never grow up, though sown in the best soil. Finally is related the idea that for the formation of seed, there is the analogy between these seeds, and the ova of animals.

Such in short is the chief contents of this little but noteworthy memoir, which at once evoked no small attention. Copies of it were made, so that it rapidly circulated amongst the students in manuscript. Presumably through the young medical student J. O. Rudbeck, it came under Professor Rudbeck's eyes, who so much liked this tract, that he went himself to Dr. Celsius, merely to ask who the student was, who had shown so much knowledge in botany. A fair written copy with scarcely varied title, was communicated to the Royal Society of Science at its meeting on the 23rd April, 1730, presumably by Rudbeck. In the minutes drawn up by Anders Celsius, it appears that the Society gladly received the author's ripe studies and experience in botany, and desired that the said dissertation might be printed and published.

Another result of this paper was that it surpassed the boldest expectations. Professor Rudbeck who was busy on his great "Lexicon harmonicum" and for its completion had received from the Chancellor many helpers subsidized by Royal stipends, wanted some one to undertake instruction in botany, that is, to hold demonstrations at the end of each spring term, concerning the plants in the botanic garden. At first he thought of the acting Adjunct, E. Preutz, who expressed himself quite willing, but Professor Rudbeck took him into the garden and noted that Preutz was not at home there. So he sent for Carl Linnæus, led him to the garden, examined him long and thoroughly, and asked him if he would undertake the duties. Though astonished at the request to a student of little more than two years standing, to lecture publicly before so great a university, he yet assented with respect, provided the Professor ventured to entrust him with the task. Professor Roberg stoutly opposed this, that a youth should be let loose on such a confidential commission, but it happened that no one else was available. It can easily be imagined that among the students it would awaken great attention, and therefore when on the 4th May Linnæus began his demonstrations, and during the whole of the following period from Easter till Midsummer the garden was filled with auditors: he wrote to Stobæus that he had almost always from two hundred to four hundred hearers, whilst the professors seldom had more than eighty, and he hoped that he would always acquit himself with credit.

What was the character of these first lectures of the young fellow? That they completely met the then wants and wishes of the students is evident, but how did he comport himself when compared with the standard of later times? It cannot be denied that according to our ideas, they must be regarded as tolerably lean and scantily scientific; they were restricted practically to the giving of the names of the plants in question and their so-called virtues, especially in medicine, some etymological remarks with a few anecdotes from classic authors. But the lectures given by the learned Rudbeck were just of the same sort, and Linnæus at the end of his first term, justified expectations. Moreover, it is very likely that youthful enthusiasm gave a fresh and brisk grace, which was appreciated by his audience.

The uncontested progress as teacher which he thus gained prepared him soon for another advantage for on the 13th June, Jubilee Day, he removed from Dr. Celsius's house to that of Professor Rudbeck, who was so kind as to permit him to act as tutor to his three youngest boys, with a salary of 50 silver dalers per annum [£3 15s.]. He was also to coach a fourth son, Johan Olof, in medicine in his leisure hours, receiving an extra 40 dalers [£3] and stipend of 60 dalers [£4 10s.], altogether amounting to £11 5s., so that Linnæus had now, as he confessed, a sufficient income through God's favour. In passing, it may be stated, that Rudbeck had been thrice married and had twenty-four children.

As regards the emoluments just mentioned, it was due to Rudbeck's appeal to the Great Consistory that a double Royal stipend should be awarded to Carl Linnæus, as one of the most spirited and promising of the young men folk, and that he should act as Docent or Assistant Professor, giving as a reason, that though in poverty he had acted with such energy and perseverance, especially in botany, that he should be encouraged by special favours. This was approved by some of the professors, who, recognizing the merits of the case, urged the Chancellor to entertain the idea. Soon afterwards the consent was given, so that Linnæus, during the period of his teaching, was regarded as the most eminent among the medical pupils.

The income now enumerated was not the only amount Linnæus possessed at that time, for he had also earned some by the instruction of students by going two or three times each week to give private lectures in the field to his colleagues. What these were during the summer term of 1730, appear from his own notes still extant, which show that not only medical students, but others also availed themselves of his guidance. What they paid is also noted; some did so in money, others, the majority, gave books, such as Caspar Bauhin's "Pinax"—the actual copy being in the library of the Linnean Society with the autograph of the new owner and the date when acquired; others again gave useful articles, as hats, stockings, hair-purses, gloves, etc.

Of still greater economic gain was, that Linnæus by living in Rudbeck's house, had daily access to the excellent library there, and also opportunities for counsel and explanations from his learned and benevolent principal. With ardent zeal he flung himself into his botanic and zoologic studies. Doubtless his duties as teacher came into conflict with his zeal for research, so that he experienced how precious time became when it must be applied to others, in order that the question of bodily requirements should be met. All went well however, the days were given to work with his pupils, and the nights to working out his new system and reform, which he was beginning in botany.

To give a detailed account of all the botanic essays which at this and succeeding periods were written is hardly appropriate to our task. Enough to say that he now began his "Bibliotheca botanica," "Classes plantarum," "Critica botanica" and "Genera plantarum," every moment being thus spent so long as he was at Uppsala. Zoology also was not overlooked; the opportunity of steadily going through Rudbeck's incomparably beautiful drawings of Swedish birds, gave him grounds for drawing up a new "Methodus Avium sueticarum" as well as "Insecta Uplandica methodice digesta," as objects for his labours. There is doubtless much truth in the statement made in his old age, that he had placed before his mind certain objectives before he was twenty-three years old, and had executed all before he came back to Sweden in 1738.

Linnæus's early effort, "Sponsalia plantarum," has already been mentioned. Partly due to his innate perception, partly perhaps to his great observation (called into being by his rapid work), his mental disposition towards flowers was strengthened, especially with regard to their reproductive parts which demanded an accurate investigation. Gradually he was led to think about the subject of systematic arrangement—something which in his earliest efforts had hardly come into the question. He then began to doubt whether Tournefort's method was sufficient, so that he set himself to describing accurately all flowers, referring them to new classes, and reforming their names and genera in a new fashion. This work occupied all his time during the summer. These systematic speculations were combined with another competitive subject, to which also he earnestly devoted himself. In his demonstrations in the botanic garden, he had been asked to compile a catalogue of the plants, so that his hearers could avail themselves of it when plants were named, and thus spare themselves the inconvenience of copying all the names quickly in the open air of the garden, which might result in mistakes in names or citations. On the ground of this modest request, Linnæus quickly drew up a "Hortus Uplandicus," which he revised and enlarged time after time after he had visited many gardens. At the beginning, he made use of Tournefort's system, but as early as 29th July, 1730, the plants were taken in "methodo propria in classes distributæ" [arranged in classes according to his own plan].

Concerning this attempt, his German pupil J. C. D. Schreber remarks, "that it was only a rough sketch consisting of twenty-one classes, and as to names, it was very different from his later efforts." This judgment is hardly sound, for even a hasty glance shows that the guiding principles upon which the classification rested, are the same as those printed later. Small improvements and changes of names meanwhile were made in the various versions which succeeded, leading to his "Systema Naturæ" in 1735. Still more so is the case with his "Adonis Uplandicus" which Professor Rudbeck in the author's presence presented to the Royal Society of Science on the 11th May, 1731. This paper was greeted by the members with unstinted applause. Concerning this, Linnæus wrote to Stobæus, "The Society at first thought I was mad, but when I explained my meaning, they ceased laughing and promised to promote my design." Thus at twenty-four years of age, Linnæus had completed his sexual system, and by a lucky chance, solved the problem, which hitherto all other botanists had failed to solve, namely the promulgation of a clear and easy scheme, by which the many productions of Nature could be arranged and found again. This simplicity formed its strength; at once it thrust aside all the older, perplexing systems, and to this day it is recognized as occupying the first place amongst artificial systems, however many there may be.

Still another production of his authorship at this time may be mentioned, less for its scientific weight and importance, than for the circumstances which were connected with its origin. It has already been related how Celsius and Linnæus took two journeys in 1729 to Börje to see a single plant, *Sceptrum Carolinum.* In September, 1730, Linnæus travelled to the same place and for the same purpose, this time accompanied by Johan Olof Rudbeck. It is evident that this plant was regarded with special interest, which, from its stately appearance, was well deserved. Discovered by the younger Rudbeck as a boy, who afterwards met with it by the Lule river during his Lapland journey in 1695, he had dedicated it to King Carl XII in verses to be found in his "Nora Samolad," with the name of "Carl's spira" [Charles's Sceptre] not long after the king's victory at Narva.

When the young J. O. Rudbeck had to put forward an academic disputation, this plant was chosen as the subject; till then it lacked scientific description, and under Rudbeck's presidency, this thesis would be brought up as a customary holiday task. J. F Bergman and Professor Roberg had produced verses on the plant, but this stately issue took on a comic colour when we read on the back of the title page of the manuscript entry, "I wrote this dissertation in one day for thirty dalers in copper (fifteen shillings), therefore another has the credit." It may be observed that when the manuscript and the printed treatise are compared, differences in the language are found, for it was first written in Swedish, but its later form was in Latin throughout. The question is valid, how far the said plant is a special genus or if it should be referred to the old genus *Pedicularis*. One suspects Professor Rudbeck's paternal hand, which here and there altered the young student's "day's labour." The whole of this little episode speaks eloquently of the intimate relation which Linnæus enjoyed in the Rudbeckian home.

That Linnæus throve there is apparent. A tempting invitation being made to him in October, 1731, to undertake tuition the following Easter at Archiater Nordenheim's in Stockholm, he accepted the proposal, but afterwards revoked it. He found himself specially drawn to his kindly host, and tried to show his gratitude to his benefactor. Not long after he had removed to the Rudbeckian household, which took place on St. Olof's Day, 29th July, 1730, he contributed to the celebration of the event by giving to O. Rudbeck, as a gift, one of the previously mentioned editions of "Hortus Uplandicus" adorned by some of his own verses. The next year, he presented an "Ödmjukt offer" [an humble offering] consisting of a description of the new genus *Rudbeckia*. He had already in his "Spolia botanica" given this name to another genus, namely that which now bears the name *Linnæa;* this was actually written so at first, then partially erased, and *Rudbeckia* substituted, but the original name is plainly to be discerned beneath it; further, in "Methodus Avium" there is a bird genus named *Rudbeckiana*, which led to the passage "Though all the world be silent, the plant *Rudbeckia* and the birds *Rudbeckianæ* sing thy noble name."

The success which had hitherto attended him became clouded in some measure by the return of Nils Rosén from his foreign travels. This man, whom

Uppsala University honoured and still honours as one of its most eminent physicians, and who is rightly termed, "the medical father in our country," was, like Linnæus, a pupil of Stobæus, who entertained of him, though young, the highest hopes. When Rudbeck obtained release from his lectures (p. 28) the medical faculty resolved with the approval of its most eminent physician, Casten Rönnow, to move the Chancellor to appoint Rosén as Adjunct, that is Assistant to the Professor, so soon as he had completed his intended journey abroad, and undergone "promotion" [graduated]. To this the Chancellor willingly agreed, and assured the said Rosén that he should enter upon his duties on his return home, and then take up his salary. He came back to Uppsala on the 4th March, 1731, and on the 16th of the same month began to lecture; soon after, it was decided that his salary should be reckoned as from the 24th December of the previous year.

Whether Rosén then wished actually to function as Adjunct, that is, to take care of the instruction (about which an agreement was made between him and the medical faculty), is doubtful, but it can hardly be characterized as improper. On the contrary he might have been justly liable to censure, had he desired to withdraw from his duties, and so disappoint the trust which was shown him by his appointment as teacher in the University. As regards the parties concerned, in this case Rudbeck and Roberg, they could not entertain the idea that a young student should displace one who for eminent knowledge and skill well deserved the post. This feeling is so entirely natural, that one would be astonished by its absence. All the charges and insults which were directed against the Uppsala professors and Rosén by thoughtless memorialists, and afterwards zealously repeated, are therefore, to use a mild expression, entirely wide of the mark.

Into the bargain too, Rosén's return at first caused

no change in Linnæus's position as a teacher. During the spring term of 1731, the former confined himself to instruction in anatomy, upon which subject he, as no other practical man was at Uppsala called upon, was much occupied. It is more than likely that he, in conformity with his duty, offered himself, or expressed a wish, in 1731, to hold each spring term botanic demonstrations, but the fact is, that the botanic instruction in 1731 was carried out by Linnæus. Presumably this was arranged by Rudbeck before Rosén's return from abroad.

That year as it happened, since Linnæus for the first time fulfilled his commission, a great change in his affairs had taken place. In the botanic garden, which he had rearranged according to his own system, and had enriched by importing rare plants from other gardens and the country, he was no longer a promising youth, but a practical scientific man, full of new ideas, and eager to put forth the same to numerous and inquisitive hearers. The exhibition, during a part of the season of blooming plants and reports of their properties, was no longer sufficient. Even the demonstrations themselves had to be conducted in a different and more scientific and comprehensive way from his well thought-out "Schema," which he followed in demonstrations of plants in the botanic garden. In connection with this came the introduction of what had never hitherto been taught at Uppsala, namely, his botanic theory, which he, later on, enlarged and reconstructed, publishing it in 1751 as "Philosophia Botanica." When on the 3rd May he began his lectures, he declared that he proposed to explain his theory on two days in each week, and on two other days to give demonstrations on the garden plants. It cannot be gainsaid that though this was an innovation in botanic instruction there was little or no opposition, and therefore Uppsala University must be congratulated on possessing the service of such a teacher in botany. But would it have benefited him had he remained as he was? It will be shown that that which happened, was the best both for him and for science.

Already when in the spring the lecture courses were published for the period from midsummer of that year to the same time next year, it was decided that Rudbeck, through Rosén as his deputy, should give botanic lectures in the botanic garden at four o'clock in the afternoon. That Linnæus by this felt mortified or depressed appears from his letters and notes at that time.

Immediately after the conclusion of his lectures, whilst Rudbeck was drinking the waters at a watering place, he arranged to take a journey to Stockholm in Rudbeck's post yacht, a means of travelling much used and enjoyed by the commonalty, in company with twenty students, who were now going down after the end of term. By much labour and exertion, they succeeded, during a calm, in reaching a small island named Kofsö, on Midsummer Day at two a.m. "The sailors and my companions," he relates, "went off to refresh themselves, and sleep, but I landed, and went to and fro, from side to side, leaving only a yard from the former path, almost as a man ploughs. I had hardly completed my course and plucked a leaf of each plant, before the captain ordered all on board, as a new breeze had sprung up, but I venture to say that hardly a single plant had escaped my hand, except mosses." A catalogue of eighty species is still extant, and bears the title "Flora Kofsöensis."

On his reaching Stockholm, where Linnæus had free lodging with Apothecary Warmholtz, he employed his time chiefly in visiting the gardens in the capital and its neighbourhood. He also made an excursion to Wiksberg, a noted watering place; the result being his "Hortus" and "Adonis Uplandicus." Moreover, during this stay in Stockholm, he did not omit to become acquainted with the medical cabinet of the Medical College, nor to learn the rudiments of pharmacy in Warmholtz's laboratory. He also obtained a large number of shells of different kinds for his own collection.

By the 24th July he was back at Uppsala. How he busied himself during the rest of this year there is not much to relate, but his time was certainly much taken up by the tuition of the young Rudbecks (p. 49), private lectures, and writings for scientific journals. Hereto was observed the promise made to Stobæus that when he composed botanic works, on which he worked with burning earnestness, no contract should be made with Æsculapius, " if I am an honest fellow," but he made use of Rosén's private teaching for a glance into therapeutics. This he found less attractive than natural history, even although he lacked opportunity to visit sick-rooms and learn from patients' own mouths as formerly. At Lund, " My natural history," he says, " has now no such free entry, though sometimes it sneaks in."

During this period he began also to arrange a portion of his written discourses ready for printing. At first he thought of beginning with " Hortus Uplandicus " on which he laboured till the end of July, 1731, in Stockholm, and would have taken from a bookseller, who undertook to publish it, only a hundred copies. " My other lucubrations," he says, " I thought of selling to Germans." This attempt, like all the others which he made during succeeding years to find a publisher, met with no other result than disappointed hopes. " First, he sent ' Hortus Uplandicus,' which passed through many hands, but nobody thought it worth printing. Upon this he sent out his ' Nuptiæ Plantarum '; printing was promised, but where this manuscript now is, not even the author knows. Now he received the promise to print the ' Fundamenta Botanica,' but the medical faculty at Greifswald had condemned it as ' food for cockroaches.' Luckless offspring! And what will happen when I appear as the antagonist of all botanists in the whole world? "

Before we close this presentation of the life and activities of Linnæus at this stage, an addition must be made of his behaviour to what then was regarded as the flower of university life, namely public disputations. The first time he appeared in such, was on the 26th March, 1729, when he opposed one of Professor Roberg's theses on the circulation of the blood. Another time he did his duty as opponent on the 23rd June, 1730, when Daniel Bonge, under Roberg's presidency, set out his treatise on the nature of salmon and their capture in Österbotten. According to his own statement he was respondent on the 27th April, 1731, in a public disputation with holders of Royal stipends, again under Roberg's presidency, and on the 25th of May in the same year, he was tempted to appear as an extra opponent under the presidency of Anders Grönvall, Professor of practical philosophy, to the graduation thesis of Thomas Jerlin, "On the ungrateful cuckoo," and this in such style, "that they never forgot it, for both sides showed such vehemence and passion." Finally it may be added, though it belongs to the next spring term, that on the 17th April, 1732, "Dr. Roberg gave out as materials for disputation, 'De Libella of Leetström' [on the dragonfly] and 'De Cornus herba' of Linnæus, but, as the latter remarks, and as Dr. Celsius informed me, Dillenius having founded the genus, I would not appear, although the thesis was half-printed, but was revoked."

To sum up the first year of Linnæus's residence at Uppsala, we find that his teachers put him forward for stipends, and he was mentioned in the Consistory and supported by most of the professors as deserving of recognition, and superior in attainments to his fellows.

But Christmas of 1731 was now approaching and in consequence of a quick change in his relations to Rudbeck's family, he had much to think about. Quite unexpectedly an event occurred, which made him resolve to give up his tutorship. In Rudbeck's house a "minx G. B" [Greta Benzelia] was an inmate, who so managed matters as to make Rudbeck's wife take a dislike to Linnæus for not keeping his pupils neat. For this reason he considered himself obliged to provide himself with another situation." In another place he says, that the person who made the mischief was the unfaithful wife of the librarian Norrelius, whose conduct gave rise to much scandal: separated from her husband she went to Copenhagen, lived a loose life and afterwards died there.

On the 18th December he left Rudbeck's house and betook himself to Småland to his parents, whom he had not seen for nearly three and a half years. He longed to meet them, and they him, specially this being so with his ailing mother. She had formerly the belief that he could not become anything above a veterinary surgeon, but her previously cooled sense had now changed, and she looked forward with joy to the hour when the home should receive her first-born, who had had the honour of lecturing as a professor, though a student for only two years.

This journey had besides another object, he wanted to consult his parents about a daring and important project. How these plans shaped themselves, therefore, will be narrated in the next chapter.

CHAPTER III

LAPLAND JOURNEY (1732)

THE subject of conversations which they had held during Linnæus's stay with Rudbeck and now renewed with pleasure, was the latter's recollection of his own journey in 1695 to Torneå Lapland. Rudbeck showed the plants he found there depicted in lively colours, and often talked about the rare phenomena and objects he saw on that journey, so that in Linnæus there arose an incredible longing for the Lapland fells. As Rudbeck's collections were destroyed in the Uppsala fire of 1702, before they had been fully worked up, Lapland, from a natural history point of view, continued almost entirely an unknown land, this fact assuring Linnæus of an extensive and grateful field of work and inciting him to illustrate Lapland in the three kingdoms of nature. Presumably it was through Rudbeck that the Secretary of the Royal Society of Science, Professor Anders Celsius, became aware that Linnæus harboured an irrepressible interest in Lapland. The recently ratified royal ordinances for the said scientific society embraced a provision for the investigation of the fatherland in all respects, and for the society's members to travel round the country, so as to take due account, and to engage suitable men in Sweden to enter on such tasks, Celsius promising that the necessary funds from the society should be forthcoming.

A journey, such as this, was then attended with troubles and dangers so great that one in our time finds it difficult to realize them. It was regarded as almost an improper and hazardous project, especially because of its condition and the prevailing ignorance in Lapland, the actual dangers from the inhabitants standing out in a greatly magnified shape. Before a definite decision was reached, therefore, Linnæus felt that he must obtain his parents' advice and consent. "Mother," relates Samuel Linnæus, "dissuaded him, fearing that so adventurous a journey might cut short her newly found gladness. She wrote to him with many reasons against his design, and quoted the old saying:

> In thy country born and bred,
> By God's bounty duly fed,
> Be not lightly from it led!*

But father left it to Carl's own choice. 'Thou hast no more than one life to take care of,' he wrote, 'if thou find it advantageous to thy future call upon God for help. He is everywhere, even among the wildest fells. Trust to Him. My prayers to God shall go with thee.'"

The result of these discussions was that Linnæus, on the 15th December, 1731, handed in a long document, addressed to the President of the Royal Society of Science, State Councillor Arvid Bern. Horn and the members. Herein he set out "that as our fatherland, no less than all other flourishing states, are now provided with scientific societies (who favour and possess the greatest studies in the entire kingdom), so it is highly necessary that our Sweden, no less than other places in the whole world, may ere long pride itself in every kind of noble science and learning.

*This recalls the old invocation translated thus by Stopford Brooke from Cynewulf, about A.D. 750:—

> Hail thou, Earth Mother of Men!
> In the lap of the God be thou a-growing!
> Be filled with fodder for fare-need of men!

"I dare say this hymn was sung ten thousand years ago, by the early Aryans on the Baltic coast." (Early English Literature, i. 220.) The Swedish form seems to be a Christianized and later version.

In the same aspect no one doubts that our country, whose natural history has hitherto lain in thick darkness, may also in a short time, boast of its efficiency through its inquiring members. Lapland first deserves accurate investigation; a tract highly admired by poets and historians, both praised and blamed. The whole of Europe and nearly the whole of the world have been searched through, but this country lies wrapped in cruel barbarism, and fabulous traditions are current." Various points of view are appended, on which investigation should be made, such as the natural productions of Lapland, the diseases occurring there, the Lapps' curious diet and cooking, their domestic medicine, with numerous other things which, for the sake of brevity, could not be specified. As to the zoology of Lapland, probably many mammals may be discovered, but on the other hand, there are many of the rarer kinds such as reindeer, lemmings, gluttons, etc., concerning which so many errors are recorded, that one would be wearied in recounting them. There must be a large number of birds which annually gather together from all parts of the world, and here, free from disturbance, lay their eggs and rear their young in peace. As this country is everywhere traversed by great streams and rivers, it should follow that here occur many kinds of curious and hitherto undescribed or unknown fishes, " and as to insects, there must be in summer a superabundance when one recalls, as travellers relate, how the people are attacked and pestered by them."

The other two kingdoms of nature have also much to offer as deserving of closer observation. Plants are found in Lapland as Professor Rudbeck has already shown, particularly rare and unknown, such as the numerous species of willows, which, and with many others, need description. Similarly we may also believe that there is no lack of mosses, now so prized by botanists. As for the mineral kingdom, we already know of the costly treasures of metals which

Nature has long concealed in her distant chambers under the Pole star, and this leads one to suspect that many splendid and useful sorts are to be found when fortune and industry are allied. Therefore since these spots are full of iron ore, here should be discovered valuable mineral waters, serviceable as watering places, with perhaps other and rarer properties. Still further would be an opportunity to investigate what Nature performs in the living body by means of the especial situation of the country, weather and the exhalations of the earth, besides the bitter cold and heat in yearly change, etc.

Linnæus continues, "that one sees clearly that no tract in the world offers a more splendid field for observation in all three kingdoms of nature, so that one with a good conscience cannot hold back therefrom. It is therefore desirable that somebody should be sent to investigate Lapland, fitly chosen and as intelligent as the journey requires, else it would be in vain, if one of other views were sent. For the purposes of the investigation the following personal attributes are requisite:

α. He should be a native of Sweden, in order that foreigners should not usurp what the natives have paid for.

β. Young and light, that he may run vigorously up the steep hills, and back into the deep valleys.

γ. Healthy, that he with greater ease and comfort may carry out, each day, his appointed task.

δ. Untiring, not only in pleasant times, for he may come short of food, he must be on foot, stooping, enduring heat and thirst, with many other difficulties, it being no pleasure jaunt for a fine gentleman.

ε. Without other duties, for here is constant duty, to outdo the best.

ζ. Unmarried, so that he may venture on the waters of the rivers, etc., without thinking about possible fatherless children.

η. Skilled in the groundwork of natural history and medicine, that in all these subjects he may have better insight.

θ. Understanding all three kingdoms, perhaps more difficult to find than a bird of paradise, for amongst all our botanists there are few who are at home in the other two kingdoms, but hardly one who is competent in all.

ι. Born a naturalist, not made, for it is remarkable how they differ when they come to practise, as I can testify by countless examples.

κ. A draughtsman that he may better describe by drawings the rare things observed."

A person so endowed, according to Linnæus's views, would be the best to be sent on the journey of discovery, " but there are few here in Sweden endowed with all these attributes, a fact which I consider it difficult if not impossible to overcome." Failing a better one, he offers himself to undertake the Lapland journey, as he at least possesses the chief of the requisites mentioned. He specially wishes to show that he had sufficient knowledge of botany and zoology, and declares himself willing, if required, to submit himself for examination. In mineralogy, which he studied under Professor Stobæus, he admits he was not as expert as desirable, but this want could be supplied, as he was to stay at Lund for a month or two to study.

He applied therefore for 600 copper dalers [£15] for the said journey to Lapland during the next summer, "as I then shall be at leisure and free, for one cannot travel, and pursue my medical studies without means." For all the troubles and risks of the journey he declares that he wishes for nothing but to have a recommendatory letter from the Royal Society, securing him a travelling stipend, so that he may finish his studies.

A few days after delivery of this application, Linnæus started, as already stated, to Småland, where he spent Christmas and the January following in the home of his parents. On the 2nd February he went to Lund to see his dear Stobæus, and to inform himself in mineralogical matters. This visit proved to be negative in results, for Stobæus's method did not agree with Linnæus's ideas, and his collection, consisting mostly of fossils, was of no value for the intended journey. He therefore returned to Stenbrohult (arriving on the 14th February), and stayed there till the end of the following month. How he occupied himself is unknown, except that he wrote his "Adonis Stenbrohultensis," and devoted himself to medical practice. Amongst those he attended were his mother, of whose illness he had received intelligence before leaving Uppsala, and his youngest sister, who was attacked by small-pox. He had the pleasure of seeing both restored to health. It is presumed that he consulted his parents as to his intended journey, perhaps also employing a good deal of time in resting after the last year's forced labour.

By the end of March spring seemed to have come to Stenbrohult, whereupon Linnæus hastened to Uppsala, where he arrived on the 1st April, but found severe wintry weather prevailing. This was not the only misreckoning which there awaited him; a more serious one was that nothing had been settled as to his Lapland journey. Since his application to the Royal Society of Science had been sent in, at least five meetings had been held, at which both of Linnæus's benefactors (Rudbeck and Celsius) were present, but not a word was spoken about any support to Linnæus for his intended journey, judging from the silence of the minutes. He was therefore compelled to send in a new application on the 14th of April, wherein he recalls his previous letter for permission for a journey to Lapland, in the ensuing summer, to elucidate its natural history. Probably because of what he gleaned from private information, he reduced his request for pecuniary help to 450 or at least 400 copper dalers [£12 10s. to £10], and gave details based upon mileage.

This document produced the desired result. The next day, 15th April, a meeting was held when Linnæus's letter was read, and it was resolved that 400 dalers in copper should be granted for travelling expenses from the Society's funds. This sum was obviously too small for the intended object, but the Society did all that it could, for according to the accounts, only one daler nineteen öre (about tenpence) remained as balance in the treasury. Information of the resolution was sent a few days later to the President, State Councillor Arvid Horn, who gave his consent and approbation. Further, the Society published in the "Posttidningen" what had been decided in order to obtain information about Lapland, and that Carl Linnæus had been chosen to investigate in all three kingdoms of nature, the healthiness and inconveniences of the country, with a description of the mode of life and the inhabitants. The agreed sum was paid over by Anders Celsius on the 26th of April.

With this the journey became assured, and it only remained for the unusually long and severe winter to end. This period of suspense was employed in necessary equipment. How this was done appears from Linnæus's Diary in which he wrote how he got his things and clothes together, thus:

"My clothes consisted of a light coat of Westgothland linsey (woolsey cloth) without folds, lined with red shalloon, having small cuffs and collar of shag; leather breeches; a round wig; a green leather cap, and a pair of half-boots. I carried a small leather bag, half an ell in length, but somewhat less in breadth, furnished on one side with hooks and eyes, so that it could be opened at pleasure. This bag contained one shirt, two pairs of false sleeves and two half shirts; an inkstand, pencase, microscope and spying glass, a gauze cap to protect me occasionally from the gnats, a comb, my journal, and a parcel of paper stitched together for drying plants, both in folio, and my manuscripts on Ornithology, *Flora Uplandica* and *Characteres Generici*. I wore a hanger at my side, and carried a small fowling-piece, as well as an octangular stick, graduated for the purpose of measuring. My pocket-book contained a passport from the Governor of Uppsala, and a recommendation from the Academy."

With such simple plenishing and so scanty a travelling purse, the young student set out in good spirits on his long and perilous journey, undoubtedly the most important one ever undertaken in Sweden, and it is proper that an adequate description should here be given. The English reader will find a detailed account in Sir J E. Smith's edition of "Lachesis lapponica," cited in the Bibliography.

On Friday, the 12th May, Old Style, 1732, at eleven o'clock in the forenoon, when Linnæus was twenty-five years old less one day, he left Uppsala. He rode out by the north customs gate, along the great north road.

"It was a splendid spring day; the sky was clear and warm, while the west wind refreshed one with a delicious breath. The winter rye stood six inches high and the barley had newly come into leaf. The birch was beginning to shoot, and all trees were leafing, except the elm and aspen. Though only few of the spring flowers were in bloom, it was obvious that the whole land was smiling with the coming of spring. The lark sang in the sky:

Ecce suum tirile, tirile, suum tirile tractat.
[Lo! it exercises its "tirile, tirile," its "tirile," i.e., an imitation of its song.]

When about eight miles had been traversed, the woods began to increase. The sweet lark which had hitherto delighted our ears, left us, another bird, the redwing, taking its place, which sang its sweetest from the fir tops, emulating the nightingale, the master singer."

These extracts from the first pages of his diary show the fresh and hopeful feeling of the spring in the breast of the solitary horseman.

The first, and also the last part of the narrative, has little of interest to detain us; we therefore pass on to his arrival at Gäfle, with a fresh passport from the Governor there, and thence through Gästrikland. The most noteworthy fact met with was that at Hamränge everybody spoke of a rare tree which grew in a croft by the road, seen by many, but recognized by none. Some said it was an apple tree, cursed by a vagrant witch. Linnæus hastened next morning by sunrise to see this rarity, which to his astonishment was only an elm, showing that *Ulmus* was not a common tree in those parts.

Passing through Hälsingland and approaching Medelpad, he fell in with some of the people he expected to meet on his journey, namely seven Lapps who were driving sixty to seventy reindeer. In reply to the question how they came to be in the low country they answered, speaking good Swedish, that they had been born by the sea, and hoped to die there.

Hardly had Medelpad been entered when he received new tidings from the fells, namely that two ptarmigan were seen by the roadside, but not within range of gun, so he contented himself with a sight of them through his telescope. Next he noticed great quantities of *Aconitum septentrionale*, which seemed as abundant as heather. He was tempted here to leave the high-road for a time in order to ascend the highest hill in Medelpad, called Norbykullen. Having tethered his horse to a runic stone, and accom panied by a guide, he went up on the western side, mostly on his knees. Up here the natives had a "look-out" for the Russians, with an extensive view over the sea. Here was also a beacon ready to be lit, whenever the enemy should come in sight.

They descended by the steepest, southern side, collecting by the way all kinds of rare plants, and finding an owl's nest with three young ones.

Continuing northward the journey was not exempt from unpleasantness; not a single horse did he mount which did not fall under him, once or repeatedly, and on one occasion he hurt himself so that he could hardly mount his steed again. He found his collection of minerals too troublesome to carry, so at Hernösand he had them packed ready to be sent to Uppsala.

The road followed the coast, where Linnæus sighted the fragments of a wreck. By the evening he reached Sundsvall and then proceeded to Fjähl, where, on the 18th May, Ascension Day, he stayed, partly for the festival, partly to rest his wearied and shaken body. The next day he resumed his way and made a visit to a cave in the Brunäsberg, where it was said, a felon lodged for a year or more, without disturbance. Soon after he left Medelpad and its sandy roads, and entered Ångermanland, where big and deep streams showed themselves, and were troublesome to cross. Passing quickly by Hernösand he ferried over the great Ångermanland river, which here forms a bay.

The next day, 20th May, Linnæus encountered an event which might easily have ended his journey. The road lay alongside a steep and high mountain, the Skulberg, in which there was said to be a cavern. He wanted to go there, but the natives said it was impossible; however, with great difficulty he got a couple of fellows to go with him to show the way. They ascended chiefly by dragging themselves up by the bushes, or stones. He was following one of the men, who was scrambling up a steep cliff, but when he saw the other man getting up better, he started to join the latter. Hardly had he stepped a yard to one side, when the first fellow dislodged a stone, which fell past him, with fire and smoke; a narrow escape from death. Directly afterwards another stone came tumbling down, whether by design or accident Linnæus knew not. At last they reached the vaulted cavern, which was fourteen feet high, eighteen broad, and twenty-two deep.

The ride on the main road was continued. Immediately Schulaskog was reached winter was encountered for the third time, and when he passed over the boundary into Vesterbotten, hardly a flower was to be seen. The dwarf birch was abundant but showed no sign of leafing.

On the 24th May he reached Umeå, a little town, well rebuilt after being burnt by the enemy. Governor Grundel was visited at his house, where Linnæus was received with great politeness. He was fond of natural curiosities and had many to talk about; his garden was shown to be well furnished, though potatoes only grew here as small as walnuts, and tobacco seeded only under the greatest care. When Linnæus left Hernösand, he had with him a general order to officials commanding them to help him on his way, and to allow him to pass unhindered through their country and forests, accompanied by Lapp guides who spoke Swedish.

Until now Linnæus had followed the road without special difficulties (if one exclude heavy rain, squalls and cold, with weariness resulting from a fortnight of constant riding). At Umeå he was obliged to seek the quickest way to Lapland. It was in vain that the natives tried to dissuade him, stating that it was impossible for him to get to Lycksele in the spring, in consequence of the floods in the extensive marshes, which then form rivers too formidable to be crossed. Linnæus stood firm in his resolution and comforted himself with Solomon's word, that nothing was impossible under the sun. He admitted that he wished to explore Lapland's southern parts, having the impression that the southern fells ought to possess a greater number of rare plants than the northern.

It was, however, during his journey in 1734, two years afterwards, that he realized his error.

On the 26th May he started from Umeå to get to Lycksele Lapland. The journey now became different from its beginning, for it rained hard and continued to do so till the following midday. This however was, relatively, an inconsiderable discomfort, as also the circumstance that he could not hire a horse quite so easily. Still worse, the track became bad, and it was hazardous to life to sit his horse, which stumbled between the stones at each step. He was now travelling in byways, "where no devil could find him again." He began to lose heart for want of a companion and also tired of being shaken on horseback.

Such was the first day, the second being no better. He started at midday, but found the going so bad, that he never experienced its like, and all the elements seemed against him. The track lay over stones interlaced with tree roots; there were holes full of water, added to by rain and springs rising out of the ground. Branches of trees sodden with rain hung down into his eyes. The small birches were bowed down; the ancient pines which had kept each other up for years now lay tumbled crosswise so that he rode with great difficulty over the path. The frequent streams were very deep and the bridges over them so rotten that Linnæus kept his seat on a stumbling horse at his peril. To complete his troubles, he had to ride without a saddle, using a cushion instead; there was no bridle, but instead was a rope round the underjaw of the beast, and in this manner he travelled towards the fells. On the second day he reached Tecksnäs.

The third day he came to Granö, whence he should have gone by boat to Lycksele. This was made, however, impossible, partly because of strong winds, and partly because the watermen went to prayers till eleven a.m. This was so late in the day that he would not be able to reach Lycksele, more than thirty miles away, by daylight, also there was no inn on the way, where one could rest. Linnæus, however, was not cast down by this delay, for he had an opportunity of describing a newly shot beaver.

The next day the journey upstream from Granö began in a "håp" or Lapp boat, the voyage continuing very pleasant, past forests on each side, in delightful weather; red-shanks and other wading birds ran on the banks and swans sported with each other. By and by the Lapland boundary was passed, and rapids rendering it impossible to sail up, the characteristic fashion of overcoming these impediments was employed. Landing below the force, the rower took his knapsack on his back, took up the slight boat, balanced it on his head, and carried it above the rapids, running so that "not even the devil could overtake him." Linnæus went after, carrying his own things, but not neglecting to note the animals and plants which he saw by the wayside. A catalogue of the plants observed by the "Tuken" [Tuggan] force is specially noteworthy, as for the first time the name *Linnæa* appears, whilst in other places it was recorded under the pre-Linnean names of *Campanula serpillifolia*, *Nummularia norvegica* and the like.

On the 30th Linnæus reached Lycksele, where

he was hospitably received by the Rector, Ola Gran, and his wife. They showed themselves solicitous for Linnæus's comfort, and tried to persuade him to stay with them until a greater number of Lapps should come to Lycksele, be interviewed by the traveller and, by announcement from the pulpit, should themselves be informed of the nature and object of his journey. Without this, disaster might happen, as the Lapps were apt to shoot at anyone taking them by surprise; they had even put a gun to the priest's wife's chest, when she had on one occasion gone amongst them without warning. The following day, which was Whit Sunday, no Lapp came to church, because the pikes were spawning, these fish being their harvest. However, many Lapps came a week later, otherwise they would have been fined fifteen shillings each, with enforced attendance at church on three Sundays. A further reason was given that the spring floods were imminent with consequent difficulties.

Sunday service was attended in the church, which was built of wood, and was so " wretched that when it rained the wet came in, and the benches being low, one had to sit in a cramped position." After this, Linnæus again started, Sorsele being his destination; and at first the journey was pursued without trouble. As happened repeatedly afterwards, our traveller recorded his dissatisfaction at the way in which the forests were treated. Damage ensued from neglect, so that instead of there being big forests of pine the trees were allowed to fall and rot, being prostrated by the storms, although some were certainly of good growth. He wondered why tar and pitch were not obtained from them, so that these forests might produce some return.

The journey now became more difficult, and Linnæus's companion, a Swedish settler, had to go into the stream and drag the boat after him for miles together. It was therefore necessary to seek a Lapp who could help them. Several Lapp dwellings were found empty, and Linnæus became so tired that his companion had to go alone on his quest. At last a Lapp was found who could not speak a word of Swedish, but he took them to his "koja" or tent, and regaled them with fish and plain water, all that was to be obtained at that time of year, and then guided them to his nearest neighbour. They were thus passed on from one Lapp to another, till they reached the river Juktan, a tributary of the Umeå.

The abominable conditions now reached such a pass that the young naturalist found himself in a most deplorable state. Wading over the river up to the middle in cold water, they could not reach bottom in the deepest part with a pole, but were obliged to use it as a bridge to pass over in peril of life.

Immediately afterwards they came to extensive marshes, through which they waded a Swedish mile (nearly seven English miles) with great trouble, their clothes being saturated with icy water, though the frost was still unmelted in places. Linnæus at this point became disheartened, and began to regret having undertaken such a journey. To add to their distress, rain and strong winds set in. At six a.m. they determined to rest, and lit a fire to dry their drenched clothes; the north wind chilled them on one side, the fire roasted the other, and mosquitoes attacked them on all sides. No parson could describe a worse hell than this; no poet could paint the Styx as being so hideous. Linnæus stayed by the fire while his companion tried again to find another Lapp. He wanted nothing more than to go back downstream, but distrusted the håp, as his body was neither iron nor steel. The Lapps are born to endure evils, as birds to fly, though Linnæus pitied them. This spot was called Lycksmyren, but he thought it should be renamed Olycksmyren, i.e., Luckless Marsh.

The travellers waited till two p.m. for the Lapp, who returned utterly wearied, having searched many native dwellings in vain, but with him came a being, Linnæus doubting whether it was a man or a woman. "It turned out to be a woman, small, with face blackened with smoke, brown glittering eyes, black eyebrows, hair as black as pitch, and hanging down all round her head; a red flat cap, grey petticoat, with bosom like a frog skin, pendulous breasts, with brass rings as adornment; a girdle round her waist, and boots on her feet. She appeared like a Fury, but spoke briskly and with pity. 'Oh, you poor fellow, what unlucky fate has brought you here, where I have never seen any stranger. How did you get here and what do you want? You see our dwellings and understand how hard it is for us to get to church.' I asked her how I could go further, either forward or any way, except the way I came. 'No, you must go back the way you came, there is no other, you cannot avoid it. Nor can you proceed, for all the streams are in strong flood. We cannot help you on your journey, for my husband is ill, who otherwise might take you to our neighbour, six miles off, who might perhaps aid you, but I think it would be entirely useless.'

"I asked, how far was it to Sorsele. 'That I don't know,' she answered, 'but you could not get there in less than a week.'

"I already felt sick with fatigue, through carrying my own effects, for the Lapp only bore the boat. I was exhausted by wakeful nights, and much water drinking, for only fish, often teeming with maggots and unsalted, was afforded me for my sustenance, for I had come to an end of the dried reindeer-meat (indigestible without bread) which the priest's mother had given me. I wanted to find folk who ate broth, and cared not if salmon ran up the stream. I asked if she had any food for me. 'No, unless you will have fish.' I looked at the fresh fish, which was swarming with maggots in its mouth, which stayed my hunger at the sight, but did not strengthen me. I asked for reindeer tongue, which the natives dry and sell. 'No,' was the answer. 'Any reindeer cheese?' 'Yes, but about six miles off.' 'If you had some could I buy it?' 'Yes, I don't want you to die of hunger in my country.' On coming to their tent I found three cheeses lying under a roof without walls; I took the smallest, which I paid for.

"I was obliged to go back the way I came, which I never desired. At last we came to our boat again, and going downstream rapidly, I lay and dried myself in the sunshine. The next day our boat split in the force, my stuffed birds floated away, but my diary, happily, was in my belt. The Lapp fortunately came up, and waded to land; the axe had gone, but he got a pole and laid it over the chasm. Our clothes were first taken over, then I stripped and followed with the pole. We pursued our way hungry and tired, passing through dense forests. From another Lapp I obtained fish, and then reached a settler's house, where they broiled half-dried salmon over embers, which I found delicious."

Finally Linnæus reached the priest's house, having been without bread for four days, and allayed his hunger. His journey was continued to Umeå, which he reached on the 8th and remained till the 12th June.

His course was now set for Skellefteå and then to Piteå, which was gained late on the 15th. The only noteworthy thing which happened was that he succeeded in shooting a hawk-owl, which hitherto unknown bird was described and sketched. The way was pleasant, through splendid forests; the inhabitants obtaining their chief income from firewood and tar.

The discomforts endured by Linnæus while on his unfortunate travels in Lycksele Lapland induced him to rest a little. He stayed in Piteå till the 21st June. That which surprised him most was that he found in the borgmaster's little garden a specimen of *Hyoscyamus* carefully cherished and regarded as a rarity. Besides, he made excursions outside the town, and amongst the skerries, but the results were not great. It was the same with the Luleå new town, which he quickly quitted as there was nothing to detain him. Then he went to old Lule by sea, as there was not a horse to be had. There he remained till the 25th June, and occupied himself with examining a medicinal spring, also studying botany and zoology.

The summer in all its splendour had now come, the weather being fine and sunny. Trees and bushes rushed into leaf, amongst them being the numerous willows of the north, not hitherto seen by Linnæus. "Though the summer is shorter here than elsewhere, it is more delightful. Never in all my days have I felt livelier than now."

It was now time to start for the fells. In a big boat towed upstream by four or five fellows (to whom he had to pay as much as they demanded) by the 28th June he arrived at Storbacken on the river Luleå, where he crossed the Lapland frontier, and hence by a long and tedious way on foot to Jockmock. Here it was that Linnæus for the first time had a glimpse of snowclad fells, though still far distant. Evidently he looked at them after he left Jockmock and reached Randijaur, for at the former place he regarded them as nothing but obtuse and long high mountains, which reared themselves one over the other. Farther on, he saw, on the 1st of July, the midnight sun, which he thought was not the least of Nature's miracles. "What foreigner would not wish to see it?" After a pause at Kedkevara, where a silver vein was formerly worked, he came to Qvickjock, where the famous pastor's wife, Pastorinna Grot, received him kindly. Here he obtained a Lapp, who served him as interpreter and guide, and to whom he gave a week's food for the festival which awaited him among his fellow Lapps.

So on the 6th July, he drew towards the fells, the first he ascended being Vallivare. What he saw there surpassed his boldest expectations. He thought himself in a new world, not knowing whether he was in

Asia or Africa, for the character of the soil, the situation, and the plants were strange to him. Nearly all the plants which he had met with on his journey were here found in such abundance, that he was frightened, and it seemed more than he could manage. During the succeeding days, so long as he remained among the fells, he was constantly encountering plants he had never seen before, as his sketches in his diary testify. As to the animal world, he became acquainted with lemmings, char, alpine ptarmigan and others. Reindeer were the object of his eager interest and accurate observation; he had good opportunities for their study, as the Lapp huts near which they were to be found were as innumerable as the forests, forming a crowd like ants in an ant-heap.

Accompanied by his interpreter from Qvickjock, who was annoyed by Linnæus stopping here and there for plants, he directed his course still further among the fells, where he had to live by favour of the Lapps, subsisting on what he could obtain from them. He did not neglect to note and sketch when necessary all that pertained to the Lapps' way of living, clothing, illnesses, etc. It seems that he quite enjoyed this strange world, so full of noteworthy objects and occurrences, even though he was compelled to reconcile himself to such poor food as was set before him, and to endure the lack of cleanliness, which sometimes even took away his appetite. He also complained of the flies being excessively troublesome.

During these wanderings he paused a little now and then, and by the 10th of July, he reached the lake Virijaur. Here he met with an adventure which might have been fatal. As he passed over a snow-field, which a stream had undermined, he broke through and would have fallen to the bottom had he not luckily saved himself, but only half-way, by the grip of his hands. His companion hastened to a Lapp hut near by, and he was drawn up by a rope little injured, although wet and so severely bruised on the thigh that it was painful for a month afterwards.

Next day the frontier between Sweden and Norway was passed, all covered with snow. A violent storm set in with sleet and hail, which stuck like an icy crust on his back, and the force of the wind was such that the travellers were not only driven forward, but often blown down and driven on some distance. Once he was rolled over a distance equal to a musket-shot till he brought up against a projection. Happily the wind was easterly, but had it come from the opposite direction progress would have been impossible.

After tramping about twenty miles, a change took place. The streams from the melting snow ran westward, bare rocks stood out, and soon the sea was seen. The landscape seemed to be a miniature garden when thus viewed from the heights. When at last they came to lower regions, Linnæus welcomed the change from the frozen fells to the warm valley, where he sat down and ate wild strawberries. Instead of snow, green plants and flowers were seen; in place of violent wind there came a beautiful scent of flowering clover; Linnæus was able to drink cow's milk and to refresh himself with food, and to sit on a chair. But at the same time he could not fail to see that the two Lapps who accompanied him showed no sign of fatigue, but on the contrary, began to run and play. To explain this, he propounded the query, " Cur lappones adeo pedibus celeres?" i.e., Why are the Lapps so swift of foot? and in his diary he set down his answer, giving as reasons—" shoes without heels, exercise from childhood, meat diet, moderation in food and drink, small stature, etc. He accounted also for their good health by the pure air, well cooked food, pure water, peaceful lives, and absence of fermented liquor.

It was Sörfjorden in Norway which was reached by Linnæus, where in its innermost bay he busied himself by investigating the sea animals and plants. On the

14th July, he started to reach the celebrated Maelström, and was received by Pastor Johan Rasch, a travelled man. The next day he proceeded to the entrance of the fjord, and met with an adventure. He was creeping along after strawberries, when he noticed a Lapp with a gun presumably after birds, but took no notice, till a musket-ball struck a stone in front of him, the Lapp vanishing when Linnæus drew his hanger. This attempt on a peaceful stranger showed their fear of witchcraft and their distrust of all who were not known to them. Another specimen of this feeling is thus recounted by Linnæus: "I showed some sketches in my book to a Lapp; he seemed frightened, snatched off his cap, bowed his head, and struck his hands on his breast, mumbled some words, and seemed as if about to swoon." Other instances are named by the diarist of the difficulties in keeping the Lapps to their work, and preventing them from deserting.

The following day a return towards Lule Lapland was begun in summer heat. Up the steep fellsides they went, seeming never to gain the top, wet through with sweat, till they began to stiffen with frost. Linnæus slipped into a stream which had made its way under the snow.

The next day they traversed the glaciers on the Swedish side, but encountered a thick fog, which delayed progress, till they came upon the track of a reindeer which was followed, and they escaped from their dangerous position.

It was four days later, on the 20th of July, that the travellers reached Qvickjock and there stayed three days; Linnæus enquiring about the life of the Lapps, their habits, house-gear, care of reindeer, hunting, diseases and the like. On the 26th July, they reached the pearl-fishery of Purkijaur, which he wished to see close at hand. As no boat was available, they made a raft, but the semi-darkness prevented a clear view of their course across the river.

Further, the tree-trunks, forming the raft, began to work loose, and it was with difficulty they avoided the force; but luckily, in the end they reached the island where the fishery was established.

The fishers used rafts of five long balks each ten to twelve feet long, and fastened with birch withes, a stone being used as an anchor. The mussels lay in clear water near the force, and standing on the forepart of the raft, the fisher used a wooden twelve-foot pair of tongs to drag up the pearl-mussels. Few pearls rewarded the search, most of the animals being without any, and these were thrown away.

Resuming the march, a new adventure occurred. Coming to a place where a forest fire had raged some days before, they found the tree stumps and ant-heaps still smouldering. Without foreseeing danger they went on, but a breeze sprang up, and the burnt trees began to fall. It was impossible to stand still because of the smoke; once, a tree fell between Linnæus and his guide, but at last they passed the dangerous place. The next trouble which followed immediately, was a plague of small, blood-thirsty flies, fiercely stinging, and not to be blown away; they swarmed on their clothes like a black covering, and could not be beaten off. Finally on the evening of the 30th they reached Luleå, intending to go on at once to Torneå, but the whole of the next day they had to remain indoors, as a violent thunderstorm with heavy rain prevented their starting. It was not till the 3rd August that they reached Torneå, where Linnæus was hospitably received in his house by Rural Dean Abraham Fougt.

In these comfortable quarters Linnæus stayed till the 9th, resting after his exertions, but also investigating the vegetation round the town. He particularly tried to discover the source of the epidemic disease of cattle, by which fifty to a hundred were lost annually, doubting whether it arose from some peculiar water or grass. He found that it came from a superabundance of the poisonous water-hemlock, *Cicuta virosa*, in the meadows. As remedy he advised the digging out and uprooting of this plant, which could be done by a girl in a month's time, and the result would be the saving of a large sum each year.

On the day above mentioned, Linnæus betook himself to Kemi, but when he reached the inn, he could get neither horse nor food, so he returned to Torneå, and some days later reached Kalix by water.

In Kalix Linnæus made the acquaintance of Circuit Judge Michael Eurenius Höijer, a noble and learned gentleman who accompanied him back to Torneå and thence to certain neighbouring copper and iron mines. Meanwhile severe frost came on, showing that autumn had set in, and as a visit to Torne Lapland was desirable, Linnæus used the same boat back to Torneå and spent a further six days with the judge, making notes of all that he observed. Whilst at Kalix he became known to the mine surveyor, Seger Svanberg, with whom he had spent some days at Qvickjock. This official offered to teach Linnæus the art of assaying, an offer which was eagerly accepted, he, in his turn, standing as godfather to Svanberg's new-born son. Only a few lines in his diary record his ten days' hard work on "Ars docimastica," which proved most valuable, later on, in Uppsala.

It was now time to journey homeward, there being three ways open, by sea to Stockholm, or by land, either on the east or west coast of the Gulf of Bothnia. The road along the coast of Finland was chosen as being new country; he learned the Finnish names for such things as he thought he should want, and thus equipped himself for his journey.

In cool and rainy weather he set out, passing through Uleåborg, Gamla Karleby (Old Charlestown), Jakobstad, Nya (New) Carleby, Vasa, Christina, and Björneborg, till on the 30th September he arrived at Åbo. During this forced march, very few observations could be recorded in the Diary, beyond the conditions of the houses, such as blackened, sooty doors and holes in place of windows. There was always a nauseous smell of sour "sik" [gwyniad, a fish], and the rooms were so hot, that his nose was nearly scorched off, and so smoky, that his eyes were constantly running. He noted with regret the extensive moors, exclusively covered with heather which defied eradication. The natives declared that two plants would destroy the world, namely, heather and tobacco.

Linnæus remained in Åbo till the 5th October, and then journeyed by Åland and Grisslehamn, reaching Uppsala on the 10th at one in the afternoon.

It was not surprising that this journey thus briefly sketched, should awaken great interest in Sweden, and not the least because of the kindly reception which so greatly helped the young naturalist. Without the explanations and help of every kind, which were so generously given him, the result of his travels would not have been so rich and valuable as they proved to be. Only in one place, Jockmock, where the clergyman and the schoolmaster wondered why the Royal Society should have sent a young student to search through Lapland, when older and more competent men could be entrusted with that enterprise, was there any unpleasantness. The conversation he had with these two worthies is amusing. The chaplain held that the sky was solid; that *Nostoc* was a plant [by accident he was right here], with similar notions, and when Linnæus attempted to put the matter in a true light, they laughed derisively and considered him mad.

A report of this journey was presented to the Royal Society of Science at its meeting on the 9th November. The minute drafted by Dr. N. Rosén runs thus: "Herr Linnæus's account of his journey to and from Lapland, with its perils and labours, was read; it included the history of novelties in all three kingdoms of nature, which he supported with his

Catalogue." Presumably this "Catalogue" was the "Florula lapponica" issued in the Society's "Acta" for 1732, practically the earliest work of Linnæus to be printed, as well as the first in his Sexual System. This report embraced his account of the black ironsand in all the rivers, the pearl-fishery at Purkijaur, twenty-three new willows, a grass resistant to the greatest cold, forage plants (which coloured butter deep yellow), the Lapps' love-potions, on the moxa, zoological details, ten kinds of bread used when grain fails in Norrland and Norway, and many similar observations. Later, on the 10th February, 1733, were also presented reports on the fatality amongst the cattle in Torne, on the *Aconitum* in Medelpad which is eaten by the inhabitants, and on a certain moss in the forests. The last was *Polytrichum commune*, which Linnæus's companion stripped from the ground it covered and used as bed and coverlet, found to be most comfortable. Another proof of the value set on the Lapland journey was that when Professor Roberg was visited, the latter hastened to write a long narration on the journey from Linnæus's dictation.

Accounts of this expedition reached foreign parts; already after the departure from Uppsala a Hamburg journal, "Niedersächische Nachrichten," noticed this event, and in the same paper, after his return, appeared a brief statement of it, and the news that the "Flora lapponica" and "Lachesis lapponica" were in preparation. Also, a Nuremberg journal published a report from London that J. J Dillenius had communicated the news to C. J Trew, himself learning it from Anders Celsius, whom he had met in London; further, Celsius himself sent a letter to Trew, when he had reached Berlin. To the same effect a Copenhagen print, "Nya Tidender om laerde og curieuse Sager" [News about learned and interesting matters] alluded to the same subject, but only in October, 1734, two years after the conclusion of the journey, a striking commentary of the slow communication in those days of events in the scientific world.

If we now put the question: Did this Lapland travel really deserve the many eulogies, which even persist to this day? For answer, an emphatic "Yes" must be given. One has only to go through his diary to be astonished that it was possible for a person so poorly equipped in every respect as Linnæus was, in practically an unknown region, not only to travel so far, but also to take such good observations. These latter dispersed so many false impressions and filled so many gaps, that even at this period the diary must be regarded as a masterpiece. Surprise is awakened when one sees the warm interest which the traveller took in most varied objects, giving a constant and acute attention which included Nature's productions, the inhabitants' economic circumstances, habits and uses, the various diseases occurring in different parts, and the customary household details, the language and sports of the Lapps. The sketches also testify to a striking, sometimes playful, sometimes poetic feeling.

In the first place, must be noted the matters relating to the domain of botany. If one measures the extent of the collections, they may be treated as extremely scanty, for hardly anything not absolutely required was collected to check the accompanying notes. But these descriptions from living material possessed a great value, forming a basis upon which was written the "Flora lapponica," a work highly estimated then, and yet recognized as of permanent value. This is due to the fact that not only was much of this material new, but it was accompanied by interesting facts concerning biology and geographical distribution and their application to different objects.

This work, drawn up by Linnæus, is the only one of considerable length and entirely founded on the observations taken during the Lapland journey. A few other notes were issued in smaller publications, though many are incorporated in later works, such as "Fauna suecica," "Flora suecica," "Species plantarum," and "Systema Naturæ." Much information was also imparted in his lectures, which were diligently taken down by his hearers. In many cases we can only wonder at his sound ideas and conceptions of the future. For instance, when enjoying the pure fresh air, the pellucid delicious water, he wrote: "In the dogdays it is usual to travel to some medicinal spring for those who seem to need it. I can say, that for many years, thank God, I have enjoyed fairly good health, though sometimes oppressed and somewhat low in spirits, but as soon as I came to the fells, I gained new life, and as it were, a heavy weight was taken off me. I spent a few days in Norway, and there felt heavy; but as soon as I got to the fells, I became at once revivified. For those who have opportunity it would be better for them to come here to drink snow water, rather than to stay in thick weather, by marshy medicinal waters." Linnæus thus perceived in advance our present day high alpine sanatoria.

As regards Linnæus himself, this journey was of the greatest importance and value. It was clear that thereby his energy was increased, his scientific outlook enlarged, his powers of observation sharpened; in a word, his development took a great step forward to the end which he was ultimately to reach. It is true, that the hardships he underwent were such that he declared "he would not take such a journey again for 2,000 plåtar [£300]" but the troubles were past, and the memories remained imperishable during the rest of his life.

When Linnæus came back to Uppsala with his treasures, there was one thing which specially troubled him, namely—his economic position. It has previously been mentioned that though (p. 65) he estimated that his expenses would amount to 600 copper dalers, he only obtained 400. Moreover, he underestimated the mileage, and on his return found that he had traversed 672 Swedish miles, 150 on foot [respectively 4,793 and 996 English miles] and the time had increased from 20 weeks to 6 months. Even allowing that a part of these distances were so-called Lapp-miles (that is, according to the guide's fancy), it is evident that the money was far too little. A great advantage was the hospitality he met with among Swedes, Norwegians and Finns. In Torneå and again in Åbo, he was forced to borrow respectively fifty and fifteen shillings. Both during and after his travels, he must have been calculating how to improve his economic position. He was "highly recommended for the Piper scholarship," but as this was founded exclusively for Lund students, which University Linnæus had left, it is not easy to understand the ground of his hopes. He had sent in his statement of expenses soon after his report to the Royal Society of Science, and later on an amended return, showing a deficit of 211 silver dalers [£15 16s. 6d.] but the Society regarded this as being too heavy a sum to refund. Finding himself saddled with debt, he exclaimed "thus are Swedes rewarded," though he challenged the whole world to accomplish a similar journey at the same cost. The Society then promised to give him a recommendation to Count Horn, but the intended journey to his house did not take place. The Royal scholarship (p. 40) which he had hitherto enjoyed, now ceased (in spite of the wishes of the medical faculty), as by its constitution it could not be continued after the spring term of 1732. But another scholarship existed, the Wrede, from which he hoped to get something to help him at Uppsala. No ordinary scholarship was vacant, but it was a custom, when the income exceeded the stipend, to bestow the surplus on needy students. An appeal supported by eight other students in Latin prose and verse was made to the Consistory in November for this surplus, but a competitor was in the field, namely, Olof Murén.

The result was that Linnæus by ten votes to three, received this gift, amounting to 30 plåtar [£4 2s. 6d.] for the half-year; an attempt to upset this arrangement in favour of Murén, being resisted by the Consistory in the following February. During the remainder of 1732 he worked on his collections, and at Christmas travelled home, arriving on Christmas Eve. His parents were saying grace before meat, and received him with delight. He had plenty to tell of his adventures, but feared that this would be the last time he should see his mother. This thought was prophetic, for in the following June he recorded that "his dearest and most pious mother passed away, in his absence, to his indescribable anguish, care and loss."

CHAPTER IV

THE LAST STUDENT YEARS AT UPPSALA—TRAVELS IN DALECARLIA (1733-34)

After the year 1732, which proved so momentous for Linnæus, he continued a member of Uppsala University for two more years. There is no occasion to record how he employed himself in his professional medical instruction, except that it seems to have been of an entirely unsatisfactory character. One of the two professors, Rudbeck, was enjoying a prolonged freedom from his chair, engaged on his great philological work; the other, Roberg, was for six months in 1733, released from lecturing on account of his rectorship, and during the remainder of that year, he confined his instruction to materia medica, and (as the University lacked a chemical laboratory) to demonstrations in chemical operations, including fireworks, with useful experiments. By this time Linnæus could not feel tempted by such a programme. Adjunct Rosén continued his anatomical lectures with energy and skill, but Linnæus had already taken in all that was necessary, and he naturally did not attend the lectures in botany delivered by Rosén in the botanic garden during May and June. Clinical instruction was suspended, no doubt because of the academic hospital's miserable condition.

Under such circumstances Linnæus spent all his time and his uncommon powers of work in scientific research, and in teaching natural history to students who desired to avail themselves of such help. In the former case, he devoted himself energetically to his

" Flora lapponica," but nevertheless because of his own observation and study of earlier authors, he drew up some smaller works, hoping to find some benevolent publisher who would disseminate his new ideas throughout home and foreign countries. In the course of the year, he had finished fourteen memoirs; some already named; but others may now be mentioned, as " Systema botanica," " Philosophia botanica," with augmentation to nearly four hundred fundamental propositions; " Harmonymia botanica," correcting many errors in genera and species, " Characteres generici," " Species plantarum," and " Diæta naturalis," showing how careful diet conduces to the prolongation of life.

The teaching Linnæus imparted in the early part of the year, does not seem to have been very profitable. Probably he did not consider himself justified in spending much time herein, for at the close of the spring term, he expected a new payment of the Wrede scholarship surplus, but in this, he found himself unhappily mistaken. After the death of the Academic Treasurer, J. Landberg, in the early part of the year, much time passed before the accounts of the University could be brought into such a state as to decide whether a surplus existed or not. Thus Linnæus found himself in pecuniary embarrassment, and ultimately it was shown that no surplus would be available during that or the following year. This money trouble was, however, modified by a student, C. F Mennander, a recent acquaintance, who paid him for lectures on Natural History.

These economic difficulties spurred him on to energetic teaching. When spring approached Linnæus found twenty auditors, resulting in 49 silver dalers [about £3 14s.], and according to custom, various botanic excursions took place under his leadership, yet he was still unable to get out of debt for his Lapland journey, though he thus had a little alleviation of his poverty.

Linnæus spent the summer of 1733 at Uppsala,

and during this period he began a series of lectures on mineralogy, lasting a month. Probably this suggested to him to employ his knowledge of assaying ores and minerals, which he gained under Svanberg the previous year. He therefore drew up a brief handbook, "Vulcanus docimasticus," and with the approval of the academic authorities, issued a notice inviting attention to it. In this he maintained the importance of the subject, especially in a country rich in metals and minerals, undertaking to explain matters so that his hearers could afterwards perform the experiments. For this, including materials and implements, his fee was only thirty copper dalers [fifteen shillings] whilst other teachers charged ten times that amount. Those who wished to take part were invited to the lodgings of Linnæus in the house of the widow Rodde, whose husband, the academic dancing master, had died in 1712.

The result was entirely satisfactory; both the subject, which was new to the University, and the lectures, attracted as many as his room could hold, so that by December he had received 200 silver dalers [£15], and amongst the audience were several notable personages, including Adjunct Rosén.

After the heavy work of term, Linnæus considered himself entitled to use some of his unaccustomed wealth upon a little travel. He left Uppsala on the 20th December and stayed at Falun in the province of Dalecarlia, as the guest of his friend and fellow-student, Claes Sohlberg, who liberally entertained him in his father's house. "Dalecarlia's sirens," he wrote to Mennander, "have tempted me to forget both friends, cares, reflections, troubles, home, studies, and time! I cannot close my ears to their songs; without joking, I have enjoyed myself extremely well."

However, the main object of this excursion was not the pleasures of the place, but to gain opportunities of increasing his insight into mineralogy and mining. He stayed therefore some time at Falun, exploring the mines and studying the smelting; he also visited certain mines in the district, and in his notes names a dozen facts which appealed to him as remarkable. His impressions he summarized thus:

Nothing is more	splendid	than	Steelwork;
,,	extensive	,,	Copper smelting;
,,	rational	,,	Ironwork;
,,	speculative	,,	Stiernsund;
,,	rich	,,	Norberg;
,,	horrid	,,	Fahlund [Falun].

The last named concerns the mines exclusively. He went down into them, and describes his sensations. "The whole way was by wooden ladders, mostly in twenty steps apiece, and perpendicular. Two ladders were often fastened together, which swung about, some slanted, but most were upright. The ways [drifts] were narrow every way, so that one had to stoop or go on all fours, often striking one's head against the roof, which showed crystals of vitriol, or were entirely black; it blew cold and strong till near the bottom, so that a windmill could work. Out of the mine a constant smoke ascended. Never has a poet described a Styx, nor a theologian a hell so awful, as that seen here, for upward rises a poisonous, stinging, sulphurous smoke, which taints the air all round, and so corrodes the ground that no plants can grow in the neighbourhood. Below, it is unspeakably dark, never shone upon by the sun, chambers filled with steam, dust and heat, till at 450 ells deep [876 feet] one reaches the solid hard earth. Here are more than 1,200 sun-fugitives, condemned to metalliferous work. The drifts are dark with soot, the floor of slippery stone, the passages narrow as if burrowed by moles, on all sides incrusted with vitriol, and the roof drips corrosive vitriolic water. The miners are naked to their waists, with wool respirators over their mouths to prevent inhaling smoke and dust. Sweat pours from them like water from a bag."

How glad and satisfied was Linnæus when he came up unharmed from the mine, appears from his final paragraph thus:

"Thou, great Creator and preserver of all, who
On Lapland fells permitted us to ascend so high;
In Falun mines permitted us to descend so deep;
On Lapland fells showed me day without night;
In Falun mines showed me night without day;
On Lapland fells allowed me no surcease from cold;
In Falun mines allowed me no surcease from heat;
On Lapland fells allowed me in one place to see all seasons;
In Falun mines allowed me not one of the four seasons;
In Lapland bore me unharmed through so many vital perils;
In Falun bore me unharmed through so many perils to health.
Praise all that Thou hast created,
from beginning to end."

At the end of February, Linnæus left Falun and reached Uppsala on the 1st March, 1734, where his first work was to arrange his collection of minerals and rewrite his "Systema lapidum"; his meetings now constituted a series such as never had been got together at Uppsala, not even in the University, so that his room, which served both as dwelling and museum, attracted no small interest. "You should have seen his museum," wrote his friend, J. Browallius, "which was available for his audience, it would have struck you with surprise and delight. The ceiling he had adorned with birds' skins, one wall with a Lapp costume and other curiosities, while a second boasted medical books, with physical and chemical instruments and stones. The upper part of a corner was occupied with tree-branches, with thirty different kinds of tame fowl, and in the window stood big pots of rare plants. Besides, one had the opportunity of seeing his collection of dried plants pasted on paper, all collected in Sweden, and amounting to more than three thousand kinds, wild or cultivated, to which must be added the Lapland rarest plants also dried and pressed. Furthermore were a thousand species of insects, and about as many of Swedish minerals, placed on wide shelves and in the most pleasing way arranged after Linnæus's own system, founded in accordance with his observations."

That this magnificence should attract pupils is quite natural. This term he taught not only general botany, but also dietetics, a science which he believed he had himself found to be built hitherto on fallacious principles, and therefore needed to be reformed. As a foundation for this he drew up his "Diæta naturalis," with seventy-five rules, after an introduction in which preceding authors were severely judged. Of his predecessors he considered Sanctorius the only one worthy of note, the others being Hippocrates, Celsus, Hoffman, Boerhaave, and six hundred others whose writings on diet (with so many rules for health that they cannot be reckoned) only proved the truth that "medice vivere est pessime vivere" [to live by medicine is to live horribly]. The reason is, that doctors have taken mankind as possessing the machinery of clockwork, not recognizing that he is an intelligent animal. So if you would live long, live as an animal of your kind should, especially in our country.

A Swede builds like an Italian;
,, takes snuff like a Spaniard;
,, dresses like a Frenchman;
,, eats like an Englishman;

A Swede drinks like a German;
„ smokes like a Dutchman;
„ takes brandy like a Russian.

It cannot escape notice on reading through the introduction that it bears the stamp of dejection and bitterness. That he actually cherished such feelings, is plain from his own notes written at this time. Although his economic condition was not pressing, one may take the end of the spring term of 1734 as the most anxious period of his richly varied joys and troubles.

The reasons for this seem to have been many. One of the weightiest was certainly the doubt and unrest concerning the future which awaited him. He could not avoid the thought that the years were passing, thus diminishing his joy in study. He was now twenty-seven years of age, and had been seven years a student. It was his constant desire to remain at Uppsala as a teacher, but doubted whether to aim at this object or to earn his living as a practising physician. In this latter profession there remained the obstacle that he had not passed any examination, nor had he been promoted Doctor of Medicine. That he had latterly considered the matter, appears from his thesis written in August 1733, in which was put forward a new hypothesis as to the cause of intermittent fevers. He afterwards used this thesis in Holland in 1735 for winning the doctorate. According to law, this should have been obtainable in Sweden, where the academic statutes permitted both Uppsala and Lund to set up doctor's promotion, but of old the idea had established itself in Sweden that only in a foreign land was the doctorate valid. It was therefore an urgent necessity for him to undertake a journey abroad, perhaps lasting some years, so as to prosecute his medical studies and provide himself with a diploma as Doctor of Medicine.

There was also another reason why Linnæus

should wish to make such a journey. He had now lying ready, at least in outline, many scientific treatises on which he had spent much labour; these with glad certainty of victory, he hoped would reform natural history from the very bottom, and at the same time ensure for himself a place as a valued member of the scientific society of that time. But one thing was wanting. He had vainly endeavoured to find a publisher at home or abroad, but hearing from a fellow countryman that Dr. Christ. Nettelbladt, Professor of Law, had settled in Greifswald, he dispatched to him the first part of his "Fundamenta botanica," with the view of its being printed, though the outlook was not very favourable. Nettelbladt did his best, but beyond printing preliminary notices in certain German journals, he could do nothing more, but recommended application to the Dutch publishers.

A journey abroad was therefore needed, but that required money. Whence was it to come? It was only by denying himself that he had gained his education, helped by some little practice of medicine, but now the economic outlook was considerably darkened. True he was the supposed recipient of the Wrede scholarship surplus, but from that he had no income, no means being at hand to pay it. Attempts were made to get some other scholarship. In accordance with the prevailing custom of soliciting the powerful influence of "most noble patrons," Linnæus applied to the Governor of Umeå, Gabriel Gyllengrijp. This person made a most humble appeal to the King, that some scholarship should be given to the needy student, specifying two stipends or scholarships, but these were hampered by special restrictions, so the appeal came to nothing.

The Governor here mentioned had recently been appointed to a post in West Bothnia, and, anxious to improve the conditions in that province, consulted Linnæus upon the methods of achieving this, resulting in a report upon crops which could be cultivated there. This attempt failed, as neither the Lapps nor settlers would try any new methods. Gyllengrijp again appealed to the King asking that Linnæus might be commissioned to undertake a new journey to Norrland and Lapland, but this also was unsuccessful.

After a long period of friendship between the two pushing and competitive young men, Carl Linnæus and Nils Rosén, a cloud now arose between them. In a wordy quarrel a sharp word fell which made a deep impression on Linnæus, so that he, rightly or wrongly, now saw an adversary in Rosén, willing to damage his future and to put hindrances in his way of gaining a living by instruction. The misunderstanding did not last long, these unhappy feelings being gradually dissipated. Among those with whom Linnæus became acquainted during his visit to Falun, was thc Governor, Baron Nils Esbjörnson Reuterholm. He heard Linnæus give an account of his Lapland journey, and being charmed, suggested that he should travel through Dalecarlia at the Governor's expense, and describe it as he had done Lapland.

How delighted Linnæus was at this is hardly credible. Without delay he made arrangements for the journey and hastened to Falun, where he enjoyed the hospitality of Reuterholm, and not only received all that was wanted, but also a generous sum for the journey's prosecution.

Already the ample funds for the journey made the Dalecarlian trip entirely different from the poorly equipped Lapland one, but other differences may be given. No one had accompanied Linnæus to Lapland, but in this case, certain of the Uppsala students offered to accompany him at their own expense. He therefore formed a society with laws for orderly conduct.

The members of the " Societas itineraria Reuterholmiana " were eight in number, whose posts were assigned as follows :

Carl Linnæus, of Småland, President;
Reinhold Näsman, of Dalecarlia, Geographer;
Carl Clewberg, of Hälsingland, Physicist and Secretary;
Ingel Fahlstedt, of Dalecarlia, Master of the Horse;
Claes Sohlberg, of Dalecarlia, Quartermaster;
Erik Emporelius, of Dalecarlia, Zoologist and Forester;
Petrus Hedenblad, of Dalecarlia, Adjutant;
Benjamin Sandel, American, Treasurer.

The document relating to these official titles and functions was drawn up, and sealed with the seal of the President.

This division of labour proved singularly well adapted, and the President kept a strict watch that every member did his duty. Each evening they assembled and reported what each had noticed, after which the President dictated to the Secretary what should be recorded. He himself could scarcely be expected to bear every detail in mind, but through the travellers' eagerness and ability, which could not be sufficiently praised, he received valuable help in the task which he had undertaken to carry out.

It was on the 3rd July, 1734, that the travelling society and its attendants (in all ten men and as many horses) left Falun in youthful gaiety, to put on record a tract of country which in many respects was still *terra incognita.*

The first day's travel was to Bjursås, the country for two miles round Falun being composed of loose stones, due to the noxious vapours from the copper mines; cattle did not thrive, but the inhabitants had their rich compensation from the underground wealth. Next succeeded some undergrowth of pines, then pines and firs in the moist spots. The next day took them to Rättvik. Travelling was good, but the country folk were not at home, so their animals were in difficulties for food. Once when the horses were allowed to eat the grass near an empty house, the travellers were accosted by a peasant from a neighbouring cottage, carrying an axe, and fiercely angry; he attacked them as violators, threatening to rouse his friends. Neither reason nor money availed, but at length he gave way on payment of three dalers [four shillings and sixpence].

Kjerfsås was the next stage, examining on the way three silver mines and a marble quarry. Next day at a wayside inn they could get nothing for their wants, so they hastened onwards to Orsa. The greatest impression made upon the party was the wretched state of the people working the grindstones, who seldom attained the age of forty, due to the dust being taken into their lungs; this was so usual that no remarks were made upon it. They were surprised at seeing two old men in church, but they were told that they were not quarrymen, but a tailor and a shoemaker. Near Gullerås they were told of an extraordinary tree, which flowered on the death of Carl XII. (30th November, 1718); the country folk called it an elm-tree; they found on travelling to investigate that the smallest child knew where it grew; on reaching it, it proved to be a common lime-tree, a proof that this species was rare in the province.

The party noted with regret how miserably the forests were neglected, and how little use was made of them. Thus they found that pines were felled solely for their bark, being food for man and beast; the trunks were allowed to rot, as no use was made of the wood. This state of things was found practically throughout the tour, and Linnæus pointed out that these logs could be dragged to the Dal-Elf river and floated downstream for sale.

The next day, 7th July, was Sunday, and after service, the party was invited by the rector, Magister Schedvin, to dinner, the host proving to be a superior kind of man. Towards evening they passed on to

Mora, where they stayed over the following day, enjoying the hospitality of the seventy-five-year-old sub-dean, Johan Emporelius, father of the zoologist of the party. After looking over his fine library, and noting the local dresses and customs, they saw through the rectory window a young cuckoo in the nest of a wagtail, and the foster-parent feeding the young.

On the 9th they bade farewell to the liberal rector, climbed Gåfshusberg, passed by Vestberg (having no time to visit it), and late in the evening, after much wandering and trouble in crossing a river, they came to Prestgården. A curious custom here was observed, namely on certain pines about 120 lists were nailed up; one pine having 56, another 35, a third 14, and so on. Each "list" was 9 inches long, 3 fingers broad, and black with cut letters; they were notices of deaths; the oldest date being 1670, thence to the present day. Each village had its tree.

As usual the clergyman took them in; in Älfdalen he was Eric Näsman, father of one of the party, and the record in the diary is "*hospitalis*." He was a man who kept abreast of the time, and was the first to plant potatoes in Älfdalen, sharing them with the country folk, to vary their diet of bark-bread, or malt-dust bread. Runic letters were still in use here, the only spot where this old writing persisted. Hykieberg was climbed, described and an inscription cut on a pine, 11 July, 1734. In 1722 an academic thesis by Z. Holenius mentioned the rare plants found here. At noon the march was resumed, and a forest, more than thirty miles in extent, had to be traversed to reach Särna; the good road now ceased and a rough stony one followed. It was past midnight before they reached Särna, where they were housed at the priest's, Gabriel Floræus. Next day was Sunday. The people here (more allied to the Norwegians than to the Swedes) were not on good terms with the inhabitants of Älfdalen, whom they nicknamed "bellowers" on account of their cries at funerals; the compliment being returned by the epithet of "soft ones."

The pastorate was reckoned at 169 square miles [about 1,127 English square miles], only one being cultivated [40 square miles]; the tenants paid nothing to the crown, and Linnæus thought it would be better if Lapps were there instead, as their reindeer in summer do not harm the meadows, and in winter do not eat hay.

On the 15th an excursion was made to Städjan mountain, over rough country. After the ascent the weather turned cloudy with rain. Gnats abounded, and when a fire was lit in a cattle-shed, the heat was overpowering, while the smoke hurt the eyes and lungs. Finally the fire, which was quenched with difficulty, burned an opening in the dry boards of the roof. The unfavourable weather continued the next day, and the party took shelter in some cowsheds in Gröfveldal. Early on the morning of the 19th they started for Gröfvel Lake, where they slept. Ill provided with food, the party shot ptarmigan and caught fish, but their bread was all gone.

On the 20th July they crossed the Norwegian boundary and reached an estate, Mugga, where they had to stay over the Sunday, in bad weather. The party then started for Röraas, the quartermaster going in advance to secure accommodation. He applied to Hans Brendal, the deputy of the mine-surveyor and burgomaster, and through him, who was most obliging, they obtained good lodgings.

On the 26th, they turned homewards; after crossing the frontier, they climbed the mountain, Svuckustöt. A toilsome scramble landed them on the top, whence a splendid prospect rewarded them. Bad weather still prevailing, thunder, lightning, strong wind and rain drove them down to the Lake Gröfvel once more, where they passed the night in a shed.

On the 28th they tried to get to inhabited parts, though the track was abominable, but they saw 100 wild reindeer. That night was spent in a cattle-shed at Idre. Their fell journey was now ended, and soon they met with the first quarter-mile post [1¾ mile, English] and thus knew they had regained the haunts of civilization; next night they were at Särna rectory.

Wearied with the ceaseless difficulties, the travellers rested the next day and part of the following one. After a short journey, they stopped in the forest of Höståkällan for the night; the gnats plagued them, and at dawn heavy rain made the tent worse inside than outside, "as the raindrops were fewer outside than in." They struck south towards Lima, and in the afternoon, the rain ceasing, a boat took them down the river, to their great content. Late in the day they came to Sähla, and the next day, 3rd August, they voyaged to Transtrand, where they found a singular personage, Lars Dahl, who was the chaplain there, garbed poorly, simple in gestures, of wise discourse and learned, though neglected by the world. Passing Lima, Malung, and Äppelbo to Näs, without noting any remarkable things, they reached Falun on the 17th, in the last stage encountering a silly woman, who took them for thieves, and thus the end of the expedition resembled the beginning. The journey had taken six weeks and three days, the total distance being reckoned at 313½ quarter miles [a little more than 518 English miles].

There can be no question that Linnæus and his young comrades had well carried out their commission. As the diary shows, they had open eyes for everything and open ears for what they could hear from trustworthy persons. Naturally the credit belongs chiefly to Linnæus, whose knowledge of nature's various dominions, and his experience gained in other journeys, with his planning, his conception, accuracy, and scientific balance, were certain to lead to nothing else than a good result.

Nevertheless, all the expected results were not attained: this applies chiefly to the botanic part.

Before Linnæus's memory stood the rich alpine flora of Lapland, and he expected something even better from Dalecarlia; only one phanerogam, *Utricularia minor*, was met with in this expedition which had not been previously known as Swedish.

It does not follow that the results obtained were without value and interest. This was the first actual investigation towards a knowledge of the plant-world of this province, and at the same time reduced previous sketches to their proper value. Petrus Ugla, who responded for the thesis " De nuptiis arborum " which had a considerable influence on Linnæus's career, had in this same year, 1734, issued a memoir " De præfectura næsgardensi Dalecarliæ," in which according to Linnæus, the author shows an unheard-of example of the danger of attacking a science with insufficient knowledge (" illotis manibus "—i.e., unwashed hands, being his actual expression), and he further points out that Petrus Ugla had included many plants only occurring in glass-houses or cultivated, which have never been found wild in Sweden. In consequence of this thesis Linnæus wrote his " Flora dalecarlica," though it was not published until 1873, a hundred and thirty-nine years later.

The zoological output is somewhat similar, only one single insect and one bird, *Picus tridactylus*, as related by Linnæus himself. " In the year 1734 when we came upon the Dalecarlian fells, I heard below in the forest between the fells, an uncommon note of a bird, which was afterwards shot; it was different in plumage and shape to other woodpeckers, and was not described or named by any author." In one or two places, pearl mussels were met with; beavers were abundant in Särna; in Älfdalen they spoke of a four-footed fish, which ran up trees; he thought it might be a water lizard. He was able to complete his account of the *Oestrus* which he had met with in Lapland. Lemmings were not seen, but in the Lima churchbook was found a remark that in 1636 a day of prayer was held because of the innumerable lemmings which came down and devastated the fields and meadows in the parish.

A larger collection was made of minerals. These seem to have been specially looked for and therefore received particular attention in the first report drawn up by Linnæus on his return to Falun, namely, his "Pluto suecicus," which, besides his previously mentioned "Systema lapidum," can be taken as the first sketch of "Regnum lapidum" appearing in his later "Systema Naturæ." His note concerning the primitive method by which the peasants procured an excellent iron from the native ore, is interesting.

Besides the natural history, observations were made on the people's ways of life, costumes, dwellings, economics, cooking, farming, hunting, fishing, diseases, medicines, marriage customs, dances, and the like. Undoubtedly its value may be recognized as written at a time when the old levelling civilization had as yet hardly affected the habits and customs handed down from father to son. Linnæus seems to have got on well with the inhabitants, both educated and uneducated, which accounted for the generous hospitality accorded. But he was less pleased with some of the food, as when in spring each household salted down blood lymph, and then buried it in the forest for winter store as "grovefish." Still less did he like the chewing-gum prepared with garlic, which was chewed by the women in church as the finest aroma.

After his return to Falun, Linnæus remained some time the guest of the Governor Reuterholm, in whom he saw "a pattern to all who study to love, understand, show favour and exercise fine judgment."

Soon he was busy in authorship; his first care being to fair copy his "Flora dalecarlica," which he handed to the Governor on August 25th, "who was greatly pleased with the description of his tour and its arrangement, and invited him to stay some time with him to instruct his sons in Natural History." The impression made upon the Governor by the report of the expedition resulted in his permission to his sons and their tutor to undertake a similar journey in the following summer. The regard and affection entertained for their teacher by his pupils are shown in the warm-hearted, elegant Latin letter written during this trip, on the 1st September, 1735, from Röraas, to their teacher, then in Holland.

Immediately after the completion of the report, he began the composition of his "Pluto suecicus." What finally induced him to take up this work was that in Falun he could thereby impart instruction. He succeeded in his endeavours and gave much delight to the mining industry. This new "Vulcanus docimasticus" was the ground-work for a course of lectures on assaying in the hired mining Assay Office, to a numerous audience. Each paid a fee of 12 to 13 silver dalers [18s. to 19s. 6d.] which quickly brought in a sum of 101 dalers [£7 11s. 6d.] He also started the practice of medicine, and succeeded in making a modest addition to his income. Presumably this was the medical occupation which induced him to draw up still another work, "Najades suecicæ" concerning the Swedish medicinal springs.

It may be added that as Governor Reuterholm had a fine library of choice books on economics, travels, etc., to which Linnæus had free access, it is easy to see that he did not lack work. But this did not debar him from the social life which, as an esteemed guest, he enjoyed with the town's foremost families, so naturally he throve amazingly, and thought that in Falun he had come into a new world where everybody loved and favoured him. Then, too, the summer's pleasant journey aroused a desire to undertake similar ones to other parts of the country. He considered that each province had its peculiarity, and considered how each might be improved to the advantage to Sweden if all the provinces were investigated and each made to help the other. He therefore hastened to invoke the help of Governor Gyllengrijp, who readily undertook to further the aims of Linnæus as to cultivation in West Bothnia, by sending a humble memorial to the appropriate committee of the Riksdag, renewing his previous petition to the King, and praying that for two years an annual sum of 300 silver dalers [£22 10s.] might be assigned to Linnæus in order to carry out his plan under the inspection of himself as governor. Although a polite answer was returned, nothing was done, as the Riksdag had abruptly closed its labours.

Linnæus was thus forced to devise new plans, as he did not want to go back to Uppsala. Amongst the friends he had made in Falun was the domestic chaplain and tutor in Governor Reuterholm's household, named Magister Johan Browallius. In their familiar talk, the pastor insisted with emphasis, that there was nothing to be done unless Linnæus went abroad and obtained the degree of Doctor, when he could return home and settle down to practice. This seemed feasible, as Linnæus by his lectures and practice had now some means, and further, that the father of his comrade Cl. Sohlberg, who was State Inspector of Mines, had promised him an annual sum of 300 copper dalers [£7 10s.] if he would take his son with him and look after him abroad; Governor Reuterholm offered him pecuniary assistance "without return," an offer which Linnæus, however, could not accept. In the end he resolved to travel abroad to be promoted Doctor of Medicine, intending to return as soon as he should be able to earn his living as a medical man.

This resolution made, Linnæus left Falun, probably in November, for Uppsala, there to provide the necessaries for the journey. In the first place, he must undergo an examination in Divinity, which took place on the 24th November, with Dean Olof Celsius and Professor G. Wallin as examiners, afterwards receiving the certificate for the one single examination which Linnæus underwent in a Swedish University. This was done, as according to a Royal decree each student desirous of travelling out of the kingdom, must pass the said examination, before he could get his travelling passport.

Provided with this *attestatum* and with an academic testimonial from the rector, Professor Schyllberg, Linnæus started on the 23rd November for Stockholm, where he spent a fortnight; he then received his passport, witnessed the Riksdag's adjournment, and on the evening of the 15th left Stockholm for Uppsala, which he reached the next day.

The following days were given up to taking leave of his friends and patrons, and getting open testimonials from Rudbeck and Celsius. He then occupied himself in packing, and left the town on the 19th December, 1734.

CHAPTER V

LINNÆUS AS MEMBER OF SMÅLAND'S NATION IN UPPSALA —PRETENDED INTRIGUES AGAINST LINNÆUS— CHRISTMAS AT FALUN, 1734-35 — JOURNEY ABROAD

THE delineation of Linnæus's career and activity during his student years at Uppsala, as previously narrated, is so far incomplete in that it does not give any account of his relations to the nation club to which he belonged. It is not specially important, but for understanding his student life, a short statement of it may be given.

It has already been stated on p. 29 that Linnæus, during his stay at Lund, neglected to inscribe himself as a member of any nation, the reason being his intention to remain only a twelvemonth at Lund, and his wish to escape the continuous, mortifying, and time-wasting "penalism" (fagging) which existed in spite of a Royal decree forbidding it. At Uppsala the conditions were essentially better, so he found himself obliged to inscribe himself of some nation as he intended to study there for a series of years; for without such inscription he would not have been able to obtain any scholarship, of which he was in the greatest need. He therefore became a member of the Småland's nation, being admitted on the 25th September, 1728, as from the University of Lund, after payment of 27 copper dalers [13s. 6d.] to the nation's treasury.

By this, Linnæus not only attained membership of the Småland's nation, but also entire freedom in the privileges such as taking part in disputations, posting and delivery of letters, and so on. Because of his previous academic life at Lund, he was entitled to be styled Dominus in place of Monsieur or simply Sieur, as were newly matriculated students from Växjö. His membership was confirmed in November.

At that time the nation's activities were undeveloped, a contributing cause being the want of a common room. The few meetings had to be held in the Inspector's house, or some other place. Actually, the only material bond of the club was, that in 1645, an iron chest was obtained to contain all their records, consisting of valuable papers relating to loans, pledges, elegies, and academic theses. In the spring term of 1733, Linnæus borrowed thirty copper dalers (fifteen shillings), pledging an article of silver as security, which debt was redeemed the following spring.

The officers who conducted the business, and had oversight of the younger members, were, besides the Inspector, the seniors and the eldest of the juniors. Not long before the arrival of Linnæus it had been decided that the juniors had no vote until they had been three years in the University.

Judging from the minutes of the few meetings, the proceedings do not seem to have been extensive or interesting. Generally most of the time was devoted to an oration or disputation on a given thesis, in which the older members displayed reading and dialectic powers. These orations were usually designated as "elegant," but as much of the disputations were confined to philosophy, they did not appeal to Linnæus. He was more likely to note the exhortations and proposals, that the officers should have the charge of the younger members in church, so that no clamour should occur near the pulpit, and that none of them should be notified to the Consistory for improper behaviour. All such rules Linnæus caused to be observed when he became tutor to the young Rudbecks. After his return from Lapland, he, in common with his compatriots, had to listen to a letter from the Chancellor, exhorting students to read Wolffius and Leibnitz with caution, and to beware of new views which might be hurtful to them. The Inspector congratulated them on their dislike of such dangerous novelties. This warning, so far as it affected him, though belonging still to the young members of the nation, led in June, 1730, to a summons of the brotherhood, and during the following years, gave cause for repeated debates. As the years went on, Linnæus, like others of his province, gained regard and influence. In the spring term of 1733, he obtained his own key of the nation's hired bench in church, the second one behind the Professor's sons, an honour which at the time was highly prized, and for which a donation of 24 öre in copper [twopence] was readily paid. At the same time, he was chosen as secretary, which post he retained during the whole of that year. Next, he was pronounced Senior, an honour which nevertheless made him (each time he was absent from Uppsala) disburse two dalers in copper [a shilling] *pro felici reditu* [for happy return], and later on, to take part in the customary observances on the admission of new members. The meetings of the members were not entirely confined to grave subjects, but also embraced convivial gatherings. Linnæus shared in these relaxations, as such opportunities were rare in Uppsala. For the nation's festivals the hours were—to begin at one p.m. and to close at nine p.m.; the Inspector was to be present with the Seniors, but the Juniors were not to claim the same equality. Regulations as to disputes were in force, and if the disputants failed to observe them, they were liable to be expelled. No member was to wear a sword or other dangerous weapon. If a new member broke a glass, he must make good the damage. These extracts from the nation's statutes, may give an idea of the students' life during the residence of Linnæus.

Finally the minutes of a meeting of the 25th November, 1734, show that the Curator proposed that their countryman and senior student of medicine, Carl Linnæus, should be awarded a testimonial. The vote was carried with the good wishes of all for his success.

With this the sketch of Linnæus's student life is closed, and in some respects it would be superfluous to add more. In most biographies, an important place is given to the so-called intrigues and persecutions with which he had to contend while at Uppsala, and to the bitterness which was thereby kindled in his mind and wellnigh ruined him. It is therefore needful to dwell shortly upon these occurrences. Most biographers also supply more or less dark reports, without giving the facts on which they rely; others are so contradictory, that one refutes the other. We may take, however, what is written by E. M. C. Pontin, who took the task of unmasking these jealous persecutors of the young student. A lecture was given to an important audience in 1849, and afterwards published in the "Aftonposten," in which he states:

"It was poverty that hitherto had hindered Linnæus in his work, which now was interrupted by envy. One became aware that the young and unnoticed Linnæus threatened to surpass those of privileged merit. This was sufficient for pettiness, ill-will, and intrigues, and their instrument was Nils Rosén.

"To drive Linnæus from the teacher's chair which he so worthily filled, Rosén wove a web of cabals wherewith to entangle influential people in the capital. These means succeeding, Linnæus was forbidden to give lectures on the ground that he had not undergone valid examination; and never did outcry achieve a more brilliant triumph. Linnæus was crushed. On the first realization of the methods employed to ruin him, and under the first impulse of a righteous wrath, Linnæus tried to draw his sword on Rosén, to redress, according to the custom of the time, the wrong under which he suffered. This act of youthful effervescence enabled Rosén to give the final blow to Linnæus. Charged with breaking the regulations against duelling, the authorities in Uppsala were unanimous in banishing him from the town.

" What was this youth, sent down from the university for violent conduct, with the piety and heart of a child, to do?

" Linnæus went to hide his pain amongst Lapland's desert fells, and in spite of persecution, still laboured for science, from which people with ample means wished to exclude him.

" Returned from his travels, Linnæus sought to obtain the newly-founded Medical Adjunctship at Lund; but his enemies were not idle. Rosén was the medical man at the same health resort where Count Karl Gyllenborg, Chancellor of Lund University, drank the waters. This circumstance Rosén employed with great skill, and succeeded in snatching that morsel of bread from Linnæus. Once again, and that the last time, Linnæus tried to obtain a living, this time as Docent; but Rosén did not rest even now. Through the archbishop, whose niece Rosén had married, he managed, by means of the Chancellor of Uppsala University, Count Cronhielm, to get a prohibition, so that no Docent should be taken on to the medical faculty

" Thus it was thought that Linnæus's future would he hopelessly destroyed, and no Celsius appeared to rescue him.

" But when the need was greatest, help was nigh, and with reason this man, who constantly wandered amongst precipices, surrounded with hate and threatening dangers, took as his motto, '*Numen adest*,' God is present.

" A letter from the Governor Baron Reuterholm proved Linnæus's salvation. It summoned him to the same province which formerly saved Sweden, etc."

[an allusion to the army raised by Gustaf Vasa in Dalecarlia].

What we infer from this quotation—and herein all who speak of "persecutions" against Linnæus agree —is that the University Adjunct, Dr. Nils Rosén, is pointed out as the hateful persecutor, who did not spare even the most paltry and infamous methods to attain his object. The mainspring of his atrocious conduct may have been, as Linnæus saw, a strong feeling of pride and ambition, and a fear that Linnæus was a dangerous competitor for the professor's chair, when the aged Rudbeck should quit the same. (Rosén was born in 1706, student at Lund 1720, Adjunct at Uppsala 1728, M.D. at Harderwijk in Holland 1730, professor at Uppsala 1740, and died 1773.)

Before we proceed to examine these accusations with which some endeavoured to blacken Rosén's reputation, a glance must be given to the position which he held at the beginning of 1730 at Uppsala University. It has already been stated (p. 29) how he became Adjunct in the medical faculty, and afterwards undertook a long journey abroad, during which he enjoyed the instruction of Boerhaave and other eminent physicians and naturalists. Returned to Uppsala during the spring term of 1731, he began at once with great energy to carry on instruction in anatomy (until then neglected), to the students' benefit and satisfaction. At the same time he became the only and most eminent physician in practice at Uppsala, as well as the most distinguished teacher. From 1732 and onwards, he discharged the requirements of his time in irreproachable fashion, including lectures and demonstrations in botany. In short, he was then almost the only one who continued instruction in the medical faculty, and that in a very serviceable way.

This was generally recognized both in Uppsala and elsewhere. So much so that when Kilian

Stobæus in the early part of 1732, exchanged his chair of natural history for history pure and simple, Lund's Consistory offered him the vacant post by a heavy vote, although he had not sought the appointment. On this ground, Rosén demanded a higher salary, which the governing body readily accorded. What he ultimately became does not properly come in here, but it may be well to recall that Carl von Linné, Johan Ihre, Torbern Bergman, and Nils Rosén von Rosenstein, in the later half of the eighteenth century, were the most distinguished men of the University, the last as practising physician earning a reputation which still remains unfaded.

With these facts before us, we are justified in asking whether it is reasonable to suppose that such a man should be frightened of competition in some future opening for a professor of medicine, with an unexamined, non-graduated student, who, however, certainly possessed for his time, wonderful knowledge in natural history; that he, apart from the simplest considerations of honour, should lend himself to such paltry intrigues? This question is the more justified, as otherwise Rosén is always noted as possessing in high degree a noble and straightforward personality. It cannot be overlooked that on the same page as Rosén is painted as the most hateful of the spiteful and irritating backbiters as a true deity of Tartarus, he is praised as a noble, peaceable man, possessing great amiability.

Still another question may be put—that as in Uppsala there were two professors in the medical faculty, both being septuagenarians and past work, it might be hoped that both Rosén and Linnæus (provided the latter gained his medical doctorate) would soon become their successors. Could it be supposed that Rosén should think it necessary, by a dishonourable act, to ruin the future of a young man, whose eminent ability and capacity he had praised? Of Rosén's contemporaries, one declared that he had never heard an unkind judgment pass his lips, during many years' intimacy; another, that he was most humane, being humanity itself; a third, that he was in all respects tender-hearted, and had a special dislike of contention, persecution and slander. Being no one's enemy he himself had no enemy, and had adopted as his motto *sine spinis,* without thorns (an allusion to his own name).

The answers to these self-evident representations are clear. They ought in truth to have led Linnæus's biographers to greater care and reflection, before throwing out these accusations against his honour and reputation.

After these general remarks we now pass on to a more detailed definition of these accusations. We may first fix upon the oft related story of the "duel," or as it may better be described, as an attempt at murder. With a marvellous confusion of ideas, people have sought to lay the blame upon Rosén, whilst for Linnæus's asserted action, extenuating circumstances have been found.

We may first of all point out, that about this "duel" not a single word occurs in any of Linnæus's autobiographies, nor can any mention of it be gleaned from contemporary accounts, letters and the like, though such an occurrence would certainly have become a *cause célèbre*. It was first mooted after Linnæus's death, about sixty years later, and that in a German biography, in all other respects trustworthy, namely Stöver's "Leben des Ritters Carl von Linné," where it is given thus:

"A young man, Nils or Nicolaus Rosén, became Linné's rival. Laying a complaint before the Academic Senate he urged that according to the statutes, Linnæus should be forbidden to lecture. He was called before the Consistory, when many of the members were favourable to him, but Rosén giving strong reasons, and as the law could not be left unobserved, the desired prohibition was granted.

"This was a blow which at once blasted Linnæus's hopes. . . . No wonder therefore that he became intensely moved; his anger became fury; he forgot himself and his own welfare and all consideration. When Rosén came out of the Consistory, he rushed insanely upon him, drew his sword, and was about to strike him down, when he was happily stopped by the spectators.

"This event naturally evoked the greatest interest, and Rosén, who possessed a permanent post in the University, reported it. In conformity with the regulations, Linnæus should have been banished from Uppsala, but happily, through one of his benefactors, this was prevented. Olof Celsius exerted himself on Linnæus's behalf . and succeeded in reducing the punishment to temporary banishment, Linnæus thus obtaining forgiveness, but no help. His impetuous temperament urged him to desperation, all his thoughts turning to stabbing Rosén, if he should meet him in the street."

No confirmation of this violent attack by a student on an Academic Professor, and consequent punishment, can be gained from the Academic Minutes of the Consistory at this period. They are preserved complete, and in them are to be found full reports of all occurrences, sometimes very unimportant. But in all these minutes for the entire period of Linnæus's student life, not a single word is made known, that he became liable to any reproach or punishment in any way. Certainly the name of Linnæus often occurs in these minutes, more so than that of any other student, but always with the expression of the heartiest goodwill, and with the most flattering expressions. Just as little is found the slightest description of Linnæus being summoned to the Consistory, or of the yielding of that body to Rosén's demand. It can therefore be asserted with absolute certainty that the whole story of the duel and banishment is a complete fabrication.

The most just explanation is that this period is to be taken as 1731, when Linnæus (since Rosén had returned from his travels and began his academic teaching) was hastily and unexpectedly dismissed from his commission to lecture in Rudbeck's place, and through hasty temper, which he could not control, came nearly being sent away by the University. But against this may be recalled Linnæus's own words: That as he knew from the beginning the recommendation for a student to lecture was only a measure of emergency (to stop simply as soon as the ordinary professor should return to Uppsala) he declared his readiness to give place to Rosén. Rudbeck opposed this as he regarded Rosén as not possessing the requisite knowledge. Either he had not the opportunity of proving this, or, what is more likely, Linnæus, before Rosén's arrival, had already begun the spring lectures as Rudbeck's deputy; it is certain that it was thus stated in the Consistory. However that may be, it is a fact that previous to Rosén's return, he delivered the botanic lectures, with no reasonable grounds for bitterness, but really with contentment. As Linnæus was, in the autumn term of 1731, an interested and satisfied pupil of Rosén in practical medicine, it shows that there was no friction between them.

Naturally, some rebuke must have been given if Linnæus had really made himself guilty of violent attacks. The alleged punishment clearly belongs to other peculiar circumstances. It is certain that he received from the Royal Scientific Society as large a subsidy for his journey as it was able to give, so that his earnest wish to undertake a Lapland journey might be realized. Also, that he received an invitation to become a member of the said Society. Truly a gentle and not especially dreadful punishment!

Whilst Linnæus was absent in Lapland during the period for botanic lectures, there could be no collision between him and Rosén. The instructions for observations which the former drew up, and the honourable judgment which the minutes of the said Society record, show that no "inexpiable hate" or even a diminished friendly relation, existed between them.

We now pass to 1733, when Linnæus, during the spring term and the whole of the summer, imparted special instruction, no prohibition being made against it. Had he then, as formerly, laid himself open to the previously mentioned attacks, he would have been guilty of a manifest and impudent lie, for in a letter dated October of that year to Governor Gyllengrijp in his catalogue of merits he says: 1. "I have occupied myself at the University in a quiet, sober and Christian way, so that nobody can convict me of the smallest offence; I have never been summoned before a judge, nor have I molested anyone in the least." After writing this letter, and at the express wish of the academic authorities, he held his assay lectures, previously mentioned, when he, a student, had the satisfaction of numbering amongst his pupils Adjunct Rosén, a very gratifying occurrence. Still less can certain biographers, such as Stöver, Pulteney, Gistel and others, ascribe the "duel" to the close of this year and regard his journey to Falun and district as a kind of forced banishment from the University. Linnæus, on the contrary, gave an entirely different reason, namely that he was especially devoting himself to mineralogy and was endeavouring to devise an arrangement of minerals, which could not be better studied than in the mining districts. Furthermore, if a prohibition had been granted it is doubtful if he could have delivered private lectures in the spring of 1734, not only on topics of natural history, but also on dietetics. Also the fact remains that six months later he received a *testimonium academicum* specially noted in the University minutes as a "handsome" one, an exceptional notification. Such testimony would certainly not have been given by grace and favour to one who had recently, in dire want, undergone banishment from the University.

To leave the enquiry, one may hope that the whole story of the duel, dismissal, etc., may be looked upon as completely refuted. But it does not follow, that as already mentioned (p. 98) no dispute took place between Linnæus and Rosén, although in the earliest Linnean autobiographies there is no account of it. This is related in somewhat contradictory terms as happening in two different years, 1733 or 1734, the later date being the more probable. The more extensive report is as follows:

Rosén, who perfectly realized that the young fellow had a considerable collection of recorded observations, and saw that if he were not repressed, he would in time become a formidable competitor for the botanic chair after the aged Rudbeck, went to Linnæus and asked him to lend him his manuscripts. When he was refused, he had recourse to threats, cursing and swearing, that if Linnæus did not lend them, he would persecute him as long as he lived. Linnæus, scared, lent the first volume, but when Rosén had copied it, he refused to give it back unless he received the second. This was a thunderbolt. Linnæus, though realizing that his whole system and collections would be ruined if that single volume were missing, after long thought, decided not to consent, although Rosén begged, promising that the first volume should be given back on loan of the other.

Poverty and oppression quickly made Linnæus take a resolution. He determined that as Samson took his revenge by killing himself and his enemies, so he might thus act against his unfriendly acquaintance, whose slander he felt painfully every hour. But Professor Oelreich from Lund, at that time a pupil of Linnæus in botany, " dissuaded him, showing that no evil happened without the Lord's permission, that to endure it was the safest, and that though He punished, He comforted. Linnæus changed his mind, praised his persecutor, and rendered thanks to the Creator." From another autobiography we learn that "when Dr. Rosén married the Archbishop [Steuchius's] niece he obtained an authorization from Chancellor Cronhielm that no Docent should ever be received in the medical faculty to the prejudice of the Adjunct. The hands of Linnæus being thus tied, his only means of support were denied him." On enquiring into this statement, it is regrettable that only one side is represented. "*Audiatur et altera pars*," to hear both sides, is an old rule, and it were to be wished that it could be applied here, especially as one of the two disputants can by no means be viewed as a perfectly trustworthy witness concerning the thoughts and intentions of the other. It is the more desirable here, as the autobiographies of Linnæus drawn up in his later years, must be read with a certain amount of caution, especially one published by Ad. Afzelius and therefore the best known. On close investigation, it shows itself to contain many erroneous statements. This is partly due to the length of time between the occurrence of the events, and the narration of them, when Linné had only his failing memory to trust to, and partly because of the want of attention that the authorities showed to the public, when otherwise they might have solved guesses, reports, misconceptions and suspicions. Add to this, that Linnæus, in none of his autobiographies refrained from employing strong, sometimes too strong, words in his representations, so that one must set limits to these manifest tendencies. Written, not for publication, but for his children, they described the chief events in the father's life, giving also a timely warning to the children, that they should not allow themselves any revenge for wrong suffered because the Allwise Omnipotent God will ever give the victory to the right and good. He, by his "Nemesis divina," also showed that punishment awaited each who transgressed God's ordinance, when by harm to another, one sought worldly advantage. As a sample of extravagant language, take his statement that at Falun, Dr. Moræus became envious of the practice Linnæus obtained in a few weeks, although the latter admitted that his future father-in-law was a prosperous man and was weary of a laborious profession. This accusation of envy resembled his language concerning Rosén.

Now let us consider the above-mentioned ordinance which the Chancellor of the University is reported to have issued; it is difficult to guard against the thought that the whole statement is entirely founded upon gossip or a failing memory. Enough that it is narrated only in the most untrustworthy autobiography, without, in spite of all research, being confirmed in any way. The Chancellor's archives, preserved in the Riksarkives in Stockholm, show no trace of any such restriction of a Docent in medicine. On the other hand, in the valid academic constitution of 1665, it is laid down that only graduates possessed the *potestatem docendi* (the right to teach), and that only by permission previously obtained of the Dean of the faculty, whilst such teaching by a student was entirely forbidden. Though the commission extended to Linnæus (then only a student in the medical faculty) to instruct in botany and assaying, was thus, strictly speaking, illegal, it was excused on account of existing circumstances, and on the term ending, those rights lapsed. That he, during succeeding terms, gave private lessons was owing to the fact that the faculty or Greater Consistory willingly shut their eyes to this illegality.

In the spring term of 1734, Linnæus gave lectures, for which he never received permission, on the subject of dietetics, thereby trespassing in Rosén's domain, a proceeding which can be regarded as still more improper for a student, as the academic regulations then in force, considered payments for lectures as a contribution to an Adjunct's scanty salary. It is quite possible that Rosén and Roberg did not quite consent to the new views in this science which Linnæus advanced and afterwards partially abandoned. Thus, if one wished to prevent Linnæus from giving private instruction, there was no necessity to invoke the aid of the Archbishop or of the Chancellor, but to use a simple application of the academic regulations. That Linnæus, as reported, made some attempt to become a Docent, is also incredible, as his name as a university professor does not appear in the then existing statutes. Neither in the University transactions nor in Linnæus's own notes does one find any hint of such a desire.

If one might venture a guess how the quarrel between the two arose, it may be gathered from the following:

When in May, 1734, Rosén should have begun his teaching in botany, he applied to Linnæus, in whose assaying lectures he had taken part, with the request that he might study the principal contents of these botanic manuscripts, of which he had doubtless heard from Rudbeck, Roberg, O. Celsius and others. Linnæus, who cannot be acquitted of a certain amount of suspicion, refused, and in consequence a dispute arose between these two young men of nearly equal age. Rosén may in this have pointed out upon what slender foundations rested Linnæus's right to teach, and that it would be easy to deprive him of this means of self-support. Startled by this, Linnæus gave up one of his botanic manuscripts, but when he found that a copy had been made of it, he determined to lend no more, and to this he adhered, in spite of more or less eager attempts at persuasion from Rosén. That the latter took measures to put certain threatenings in train, after the warm contention, is not certain, and in any case they would have been in vain, as Linnæus a short time after quitted Uppsala.

That this quarrel left any bitterness with Rosén is nowhere stated, but it was otherwise with Linnæus. He was, as he himself said, *præceps in iram*, quick to anger, and grief and irritation were kindled in him when he thought of the poverty, despair, and anxiety which he believed awaited him. These feelings encouraged thoughts of a violent revenge, but these were stemmed by an old friend's persuasions, who appealed to those warm religious feelings, which from childhood had permeated the whole being of Linnæus. "God should become my avenger," said he, and added "since then all went well with me." He clearly entertained feelings of revenge, as shown in his "Nemesis divina," quoted at the close of this volume, but conquering them, he acknowledged that everything prospered with him.

Evidently it is from the confession of this flaming wrath by Linnæus that well-meaning and uncritical biographers derived their material for the whole "duel" story, and all its supposed consequences, not considering that it is a long step between hasty thoughts and violent actions. Besides it is evident from Linnæus's own words, that he did not long cherish bitter feelings against Rosén, but that *facile placabatur*, he was easily appeased. This is shown by the greetings which he sent during his residence abroad to Rosén, and their familiar mutual letters at the time when Rudbeck was about to vacate his professoriate.

Linnæus's relations with Rosén are well shown in the *Mæcenatibus et patronis* mentioned among his patrons, in the dedication to his doctoral thesis in Holland where Rosén's name occurs, and two years later in Linnæus's "Corollarium," which he dedicated as *devota mente*, *Viro illustri D. Nicolao Rosén* alone. It appears incredible that Linnæus should have acted thus, had his soul been full of bitterness and ill-will, unless he were guilty of cringing hypocrisy.

Another story accuses Rosén of having induced Count Carl Gyllenborg to prefer J G. Wallerius to

Linnæus as Adjunct at Lund. Now if Rosén dreaded Linnæus as a competitor for Rudbeck's chair when it became vacant, would he not willingly have seen him settled at Lund, awaiting old Professor Döbeln's resignation or death?

It was widely known at this period that the medical faculty at Lund was in an unsatisfactory state. The University's powerful Chancellor, Carl Gyllenborg, had, so far back as 1731 (that is, before he knew Rosén as a physician), sent in a humble memorial to the State Secret Committee, stating that Lund needed an Adjunct. This produced no result, but in the following spring, the Consistory was astonished to get a precept from the Chancellor, requiring them to appoint Johan Gottschalk Wallerius, Adjunct in Medicine at Lund. As no funds were available for stipend, the Consistory was moved to permit him to receive a double Royal Scholarship. This was effected in May, 1732, and Wallerius remained in this post at Lund till 1741. He left on account of some unpleasantness, partly due to his delay in writing a disputation, partly due to his marriage. His first thesis did not appear till 1740, and in the next year he stood as a competitor against Linnæus, as will be shown in due course.

If it be asked, what share Rosén had in instituting the Adjunctship and in nominating Wallerius, it is evident that Count Gyllenborg's decision was taken before Rosén came home. In the summer of 1731 he was drinking the waters of Wiksberg, near Södertälje, but that the newly arrived Rosén was physician there, is not certain, indeed hardly credible. It is, however, certain that when Gyllenborg was living in Stockholm, he was his medical attendant. That they conversed on medical topics and that Wallerius was recommended by Rosén may be regarded as correct.

It thus appears, (1) that any measures for the post of a medical Adjunct at Lund was not granted; (2) that no such place could be allotted until vacant;

(3) that if Linnæus in 1733 announced himself as seeking it (no such application can be found), he came too late, for Wallerius received his warrant in May, 1732; also (4) that the "morsel of bread," which Linnæus was supposed to have lost, was only the double Royal Scholarship, that is to say, less than the income from the surplus of the Wrede Stipendium, which he was counting upon.

To put another point: which of the two, Linnæus or Wallerius, was the best fitted for the post of Adjunct? It is apparent that the former was pre-eminent in botany and zoology, while the latter had greater experience in practical medicine, owing to his long enjoyment of Rosén's instruction. Add to this, that Linnæus was only a student, while Wallerius was a qualified Master in Philosophy. The former had never held a disputation, but the latter had done so on three occasions, the last time being on a medical subject, wherein he had showed high merit. Thus no thought of injustice can be imputed to Linnæus. The future also showed the high attainments of Wallerius, as he became one of Uppsala's most celebrated and eminent professors.

Rosén has been reproached for his proposal put forward in 1730 concerning the promotion of doctors in medicine. Linnæus mistakenly took this to be directed against him personally, and regarded Rosén as the originator of an intrigue. In this respect it is easy to perceive that the latter was innocent. Probably Linnæus was misled by loose and distorted reports. The actual state of things was thus: In the academic constitutions the faculty was recognized as having the right to promote to the doctorate. On this ground, Lund had once, in 1689, granted this dignity. At Uppsala, in 1680, the Chancellor had submitted—that a Licentiate (graduate) should be promoted to Doctor, with the sanction of the authorities, but the faculty raised difficulties, and the project fell through. In 1697, however, the day for such promotion was fixed, but difficulties arose. The consequences were, that an understanding was formed, that the degree of doctor of medicine should only be obtained abroad.

At this time the Medical College at Stockholm suggested an alteration, asserting that many Swedes who received the doctor's diploma abroad, went preferentially to such universities. The college now therefore urged that the Swedish Universities should be empowered to grant degrees, after due examination. Rosén had the slenderest share in this suggestion, as he was engaged in a lively conflict with the said college, which was decided, after examination by the King, in Rosén's favour.

The academic Consistory was blamed, as well as Rosén, for its action against Linnæus when even his benefactors, Olof Celsius and Rudbeck, were thought in the end to have failed him and his just cause. The chief charges against it were, that the professors had met Rosén's desire, to apply both old and new regulations, thereby rendering relief to his wounded vanity. In other words, the members of the academic Consistory were thought to have gladly seized the opportunity to rob Linnæus of his right to teach, whereas they only vindicated the annoyance they felt at seeing their auditorium empty, while a young man, raised up and idolized by the academic youths, received their vociferous confidence and "conducted them round Flora's delightful flowery field as the interpreter of Nature, unmindful of interpreting Cicero and Demosthenes."

It may at once be pointed out, that not a single word in the slightest degree can give support to these accusations in any of Linnæus's notes, letters, or printed writings; they are exclusively the unrestrained fantasy of the author here cited. Thoughtless repetitions have not been wanting, but even these, on reflection, should have been discarded, when they wrote such statements as "The fathers in Uppsala were at one to drive Linnæus from the University," when as a fact they did not even condemn him to the most modest punishment.

What was the reason of the empty audience chambers? Linnæus, as Rudbeck's deputy, gave instruction by lectures on plant demonstrations, from two to four times a week in the afternoon from the beginning of May to Midsummer, and arranged botanic excursions into the country twice a week during the same period with the addition of a few paying members. It is hard to see that this should have resulted in all the professors having to lecture to bare walls, especially as according to that year's list of lectures, none were held at the same time as those of Linnæus, and nearly all in the forenoon. Far from any ill-will being shown to Linnæus in the academic records, he was mentioned more often with commendation than any other student, and his services as Rudbeck's deputy were specially acknowledged. The Consistory even tried to support him by stipends as already mentioned. After he had quitted the University, an applicant for the surplus of the Wrede stipend had for answer, that he must wait until the said scholarship was vacant. That Linnæus did not receive the travelling scholarship did not depend upon any ill-will, but simply that it was not vacant, and moreover he was not eligible for it.

The accusations, calumnies, and spite, which for more than a century have been lavished on Linnæus from named or nameless "enemies, envious persons and persecutors," have now been subjected to close scrutiny. It seems to us high time to reduce these accusations to their true value, so that future biographies may be spared the erroneous statements which have too long been taken for confirmed truths. Linnæus's student days offer so much instruction, uplifting and wonderful, that one is not obliged to illuminate them with an invented martyrdom. It must be a duty and pleasure for his biographers to remove the ugly blemishes, with which unreflecting people have tried to defile his character and actions. If during this period he really had enemies, and if this were the real reason for his travels in Lapland and Dalecarlia, his stay at Falun and so on, then assuredly they directed his progress in science in so fruitful a manner, that not even his best friends and helpers could have devised a better or happier result.

With what feelings Linnæus himself looked back upon his student years at Uppsala appears from the words which, on his hour of departure, he wrote in his diary:

"1734, Dec. 19. At eight in the morning I said good-bye to Uppsala Academy, to which Almighty God so marvellously conducted me, living now in difficulty, now in enjoyment, now in poverty, now in abundance; now in blame, now in honour. To Thee, Great God, be thanks."

The journey which was now entered upon was not to foreign parts, many reasons inducing Linnæus to go to Falun. Through rain and sleet the way lay to Sala silver workings, which he reached on the 20th, quite cold and frozen, where he received the hospitality of Assayer Stockenström for a couple of days. Next he arrived at Hedemora and lodged with subdean Anders Sandel, resuming his travels the next day, and arriving at Falun on Christmas Eve. Here he was a guest in Inspector Sohlberg's house, where the whole of Christmas passed in the greatest pleasure, the festivities being continued till past Twelfth Day.

It must not be supposed that all his time was given up to pleasure. On the contrary, he devoted himself with great energy to the revision of certain of his writings, which he was taking with him. He has recorded that he made a new edition of "Systema mineral.," began "Sponsalia plantarum," and completed his "Flora dalecarlica"; besides this, he wrote letters, visited the sick, and investigated a mineral spring.

However, it was something quite different which

took the most important place in Linnæus's thoughts. During the preceding autumn in conversation with Johan Browallius, the latter pointed out that to provide means for foreign travel, the best way would be to marry some rich girl, who would make him happy. This theoretically pleased Linnæus, but he made no attempt to carry out the plan. Now during this Christmas, he met with the eighteen-year-old maiden, Sara Elizabeth Moræa, who seemed very attractive, and he soon began assiduously to wait upon her. Thus in his jottings in an almanack, he has noted that on the 2nd January, 1735, he called on her in his Lapp costume, and again on the 3rd, when the parents were out; on the 10th and other visits, and on the 15th he was a guest of the Assay Master in Falun, with his sweetheart, and finally on 16th he records a delightful day spent with her. His feelings for the chosen one apparently did not escape the notice of others, for when he on the 19th was the guest of the artist Trygg, it went so far that the host or some of the company wagered two cans of Rhine wine, if a christening did not happen in four years.

Now began unrest and trouble. The young girl and Linnæus noted that if he had first gained the parents' esteem, no one else would have become his affianced. But he now had to encounter difficulties. Her father was the town physician in Falun, Assessor Dr. Johan Moræus, learned, experienced, and well to do. He liked Linnæus extremely, and so was often visited by the latter. Moræus had repeatedly declared that the practice of medicine, with regard to income was more precarious than any other profession. therefore he had resolved that none of his children should follow it. This, with the consciousness of his own narrow economic position, could only awaken despair in Linnæus, a simple student. He realized that he was a poor man who could not maintain a wife, while she was wealthy. He knew that she was courted by many eligible suitors, but to cast her out of his thoughts was impossible. A decision must be made. After long thought he made a declaration to her parents asking for the love of their dearest child. This happened on the 20th January, but the answer was withheld till the 27th, for though the father, who entertained the best hopes for Linnæus, had not the heart to refuse him, the mother cherished other ideas. When Linnæus at last received the desired yet dreaded answer, he found to his surprise, that it surpassed his boldest expectations, but the wedding was not to take place till three years after his departure on his foreign travels. During the suspense he had not failed to visit his beloved, and on the 22nd he gave her the betrothal ring.

The month which followed seems to have been taken up with a lover's usual thoughts and occupation, exchanging visits with friends and acquaintances. Visits to the father and mother-in-law elect, did not become fewer, and on the 3rd February he gave a written declaration of fidelity. Meanwhile the hour of parting drew near. The 18th and 19th were employed in leave-taking and in receiving congratulations and presents. He had agreed with his intimate friend Browallius to take charge of his sweetheart's letters, while he was living abroad.

It was on the 20th February that Linnæus and his travelling companion, Claes Sohlberg, set out from Falun, provided with the usual passports from the Royal Council. As to their equipment, we only know that Linnæus had at least the chief of his manuscripts, and his Lapp costume, which in Holland afterwards occasioned much attention. His means, he himself reported, consisted of 260 silver dalers (less than £20), an amount less than he had reckoned on beforehand. He had expected to receive from Inspector Sohlberg an annual allowance of 300 copper dalers [£7 10s.] and he had a claim on the two Sohlbergs of 30 plåtar [£4 10s.] due on their agreement for assaying. But when the journey was begun, these 300 dalers were forgotten; and he received only 12 plåtar [£1 16s.] as remuneration. He could not draw back from the journey as arrangements had been made, nor could he reproach the old man, the Inspector, who had boarded and lodged him for half a year. He therefore committed himself unto God's hands, who had directed him so wonderfully hitherto, and determined to serve his travelling comrade with all fidelity, knowing that God repays according to one's deserving. He had his savings, the income from his medical practice, a contribution from Moræus as a token of affection, and a little purse from his betrothed. He reflected on his father's assurance of the help of the Almighty when he went to Lund, and now he committed himself to Providence.

Journeying south, progress was slow, as they stopped to inspect mines and works, but they reached Jönköping at last, where they paused for four days. The next stopping place was Växjö, where Linnæus was entertained for five days by his old teacher, Rothman, and the Governor; then he reached his old home at Stenbrohult, where he found his old father, brother and sisters. The mother had died since his last visit, and the house was in confusion.

Here Linnæus and Sohlberg stayed for a whole month, recording few notes. His father since his wife's death was much depressed, and, dreading the future, committed the youngest daughter and his library to the care of Linnæus, in case of his own demise. He also lamented that he could not add to the travelling purse of his son, though fearing that Carl, on account of his scanty means, might have to remain abroad. All that Linnæus asked of his father was a soft skin, which he made into a money belt.

On the 15th April he bade his sixty-year-old father and the family farewell, setting out in splendid weather, and amid all the signs of spring; they reached Helsingborg on the Sound two days later, procured passports to cross the channel, and went on board at half-past five of the 19th. Thus was begun the journey abroad from which Linnæus returned three years later, not as an insignificant student, but as one of that period's eminent naturalists, a celebrated and esteemed man of science.

CHAPTER VI

RESIDENCE ABROAD—RETURN TO SWEDEN
(APRIL, 1735—JUNE, 1738)

AT that time Helsingör was the central point for Scandinavian transit outwards. This took place exclusively by sailing boats, and as the traffic could not be regulated as are modern tourist passages, it was necessary to take advantage of any vessel sailing to the desired port of arrival. In this respect Helsingör offered the best opportunities, as all sailing craft had to call there for the local customs.

To await events Linnæus and his companion took lodgings for several days. They employed their time in seeing the town and environs, having as guide the Swedish Consul Slyter. A few plants were noted, but in other respects there was nothing noteworthy. The town, though well built, had houses of brick-nogging with tiled roofs and pumps everywhere. They particularly noticed the soldiers in red uniform. Linnæus was least pleased with the inhabitants, and recalling the kindness and generosity enjoyed in Norway, he recorded that the people here were entirely different from those in the north.

Their intention was to sail direct to Holland, but hearing of no vessel bound thither the plan had to be altered. Both travellers eventually embarked on the Lübeck " The Travelling Tobias," which with sixty other vessels of different nationalities was waiting for a favourable wind. The food was worse than in the town, though as costly. Rye bread as white as wheat bread and chestnuts were good, and they had porridge for dinner and supper. Only French wine was drunk, and although of fair quality Linnæus wearied of it, and longed for pure water.

At last at dawn of the 24th, the wind came from the north-west, and it was amusing to see how all the vessels in the Sound in one instant hoisted sail and raised anchor. The journey was now south, and land was soon out of sight. Linnæus himself escaped sea-sickness, to which fact he ascribed his use of the sailors' customs to lying fore and aft (not athwart) and of drinking sea-water. His companion was ill, and there was little pleasure in sailing, as the ship lay over on her side, and one became afraid of every lurch. Meanwhile all went well, and on the 26th the vessel anchored at Travemünde. The journey was continued by road to Lübeck, which was reached at noon. "Here it was most splendid summer, the country was a paradise consisting of level fields and splendid cornfields, with beech and oak woods in the valleys. The heaths were adorned with gorse with its fine yellow flowers."

A couple of days were spent in Lübeck, where, however, there was nothing to be gathered of medical or biologic interest, the doctors there being of small repute as scientific men. There were, however, a few things to attract his attention, as for instance in the streets there were four-sided lanterns on poles, which burned all night, and at every street corner there were pails of water to extinguish fires. On Sunday they went to church, but complained of the long psalms which were sung. The men were garbed in black, with black capes, though it was not raining.

At six a.m. on the 28th, the start was made from Lübeck in a diligence drawn by six of the biggest riding horses, and at six p.m. they gained Hamburg, though not without a little adventure. The driver had taken his team so near a cornfield, that the farmer was moved to strike him and knock him off his horse. When the quarrel had lasted an hour, Linnæus advised the peasant to sue the driver at law, and not to delay the post. The red-faced fellow turning upon him with his axe, Linnæus would have tackled him, but was prevented by the other passengers.

The travellers halted at Hamburg till the 16th May, and here Linnæus enjoyed himself, visiting the pretty gardens and other noteworthy places. At the same time he made acquaintance with the resident naturalists, who showed him much politeness, entertained him well, lent him books, showed him their collections, and drove him round in and outside the town to see libraries, museums and gardens. The first visit was naturally to Johann Peter Kohl, who in his "Hamburgische Berichte" had already made known the name of Linnæus among the learned. He was remarkably polite and showed him every attention: Linnæus had also the pleasure of reading his own name many times in the said "Berichte," and always mentioned with respect. Among others who entertained him may be named, Gottfried Jacob Jænisch and Johann Heinrich von Spreckelsen, Licentiate in Law, in whose beautiful garden were many exotics and orange trees. He had a large number of books on botany in his library, and also possessed so many fossils that Linnæus had never before seen so large a collection. Johann Alb. Fabricius, Doctor of Theology, showed him his extraordinarily extensive library, many rooms being lined with books in place of tapestry. The great drug merchant, Natorp, took him to his house, where he saw numerous preserved lizards and snakes and many other rare things. But he did not omit to take a survey of notable buildings in the town, such as the Exchange, the Synagogue, and the old Reformed Church which was then turned into a vast wine-cellar, etc.

It is plain that Linnæus found himself very happily situated in the splendid town of Hamburg, with its fortifications, fine houses, handsome people, pleasant, lively, and French in manner. The reverse side was the disagreeable pervading smells, or rather stinks, the result of neglected cleansing, and the immorality which was not restrained but openly practised in the disorderly houses between Hamburg and Altona, "where flutes, oboes, dulcimers, trumpets and waltzes were constantly heard."

What was the impression made by the young Swedish student, hardly of mature years, on the noted naturalists in the foreign town? The answer to this question is found in a long article in the "Hamburgische Berichte." Here are given some titles of the works which Linnæus had brought with him to get printed, also a description of his collection of nearly one thousand rare insects, found in Lapland and Dalecarlia; a picture of his complete Lapp costume with the magic drum belonging to it, whose use was described, and so on, which awakened the liveliest interest and the greatest surprise.

His thoughts and deductions were methodically recorded, and that he possessed an uncommon judgment in conjunction with an inborn power of observation, is certain. His ardour, endurance, and energy were unusual. In the desire to search out and discover such things as had hitherto remained hidden from the sharpest eyes, in all that appertained to the three kingdoms, he had few equals. Moreover he was active in reading and noting, and thus had acquired great experience and such well-founded insight in many directions, that he, though only twenty-eight, in this respect stood out from many older persons. His intellectual precedence was adorned by an equal excellent temper, for amongst learned men he was distinguished for modesty, together with a natural straightforwardness, love of truth, genuine piety, readiness to oblige, free also from envy or jealousy, and possessing a constant great love for mankind.

Everyone was pleased, and probably Linnæus would have stayed longer in Hamburg, where the libraries and museums had so much of novelty to show him, but he felt himself obliged to continue his journey.

Among the chief objects in natural history in Hamburg was a seven-headed snake or "hydra" which had in the year previous been drawn and described in Seba's "Thesaurus." This monster was stated to have had its place on the altar of a church in Prague, which in 1648 was Königsmark's share of booty; after his death it was inherited by Count Bjelke, and after changing fortunes, at last came to Hamburg, where it was kept in the collection of Burgomaster Johann Anderson and his brother. Many said it was the only one of its kind in the world, and thanked God that it had not multiplied. It was related that the Danish King Frederick IV vainly offered 30,000 thalers [£4,500], but since then the price had sunk to 10,000 florins and 4,000 rixdalers [nearly £900]. At the time when Linnæus was in Hamburg, negotiations were being carried on to sell it for 2,000 thalers [£300]. Naturally, Linnæus was particularly anxious to inspect this marvel, and by Kohl's help succeeded. It only needed a short examination of the beast, whose movements were ostensibly actuated by seven different brains, for Linnæus to exclaim, "Great God, who never put more than one clear brain in one of thy created bodies." He perceived at once that the heads with their gaping jaws and the two feet provided with claws, belonged to weasels, and that the whole covering of the body consisted of snake-skin pasted thereon. Evidently this hydra was just the opposite of that certified in Seba's work as "Nullement l'ouvrage de l'art, mais véritablement celui de la nature." Linnæus saw that it was doubtless made by the monks as a representation of the dragon in the Apocalypse, and that the learned people, both old and young, in their credulity, had been deceived. This conception was not disturbed by the second visit which he made on the day before he left Hamburg.

That he did not refrain from telling of his discovery was natural, and equally natural was it that thereby the outrageous price set upon it fell at once to nothing. Linnæus and his friends feared that the Burgomaster on his part would make trouble about it. So, probably on the advice of Dr. Jænisch, who Linnæus afterwards declared was his only true friend in Hamburg, he decided to continue on his journey. His fear of Anderson's "revenge" may, however, have been superfluous, as the unveiling of the deceit does not seem to have occasioned great remark. At least Kohl relates shortly after, in a letter to Linnæus, that since his departure, nothing more was heard about the *Hydra;* for it might be in the owners' interest for the unpleasant discovery to be buried in silence.

It was the 16th May when Linnæus and his travelling companion bade farewell to their friends in Hamburg and prepared to depart. They took their way to an inn in Altona, then a Danish town close by, and the next day early they stepped on board the Hamburg vessel, paying one ducat [9s. 2d.] each for passage to Amsterdam.

The voyage began in a grievous storm of rain, succeeded by adverse wind which forced them to anchor during Whitsuntide; Whit-Monday they went to church, which was decked with leaves like a forest, hiding even the pulpit and altar. Naturally, opportunity was taken to make observations in natural history, noting that the frogs croaked far louder than in Sweden, and each had its own voice. Some sang as if they were lately fed, and some so badly, that one might die of melancholy.

Sailing was resumed on the 20th, but the wind was unfavourable and progress slow, and more than once they were able to go ashore and obtain some objects of interest. Passing the coast of Groningen and West Friesland on the 30th, they practically left the sea, but, having experienced a violent thunderstorm, it was not till the 2nd June that they reached Amsterdam. Here Linnæus spent a few days sightseeing and visiting Jan Burman and Seba; he was surprised at Burman's extensive library and Seba's incomparable apothecary's establishment. Then they travelled across the Zuyder Zee to Harderwijk, arriving early the next morning.

Harderwijk, a small town in Gelderland, then boasted a university, which was greatly frequented by foreigners, especially Swedes, wishing to obtain the degree of Doctor of Medicine. Here not only Linnæus's teacher and benefactor, Rothman, but Rosén, and most of the then members of the Collegium medicum in Stockholm, had obtained the same dignity. This may have been the reason that made Linnæus prefer this university to Leyden, which undoubtedly could claim to possess more eminent professors, both in medicine and natural history. A contributory reason may have been that Harderwijk did not require long residence, and thus demanded less economic sacrifice.

On the day of arrival, Linnæus looked over the University and called on the Professor of Mathematics, J. H. van Loms. On the 7th of June O.S.—18th N.S.—his name was inscribed in the Album studiosum of the University, and on the same day he was sufficiently and sedulously examined in general medicine. He took as his subjects two aphorisms of Hippocrates, upon a diagnosis, prescribed for the treatment of a case of jaundice, when he, with deep learning, encountered all the questions concerning doubtful points and arguments; the result being that he was declared Candidate of Medicine. His thesis already prepared in Sweden, "Hypothesis nova de febrium intermittentium causa" [New hypothesis as to the cause of intermittent fevers] in which he sought for the causes of ague in certain parts of Sweden being so frequent, by assigning the drinking of clayey water as the determining act—he had previously left for inspection in readiness for his examination, with Professor Jan De Gorter, who returned it on the 19th, marked as usual, "imprimatur." The same day, though Sunday, it was handed to the printer, who so hastened his labours, that it was presented on the 24th, when the author calmly responded to the remarks of the official opponent. When thus, in every respect, he made manifest "a praiseworthy education and distinguished medical knowledge," he was on the same day invested by the rector of the University, the previously mentioned Professor De Gorter, with delivery of a gold ring, a silk hat, and a diploma, promoted to the position of Doctor of Medicine, and thereby recognized as having, after the usual declaration, the right to advance to the upper (or doctor's) chair, publicly to justify medical treatises, to teach the craft of a physician, to visit the sick, to prescribe for them, to hold disputations, to promote for graduation with all other kindred matters, to exercise the duties of a physician. In addition there were committed to him all powers, privileges, dignities, prerogatives, and doctoral insignia, which by law and custom in any university whatever, were extended to an actual lawfully-promoted Doctor of Medicine. The same day his name was inscribed in the Album doctorum; following this De Gorter signed his name, also expressing his good wishes in his "brief," and Linnæus departed from Harderwijk. Naturally he did not omit to inform his friends at home of the rapid progress made.

Until this time Linnæus had done particularly well on his journey, but now he had spent all his money. He would gladly have gone home direct, but money was wanting, so he accompanied his comrade, Cl. Sohlberg, to Leyden, where he intended to pursue his medical studies, inasmuch as he did not care to apply to his prospective father-in-law, full well knowing his disposition. Their course was through Amsterdam, a short halt being made there to visit Professor Burman again and look at the nurseries. The way continued by Haarlem to Leyden, where they arrived on the 29th June. Linnæus inscribed his name as studying in the University, in whose album it remained as late as 1739.

Unquestionably it now looked black for Linnæus, so far as the realization of his hopes to get the books he carried with him printed. What helped him was the same quality which had already stood him in good stead in Lund, Uppsala and Falun, namely his great power of winning confidence. To push forward his new, and for that time, daring, almost revolutionary scientific views, his devoted friends did not hesitate, though with no small trouble and sacrifice, to try and smooth his path. He had not been long in Holland, but he had had time to secure many such friends, some of whom may now be mentioned, because they came to exercise a considerable influence on Linnæus's career.

One of these was a senator in Leyden, Dr. Jan Fredrick Gronovius, whose keen interest in natural history gained for him the testimony that "he was the most inquiring man Linnæus had met in Holland, and his herbarium had not its equal." On returning Linnæus's visit, Gronovius saw his "Systema Naturæ" in manuscript, with great astonishment. Being well to do, he wished to publish the same at his own expense, and a common friend, Isaac Lawson, a learned Scot, who had travelled much but was then in Leyden, joined in the same request. Linnæus thankfully accepted this offer; it was put in hand on the 10th July, but the printing progressed very slowly, lasting into December. Thus this celebrated work, the naturalist's golden book, saw the light, and at once displayed the author to be of sharp intelligence, with insight, and a courageous reformer of system in all three kingdoms of nature. As to its extent, it was extremely modest, only eleven printed folio pages, but its great worth appears from this circumstance, that during Linnæus's lifetime, no fewer than sixteen ever-increasing editions or reprints came out. In the preface, dated 23rd July, 1735, Linnæus openly declares that if an interested reader should gain profit from this little work, he had exclusively to thank Gronovius and Lawson; and these thereby not only laid the foundation of Linnæus's world-wide reputation, but secured for themselves the warm gratitude of all naturalists.

A still more eminent and influential helper was gained by Linnæus. This was no other than the venerable old man, Herman Boerhaave (1668-1738), regarded in the whole of Europe as the chief medical oracle of the time, "Hippocrates redivivus." The position which he had filled with so great renown in medicine, botany and chemistry, he had, it is true, relinquished in 1729, but he still continued (till a short time before his death) to impart medical instruction, for which advantage pupils came from all countries to Leyden. Also for botany, in which science he had engaged as author, he retained a warm interest, and great was his pleasure when once a week he betook himself to his country seat outside Leyden, and there, free from other vexations, he could refresh his mind in working in his "arboretum" or park, where grew every kind of tree that could bear the climate.

Clearly Linnæus would not willingly have left Holland before meeting with this great man of science. His immense reputation as practising physician caused him to be overwhelmed by patients seeking advice; his abundant wealth made it possible for him to confine the number of his visitors within measurable limits, without regard to their condition or means, and even these were admitted in turn, a measure which his age and decreasing strength demanded. Unfortunately it was generally reported that his servant for his own profit only admitted those who could or would employ jingling methods of persuasion.

Conscious of his own weakness in this respect, Linnæus despaired of getting his wish fulfilled, but by advice of Gronovius, he decided to make an attempt by letter, craving the favour of an audience. This succeeded beyond expectation; after a week's waiting, the great Boerhaave received him with great kindness and wrote in his "brief" as follows:

SIMPLEX VERI SIGILLUM.
EX VOTO CLARI NOBILISQUE
VIRI
CAROLI LINNÆI
SCRIPSI
FAUSTISSIMOS EIDEM LABORUM DURISSIMORUM EXITUS
PRECATUS
H. BOERHAAVE.

LEIJDÆ, 17 $\frac{5}{7}$ 35.

A fabulous account was published that Linnæus gained admission by sending a copy of his "Systema,' but as the above was written on the 5th July, and the printing of it did not begin till the 11th of the same month, and ended on the 13th December, it is manifest that this account is entirely false.

A few days after this, they met again in Boerhaave's arboretum, "a paradise, Holland's miracle, whose like no mortal can imagine," and in the walks round it, Linnæus was able to show his insight in botany and its literary history. An old tradition, which is not intrinsically unlikely, and is to some extent supported by Linnæus's own words, exemplifies an episode from this remarkable meeting. In the garden stood, it is said, a tree which Boerhaave regarded as a rarity, that had not yet been described by any naturalist. To his complete astonishment Linnæus declared that it was well known to him, and that it grew abundantly in Sweden; further, it was not only not undescribed, but was included by Vaillant in his great "Botanicon Parisiense" as "Cratægus folio subrotundo laciniato et serrato." This Boerhaave himself disputed, as he had in 1727, on the suggestion of the English botanist, William Sherard, obtained the said work. Nevertheless, as Linnæus stood to his statement, the book was sent for, and the description of the tree was found and confirmed. It was the White Beam, which in the greater part of Europe does not occur in a wild state.

It is certain that Linnæus at his first meeting with Boerhaave made so good an impression, that he found in him a friend whose benevolence was not evanescent, but on the contrary, remained warm and unchanged to Boerhaave's last hour. After his death, Linnæus was able to say, " With Boerhaave I have lost the most devoted friend, the most obliging teacher, the best benefactor. The memory of my medical father Boerhaave, I shall ever hold constantly in honour."

The first testimony of Boerhaave's estimation of Linnæus's work in botany, and his unquenchable love for it, was his offer to send him to the Cape, there to collect plants for two years, for the University garden in Leyden, and afterwards proceed to America. He promised that he should travel free, and on his return should enjoy the status of a Professor with suitable emolument. The offer was unquestionably tempting, but before deciding, Linnæus consulted his friends in Sweden, particularly Olof Celsius. The latter suggested a plan for a visit to his friend Dillenius in England, and as Linnæus was well versed in assaying, he urged an application to Peter Collinson in London, who would provide a suitable post in America, where the English had mines, and added, that if this project fell through, he ought to conduct a diligent correspondence with Dillenius. Celsius seems to have discouraged the projected journey, as he well knew it would not suit him. But before this letter came, Linnæus had determined to decline the offer, with the excuse that he could not bear hot climates, having been brought up in a cold one; the true reason, however, was that his affianced bride in Sweden held him back. (" Sweden " was the name of Moræus's country house outside Falun.)

During this period of uncertainty, Linnæus found himself tolerably active. At one time he went to the seashore and botanized, then to Amsterdam, back to Leyden, to Utrecht to view the University garden, and next day once more to Leyden. He then decided to go back to Sweden without delay, but when he went to take leave of Boerhaave the latter advised him by no means to quit Holland at once, as he intended, but to settle down and live in the Netherlands. However, as Linnæus meant to pass through Amsterdam to Sweden, he begged him to call on Burman. The latter was born in the same year as Linnæus, at twenty-one years of age became Professor of Botany and manager of the Botanic Garden at Amsterdam, and died in 1780.

The next day, 2nd August, he started for that town, and hastened to discharge Boerhaave's instructions. During their conversation Burman asked what plants in his herbarium Linnæus wished to see. " I should like to see many, perhaps all," was the answer, " but I do not know what you possess." On which Burman handed him a dried plant with the remark that it was rare. Linnæus took a flower, moistened it in his mouth, examined its parts, and declared it to be a *Laurus*. " That is no *Laurus*," said Burman. " Yes," answered Linnæus, " and a *Cinnamomum* into the bargain." " True, it is a *Cinnamomum*," admitted Burman, and then Linnæus gave reasons for uniting both genera. So the talk went on, the result being that Burman asked Linnæus if he was willing to help him in working up Ceylon plants on which he was then engaged. He also offered him a fine room, service and board, and Linnæus closed with the offer till the following year. As he at the same time received a bill of exchange for 200 silver dalers [£15] from Sohlberg, he could view in peace at least the near future, and further found to his delight, the opportunity of printing in Holland the treatises he had ready. On the 9th August he handed his "Bibliotheca botanica" to the press, and soon after his "Fundamenta botanica"; the former being dedicated to J. Burman, "as a lasting remembrance of the special friendship and kindness with which he treated me during the time when this work was in preparation." Recreation during this strenuous period was provided by his enjoying himself in looking over Burman's work on Ceylon plants, and diligently visiting the *Hortus medicus* at Amsterdam.

During one of these visits, there happened an occurrence which in high degree recalls the first meeting of Linnæus and Olof Celsius in the Uppsala botanic garden in the spring of 1729. While he was living in Holland he had certainly heard of the Director of the Dutch East India Company, Georg Clifford, LL.D., a very wealthy man, who took a particular delight in botany, and had set up an incomparable botanic garden on his estate between Leyden and Haarlem, the maintenance of which costing him annually 12,000 gulden [£1,000]. He believed that this said garden could not be materially different from many others, "which cover the highly cultivated Holland," so that he neglected to visit the same. When one day he was wandering round the medical botanic garden, he was accosted by an entirely unknown person. This happened to be the said Clifford, who invited Linnæus to pay a visit in the company of Burman to his country house at Hartecamp, to take the foreign plants and rare animals there into closer observation.

For the kindness thus displayed to an unknown foreigner, Linnæus had to thank Boerhaave. The latter was Clifford's physician, and on one of his visits to obtain relief for his hypochondriac trouble, Boerhaave declared: "You cannot live a happy life unless a physician is daily with you to watch over your meals and the rest of your diet, etc., and to carry out my advice, if anything more serious should happen." On Clifford declaring that he would willingly take that advice if only such a physician could be obtained, Boerhaave answered, "There is a Swede whom I can recommend, and being a botanist as well, he can look after your garden also."

The visit to Hartecamp took place on the 13th—14th August, and Linnæus's boldest anticipations were surpassed. "My eyes," he says in a dedication to Clifford, "were enchanted by so many natural objects, of masterpieces supported by art, alleys, plant-beds, statues, ponds and artificial mounts and labyrinths. Your menageries delighted me, full of tigers, apes, wild hounds, Indian deer and goats, South American and African swine; with these mingled flocks of birds, American hawks, various kinds of parrots, pheasants, peafowl, American capercailzies, Indian hens, swans, many sorts of ducks and geese, waders and other swimming birds, snipe, American crossbills, sparrows of diverse kinds, turtle-doves, with innumerable other species which made the garden re-echo with their noise.

"I was astounded when I stepped into the plant-houses, full as they were of so many plants, that a son of the north must feel himself bewitched and struck with wonder when he thought of the distant lands from which they were brought. In the first house were kept many kinds of flowers from southern Europe, such as Spain, south of France, Italy, Sicily and Greece. In the second, treasures were found from Asia, such as cloves, *Poinciana*, mangosteen, coco-palms and other palms as well. In the third, Africa was represented with its peculiar, not to say scientific plants, such as *Aloë* and *Mesembryanthemum* with their numerous forms, carrion-flowers, euphorbias, *Crassula* and *Protea* species, etc. Finally in the fourth house were cultivated the delightful natives from America and other parts of the New World;

great numbers of *Cactus*, orchids, crucifers, yams, magnolias, tulip-trees, calabash-tree, arrowroot, *Cassia*, acacias, tamarinds, peppers, *Anona*, manchineel, cucumber-tree, and many others, and encompassed by these, pisang (*Musa*), the stateliest amongst all the plants of the world, the magnificent *Hernandia*, silver-leaved *Protea* and the valuable camphor-tree. When I afterwards came into the truly royal residence and into the most instructive museum, whose collections spoke no less of their possessor's renown, I felt, as a foreigner, quite transported, because I had never seen their equal. My earnest wish was that I might lend a helping hand to their preservation."

Linnæus's feelings and wishes being as stated, Clifford was no less eager to see Boerhaave's plan realized, and all the more, as he had received a letter from Gronovius candidly and sincerely setting forth the great advantage he would obtain by Linnæus taking charge and regulating the herbarium, and the rest of the natural productions, as well as drawing up a *Hortus hartecampensis*. He remarked the peculiarity about Linnæus, that he knew the Indian plants, which he had never seen, as soon as he had opened a flower and counted its parts. This set him wondering, and he therefore came forward with his proposition, that Linnæus should exchange Amsterdam for Hartecamp. Notwithstanding the great desire Linnæus had for this change he considered himself bound by his engagement with Burman, and the latter seems to have been little disposed to relinquish his newly acquired, helpful coadjutor. In a visit to Clifford's rich library, Burman, with delighted astonishment, beheld the second volume of Sloane's great work, "A Voyage to the Islands Madera . . . and Jamaica," and gave utterance to his great pleasure. "I have two copies of it," Clifford hastened to say, "and I will give you one, if you will give up Linnæus to me." The decision was now left to the latter. After a journey to Leyden on the 18th of

August, probably to take the advice of Boerhaave, he accepted the offer, and bound himself to remain at Hartecamp over the winter, on condition that he should have free board, housing and a salary of 1,000 florins per annum [£83 6s. 8d.]. The next day he made haste to inform his friends at Falun, Inspector Sohlberg, Magister Browallius, and naturally, Sara Lisa Moræa. On the 13th September he removed to Hartecamp and took up his duties. His principal conducted him into the plant-houses in which were certain plants unknown to Linnæus, especially some from the Cape. After investigation he was able to assign names to some, but others he declared were still undescribed, which highly pleased Clifford. His own contentment Linnæus expressed in a letter to Gronovius, describing himself as being in a paradise.

Burman and Clifford were not the only persons who wished to make use of the young Swede's knowledge and industry. It has already been mentioned (p. 138) that the rich apothecary, Albert Seba, was living in Amsterdam, where he had a fine and valuable collection of illustrated works. He was now engaged in prosecuting the same, but old age, ill-health, and perhaps inaccurate views, put hindrances in the way, causing him to turn to Linnæus for help. But Linnæus, now engaged by Clifford, could not undertake the commission, and besides, the third volume next to be printed, was on the subject of fishes, least liked by Linnæus. Circumstances, however, enabled him to make use of one of his friends, an Uppsala comrade, Petrus Artedi. He had left Sweden about the same time as Linnæus to pursue his studies abroad. He first went to England, where the naturalists, especially the celebrated Sir Hans Sloane, received him with the greatest kindness and gave him opportunity in his own and other museums and collections to add to his already considerable knowledge concerning fishes, then a scarcely known class. He stayed in England until he found himself obliged by his diminishing funds to go to Holland. During a visit by Linnæus to Leyden, the two friends met unexpectedly on the 8th July, and tears in the eyes of both testified to their emotion at the unforeseen meeting. Each had much to impart to the other since they last met, about works and future plans. Artedi's greatest wish was to obtain the degree of Doctor in Medicine, but as his means were nearly exhausted, he saw no possibility in a foreign land of obtaining food, clothes and needed books, so was thinking of returning to Sweden. "But Linnæus comforted him, that he was no longer at Uppsala under restriction and persecution, and prophesied that if he would be careful all would be well." He persuaded him to go with him to Amsterdam, where they visited Seba, with the happy result that Artedi was installed as Seba's helper, and he so diligently devoted himself to describing the fishes, which were to be included in the volume of Seba's work now in course of preparation, that only a few remained undescribed.

Linnæus thereupon returned to Hartecamp, but as soon as the preliminaries to his "Fundamenta botanica" were finished, he hastened to Amsterdam to meet his friend. Artedi, generally of few words, produced all his manuscripts which he had never shown before, went through them and said that as soon as Seba's work was finished, he would take up the final revision and polish the same, so that they could be printed before his homeward journey Although Linnæus was much engrossed by other objects, Artedi was unwilling to part from his friend, "but," he says in the preface to Artedi's "Ichthyologia," "had I known that this was to be our last talk, I could have wished it had lasted longer."

A few days later, on the 27th September, Artedi was invited to Seba's house to supper, where he stayed in happy conversation with many fellow guests till late at night. On his homeward way in the darkness, and not being well acquainted with the neighbourhood, he fell into the "gracht" (canal) at one a.m. and was drowned.

Not till three days later did Linnæus learn through Sohlberg of his friend's sad fate, when he hastened to Amsterdam to be at the funeral. Seba, whom he visited, paid 50 florins [£4 3s. 4d.] towards the expenses. "When I saw," he relates, "the lifeless, stiffened body, and the froth upon the pale blue lips; when I recalled my oldest and best friend's unhappy fate; when I remembered how many sleepless nights, wearisome hours, journeys and expenses, the departed had undergone, before he attained such a measure of knowledge as to be able to compete with any, I burst into tears, when I foresaw that all this learning, which should have secured for him and his country immortal honour, threatened with annihilation; I felt that the love I cherished for my friend compelled me to fulfil my promise that we once mutually exchanged, namely, that the survivor should publish the other's observations."

But here arose difficulties. Artedi's relatives in Norrland, to whom Linnæus applied, gave him full right to take over the manuscripts left, but the landlord in whose house Artedi lodged, definitely refused to give them up, until his preferential claim was fully paid. An attempt to induce Seba to liquidate the debt, did not succeed, and a public auction was arranged, threatening the dispersal of the collections. In his vexation Linnæus applied to Clifford, who willingly paid the requisite sum. Thus it became possible for Linnæus, though with much trouble, to give to the world the fruit of his friend's many years' work, and at the same time by publishing Artedi's "Ichthyologia sive opera omnia de piscibus" to ascribe to him the honour of being the actual founder of a scientific system of fishes on a large scale. In later years he founded the genus *Artedia* in his friend's memory, upon an umbelliferous plant, one of a group to which Artedi had devoted much attention.

Linnæus was now living with Clifford, forgetting his native land, friends and kindred, oblivious of past and future cares, for two untroubled years, which he himself was accustomed to denote as his most pleasant years. Formerly living in narrow or straitened circumstances, he could not at first realize it as both unusual and delightful to be able to live like a prince, having everything he wanted, splendid lodging, grand gardens and glasshouses, a fine library, with liberty to order all plants which were wanting in the garden, and to buy any books which were not in the library. He was enabled to say " Others must travel home for the sake of money. I am afraid to do so for the same reason; here I can do as I like, but not so at home." His relations with Clifford and his family being most cordial, he was like a son in the house.

He now applied himself with ardour to arranging the herbarium at Hartecamp and added to it many dried plants. To increase its contents, each month he visited the gardens at Amsterdam, Utrecht and Leyden, the last especially yielding him many rarities. Supported by the skilful gardener, Dietrich Nietzel, he had great success in his efforts. Already in January, 1736, he succeeded by clever management in getting the pisang (*Musa*) to flower for the first time in Holland. This was inspected by practically all in the land, even the most distinguished, and Boerhaave himself came to receive a demonstration by Linnæus of this, then held to be the finest of all plants. He drew up a small volume on its cultivation, entitled " Musa Cliffortiana," by which every gardener afterwards was able to induce it to flower. By his direction the plant was established in rich soil, water being withheld for many weeks, after which it was deluged as if by tropical rains, to which treatment it responded. The next year it flowered twice as freely at Hartecamp, and produced fruit.

Linnæus did not spare his activities this year, for knowing how to use his time, worked night and day to amplify the notes which he had sketched out at Uppsala. Quite early in his stay with Clifford, his "Systema Naturæ," "Fundamenta" and "Bibliotheca botanica" were in the press, but before these were completed, he began the final revision and printing of two other works (both of which are to-day considered classics) and thoroughly confirmed the great reputation the young author had already secured. One was "Genera plantarum" dedicated to Boerhaave, who afterwards expressed himself in the most flattering terms about this work, "as displaying to the astonished reader unceasing industry, uncommon consistency, and unequalled learning," and "Flora lapponica" which was admittedly that whose publication both in and outside Holland was awaited with the greatest eagerness, as on it he had bestowed especial care and attention. It has already been mentioned (p. 86) that after his Lapland journey he was busily occupied at Uppsala with drafting this flora, but the manuscript thus prepared in Holland underwent a thorough and time-consuming revision, by which not only its extent, but its scientific value was considerably increased, which can be proved by comparison of the manuscript, now at Uppsala, with the printed volume. Its publication was effected with special difficulties and very great expense, but willing helpers were found, and even a society formed in Amsterdam, to undertake the plates. Linnæus hoped that it would be issued early in 1737, but he feared it might be late in the year, before that happened. Both "Flora" and "Genera plantarum" were dated 1737.

During this work, and probably through consultation with Artedi, there arose in Linnæus a great desire to visit England to see the large museums there and to make acquaintance with its most eminent naturalists. He hoped also to find plants there for Clifford's garden. The latter readily gave permission and means to undertake this journey, but did not wish Linnæus to stay away too long. The agreement was made that the journey should not take more than eight weeks, of which two were reckoned for outward and homeward voyages. This agreement was soon found impossible to keep. On the 21st July, Amsterdam was left, and after a visit to Boerhaave's country seat and Leyden, Linnæus arrived at Rotterdam on the following day. On his departure the wind was so unfavourable, that it required nearly a whole week to get to London. On landing, he took up his abode with the pastor of the Swedish church, Tobias Björk, in Princes Square, near Ratcliff Highway.

Amongst those first visited by Linnæus, was the President of the Royal Society, Sir Hans Sloane, who, for bringing together his world-renowned natural history museum, had of his own private means spent no less than £50,000, and so more than any other mortal had gathered a museum whose equal was not extant. To this protagonist amongst English naturalists, Linnæus brought with him a recommendary letter from Boerhaave, in which he testified to his great appreciation of Linnæus, and ascribed to him a more than prophetic power. "Linnæus, who brings you this letter, is particularly worthy of seeing you, and of being seen by you. He who sees you together, will look upon a pair of men, whose like can hardly be found in the world." Sloane invited Linnæus to go through certain "herbaria viva" on the 27th July, amongst them being those which belonged to Plukenet, Petiver and Camell. On the day after, he visited the museum of the Royal Society in Dr. Cromwell Mortimer's company.

Another acquaintance which Linnæus especially longed to make, was with the administrator of the Apothecaries' garden at Chelsea, Philip Miller; who conducted him at once through the garden and showed him its rarest plants, employing the then long-syllabled names. Linnæus, who had come to recognize the unsuitability of these names, and who probably found it difficult to speak Latin with an Englishman, maintained a reserved silence, which Miller conceived as due to ignorance. "This Clifford's botanist does not know a single plant" was the judgment he formed and expressed to an expert, who reported it to Linnæus. When on the next day, Miller employed the same kind of names, Linnæus thought he ought to point out the existence of other names, more accurate and shorter, which should be used, and gave an instance, whereupon Miller took offence and became unfriendly, but this ill-humour soon passed away.

From London, Linnæus travelled to Oxford, where he saw Sherard's herbarium, library and the University garden. His judgment on the collection of dried plants was that it excelled all others in European, but he was less impressed by the exotics. He was kindly received by Dr. Thomas Shaw, Divinity Professor, who had travelled in Barbary and was especially charming, as he considered himself a disciple of Linnæus, after reading his system with much enjoyment. Linnæus longed most to become acquainted with Dr. Dillenius, the then Sherardian Professor of Botany. His surprise therefore was great, when on his visit to him he was received so haughtily, that he was scarcely invited to step in. Linnæus heard also, before he was admitted, that the professor remarked to James Sherard, who was present: "This is he who is bringing all botany into confusion." A walk in the garden, however, was taken, and then Dillenius burst out with little angry remarks and contemptuous gestures. Linnæus, not allowing this behaviour to frighten him, stayed with him three days, but was hardly permitted to see a single plant.

During this walk in the garden Linnæus noticed *Antirrhinum minus* which was unknown to him, so he asked what it was. "Don't you know that?" answered Dillenius. "Yes, if I may take a flower, I will say at once." "Take it," said Dillenius, and the answer was at once forthcoming. As no improvement in his entertainment showed itself, and Linnæus's travelling expenses beginning to dry up, he determined to betake himself homeward. As he was unversed in the English language, he asked Dillenius to let his servant order a carriage for the following day, wishing to pay for it in his presence. "I could then no longer put up with it," said Linnæus, "but desired as the only favour from him, to explain why he thought I was bringing botany into confusion." Dillenius refused, but when Linnæus became pertinacious and continued, "Why was he now so angry with me when formerly he was polite?" "Step in with me," he cried suddenly, and took up the first printed sheets of Linnæus's "Genera" which Gronovius, in his innocence, had forwarded to him. On almost every page N.B. (*nota bene*) had been written, and on Linnæus asking why, the answer came, "Yes, in your book there are many erroneous genera like N.B." Linnæus disputed this, but declared that if the opposite were shown, he would gladly correct it. Instead of arguing, they began to examine the flowers and by dissection to judge which was right; finally coming to a perfect agreement when Linnæus's statements were found to accord with nature, though not with the old writers. As a consequence Dillenius admitted that "Genera plantarum" was not written in order to oppose him, the result being that Linnæus had to leave his travelling companion, and remain for a whole month. From that time forward they were hardly apart for two hours while Linnæus was at Oxford, and when he at last left that city, Dillenius embraced him and parted from him with tears, having before that invited him to live and die there, as the professorial salary was sufficient for both. On parting, he presented him with a copy of his "Hortus elthamensis"; also all the living plants which Linnæus wished to have for Clifford's garden.

He wanted also to have some plants from the Chelsea garden, so on his return to London, he again paid a visit to Miller. The latter, however, kept away, and it was not till the evening that Linnæus met with him. He was then in a good temper and willingly agreed to give him plants. It was therefore with great satisfaction that Linnæus prepared for his return to Holland. Others in England he records having met, were G. D. Ehret, Professor John Martyn, and Peter Collinson. Sir J. E. Smith mentions as a tradition that Linnæus was so enchanted with the gorse in full flower on Putney Heath, that he flung himself on his knees before it, but as the gorse is a spring flowering plant, and Linnæus was only in England in late summer, the tradition is unfounded. Gorse, *Ulex europæus*, is unable to stand the Swedish winter, and already on p. 135 has been mentioned his record of noting it near Hamburg.

Now came a time of strenuous labour, for it was necessary to begin that work which more than any other should preserve the remembrance of his activity at Hartecamp and Clifford's generous love for botany. So "Hortus Cliffortianus" was attacked as Clifford ardently desired, for he knew that Linnæus would stay to finish it, and guessed that otherwise he would soon go back to Sweden. For its beautiful appearance Clifford spared no expense, Linnæus on his part displaying a marvellous power of work. Almost alone he wrote this sumptuous volume, correcting the press himself, all within nine months, which might have taken others years to do. This was not all, for realizing that when it came out, many would wonder at the new names employed, he applied himself in the evenings, when tired of the "Hortus," to writing and publishing his "Critica botanica." He lamented that he could not devote as much care to its latinity as he wished, as he had been persuaded by Boerhaave to issue it speedily. Further came the authorship of his works "Corollarium," "Methodus sexualis," and "Viridarium Cliffortianum," the translation into Latin of his friend Browallius's " Thoughts on Natural History," with a preface by the translator, and the completion of his " Flora lapponica " and " Genera plantarum " which were begun before he visited England. The result was that his books, which bear the date of 1737, consist of nearly 500 pages in large folio, and more than 1,350 pages in octavo with 46 plates. If that be marvellous, it is still more so if one takes into account the scientific value of the said works and the great influence which they exercised on reforming and developing botany. It will further be seen that the said year 1737 did not end before he was busily engaged with writing and printing other and important productions.

Linnæus was fully justified in declaring that he had carried through an amount of work, which before had hardly been witnessed. Although we cannot definitely adduce all the objects he undertook at that time, it is possible to set forth the extensive, constant and varied correspondence which he conducted with naturalists in many countries. Among those with whom he interchanged letters, we may note such men as Johann Ammann, botanical professor at St. Petersburg, A. E. Büchner at Halle, Dillenius, Fahrenheit, the celebrated physicist, J Gesner, professor at Zürich, Albert von Haller, the world renowned physician and naturalist at Göttingen, J E. Hebenstreit, professor of medicine at Leipzig, L. Heister, professor of botany at Helmstedt, F. C. Lesserus, C. T. Ludwig at Leipzig, F. O. Mencken, a Polish Court Councillor, P H. G. Moehring, doctor of medicine in Jena, Fr. Boissier de la Croix de Sauvages, medical professor at Montpellier, J Scheuchzer, professor and musician at Zürich, J. G. Siegesbeck, botanic demonstrator at St. Petersburg, Sir Hans Sloane, C. J. Trew, medical doctor at Nürnberg, and many others, not to mention his Dutch friends with whom his correspondence was specially animated. It is true that sometimes these letters only contained notes about a few plants, or well-turned compliments, with which letters then abounded, but usually they contained many scientific details, accurate explanations which were asked for, and extensive inquiries and sources of information. In whatever case it was, it demanded considerable time, especially as Latin was the only language, accept in letters to Sweden, which Linnæus could employ.

During this hurried, nervous work the only recreation which Linnæus allowed himself, was a visit occasionally to Amsterdam, where Burman lived, and there he was always welcome, and where he interested himself with Ceylon and African plants: also to Leyden, where he sometimes listened to Boerhaave's lectures, or took part in a disputation, whether Linnæus's method was the best. Liberal hospitality was afforded by Botanical Professor Adrian van Roijen, by Lawson and particularly Gronovius, in whose household he was an intimate friend. There was always a room placed at his service, and he was invited to celebrate the host's birthday and Christmastide. Friendly reproaches were sent to him from Gronovius's wife, when he once failed in his promise to spend several days at their house, and when there to add to their amusement by donning his Lapland dress, an offence which could only be atoned for by a speedy visit. After such an event, Gronovius could not refrain from telling, with a certain pride, how all his friends talked about his guest, and how he could hardly go out without meeting someone, who in the choicest language, would ask after Linnæus, his travels, and his Lapland journey. It is quite evident that what Linnæus had to relate about Lapland and the Lapps (about whom at that time the most fantastic and laughable representations were current) attracted eminently flattering attention. A journey to Lapland then seemed uncommon, and was regarded as united to great dangers and fatigue, such as, in our days, a voyage of

LINNÆUS IN LAPP COSTUME
(Portrait by M. Hoffman, painted about 1737).

discovery to the world's most inaccessible regions would entail. A testimony to this is the portrait, the oldest one extant, of "C. Linnæus e Lapponia redux" in Lapp costume, painted by the eminent artist Martin Hoffmann, during his residence in Holland.

At Hartecamp he lived "in the best circumstances a mortal could wish for"; he had all the services of cook and other servants, and could entertain those who paid him visits, with all festive liberality. He could also undertake what excursions he wished, and could drive through Amsterdam streets with a carriage and pair of horses. Nevertheless he began to be troubled with home sickness, principally induced by over-strain and the feeling of being a foreigner, from which he could not free himself, for "his genius was so little for speech that he never learned Dutch, though he lived for nearly three years in Holland." Especially during the period when Clifford and his family were away from Hartecamp, and consequently he was alone with servants and his work, he felt himself "a solitary monk, penned up within two walls." Dutch habits and customs and the whole temper of the people did not seem to suit him, so that his health suffered. "From all this occupation he was so worn out in the autumn of the year, that he could no longer endure the Dutch air," and found "that the Dutch climate is not long wholesome for a Swede." Though he saw that he was not justified in stopping his strenuous work in such desirable circumstances, as rich collections, garden and library at Hartecamp, his decision ripened to leave Clifford and betake himself home to his expectant bride.

This decision he imparted to his host, who would not allow him to leave for some weeks, suggesting that he should remain at his expense at Leyden, to hear Boerhaave before the botanic chair at Utrecht became vacant. After the anticipated death or resignation of the aged Serrurier, who was past eighty, Linnæus would be certain to succeed him, and Clifford meanwhile would pay him a salary. All persuasion, how-

L

ever, was in vain; he determined to go back in spite of all offers, all comforts and honours in the place where all botanists sought him as an oracle, for Linnæus's sweetheart drew his mind towards Sweden.

With feelings of the warmest gratitude, to which he gave handsome expression in the preface to "Hortus Cliffortianus," he took farewell on the 7th October, 1737, of his "one botanic Mæcenas" Clifford from whom he had received, besides his agreed salary, a sum of 100 ducats [£23 10s. 10d.] on account of the "Hortus Cliffortianus." His intention was to go to Paris and stay a short time there, after that to travel into Germany and visit Ludwig at Leipzig, and Haller at Göttingen, especially to work up mosses with the latter, and then to go straight home. In writing to Haller, he expresses his fear that he would not be able to come to him very soon, as he wished to stay at Leyden, and there bid adieu to his friends and acquaintances.

This fear was justified. Professor van Roijen was aghast that Linnæus should so soon leave the place, and offered him every advantage if he would remain with him for half a year, to put the academic garden in order, to help him in his work, and demonstrate "Fundamenta botanica." Boerhaave, Gronovius and others, who wanted to keep Linnæus permanently in Holland, did their utmost to upset his plans of travel, and as a result, he decided to remain in Leyden to the end of February, 1738. What caused so much hesitation was the fear that Clifford would probably be hurt, as indeed he was, but Linnæus endeavoured to excuse himself by stating that he remained for no other reason than to honour himself and his worthy friend Mijnheer Clifford. This was in a certain degree consonant with truth, as the relations between Clifford and Linnæus would be publicly made known in so brilliant a University, and would further honour the name of Linnæus, as he had in "Hortus Cliffortianus" and his other writings, shown his connection with so famous a garden as that of Clifford.

The administrator of Leyden's botanic garden was Boerhaave, who arranged it after his own system. This was ignored by others, and Professor van Roijen had quite decided to abolish it and take up the Linnean. It was for this that Linnæus's help was wanted, and 800 florins [£66 13s. 4d.] assigned for it. There was no complaint to bring against van Roijen, and there was no cause to seek a motive for this resolve. There was a report that Boerhaave had suggested the marriage, with a dowry of a million gulden [more than £85,000], of his only daughter and Linnæus; but it was only a mere rumour, and without any support; she subsequently married Count Toms. Meanwhile Linnæus would not permit of this shock to one who had been so good to him, but as Boerhaave's method was not to remain, he helped van Roijen to work out a special one. The plants were reviewed by van Roijen and Linnæus, who also contrived new names, and arranged the details of instruction in the new philosophy To Linnæus's delight he found that the pupils at Leyden would hear his "Fundamenta" publicly explained.

The reasons why Linnæus considerately tried to avoid wounding Boerhaave's feelings are easy to see. As already set forth he was indebted to him for much kindness, benefiting by his public and private lectures, and receiving from him clinical instruction at the hospitals. Not long before also Boerhaave had given him a new and striking testimony as to his confidence in him. In the year 1737 there was a vacancy in the medical service in Surinam, which could be filled by Boerhaave, who, however, offered it to Linnæus, pointing out that his predecessor had within five years amassed several "tons of gold" [a ton = £1,400], as the only doctor in the place. He also tried to lure him by an account of the splendid plants that could be found in so fine a climate, but love for his affianced bride held Linnæus back. Then Boerhaave allowed him to propose the most suitable man for the post, because no one knew the young Doctor better than himself, and thus Johann Bartsch of Königsberg became the possessor of the post, a small, handsome, quick, learned and methodical young man, an intimate friend of Linnæus, who had taught him not only botany but entomology. He started on the 2nd October for Surinam, but most unfortunately sickened and died a few months after his arrival. Linnæus, who called the genus *Bartsia* after him, sharply complained of the brutal conduct of the Colonial Government, which caused the death of this amiable young man. Another Swede, Tiburtius Kiellman, befriended both by Sohlberg and Linnæus, became Bartsch's successor, but he too soon died in that unhealthy climate.

The good relations which prevailed from the first meeting between Boerhaave and Linnæus were never clouded, but continually became more intimate. This was shown when Linnæus, shortly before his departure, came to bid farewell to his old teacher. The latter was seriously ill of dropsy, accompanied by asthma, so that he could not lie down in bed, but had to sit propped up. For some time, no one was admitted to his room; Linnæus being the only exception. He came to kiss his great teacher's hand, with a sorrowful "Vale" [Good-bye], but the sick old man had still so much strength that he carried Linnæus's hand to his lips, and kissed it in return, saying: "I have lived my time and my years are done, what I could do, I have done. God preserve thee for what remains. What the world asked of me it had, but it asks far more yet of thee. Farewell, my dear Linnæus." Tears prevented more, but when Linnæus went home to his lodgings, Boerhaave sent him a fine copy of his Chemistry. He fell asleep soon after, on the 23rd September, 1738.

In Leyden, with its brisk life and busy interchange with many educated persons, Linnæus prospered. He was inducted into a club, where Dr. J. F Gronovius, Dr. van Swieten, himself, Isaac Lawson, Lieberkuhn, a big and stout Prussian, who had matchless microscopes, J. Kramer, a fast and careless German, a student in all the faculties (with incomparable genius for remembering everything which he had heard or read), with J. Bartsch, were members. When they were gathered together, it was the duty of the host to demonstrate something in his province, as Gronovius in botany, van Swieten in practical medicine, Linnæus in natural history, Lawson in history and antiquities, Lieberkuhn in microscopical subjects, Kramer in chemistry, and Bartsch in physics. That Linnæus was the active spirit in this society, appears from a letter of Gronovius to his friend Dr. R. Richardson of North Bierley, in Yorkshire, in which he states: "Last winter we had a most excellent club or union which met each Saturday, with Linnæus as President. Sometimes we examined minerals, other days flowers, insects or fishes. We made such progress that with the help of his tables [Systema Naturæ] we could refer each fish, plant or mineral to its genus, and subsequently to its species, although none of us had seen it before. I consider these tables to be of the highest value, and everybody ought to have them, hanging up in his study, like maps. Boerhaave values this work highly, and they are his daily recreation."

Other work of Linnæus in Leyden consisted chiefly in drawing up and printing his "Classes plantarum" and his deceased friend Artedi's "Ichthyologia," both of which came out in 1738. He also helped Gronovius with his "Flora virginica" in which Linnæus's principles were embodied. According to his custom he was very busy here, but not to that degree as in his last days at Hartecamp, with soul and body in ceaseless toil. From this it resulted that at Leyden, though he no longer lived as splendidly as a king, he became stouter and lively, though he still laboured abundantly. His economic condition was excellent, as he lived well at Professor van Roijen's expense, receiving money from him as he had from Clifford. Besides, he had in the wealthy Lawson a liberal friend, who often asked Linnæus if he wanted money, and when he answered "No," he would press 60, 80, 100 gulden [£5, £6 13s. 4., £8 6s. 8d.] on Linnæus, saying that he had enough for himself, for he had had the forethought to save.

So the time passed till the spring, when at Easter he received unwelcome news from home. His friend Browallius, appointed professor at Åbo, was asserted by another friend (presumably Mennander) to cast amorous glances at Sara Lisa Moræa, and when the time fixed by Assessor Moræus for Linnæus's stay abroad had passed, Moræus considered himself no longer bound by his promise concerning his daughter. Linnæus would have hurried back, but was held back by a bad ague. As soon as he recovered, he was invited by Lawson and some Englishmen to a little oyster feast, at which he ate only a single oyster and drank one cup of good wine. The following day, however, he was down with cholera, and Dr. van Swieten had to employ all his skill to save him. Hardly had he somewhat recovered, though still tottering, than he received a visit from Clifford, who still felt somewhat offended, because, according to his account, if Linnæus wished to stay in Holland with a salary, he, as willingly as anyone else, would have given it. That his dissatisfaction was not deeply rooted, appears from this, that on seeing Linnæus so weak, he invited him to go back with him to Hartecamp (pointing out the risk of travelling in his feeble condition), and resume his former happy life, to walk about as he pleased, and he would give him a ducat a day [nine shillings and twopence] as long as he liked. This generous offer was gratefully accepted, and about two months were quietly spent, when Linnæus had so far recovered, that he could once more think about his Paris journey. "He did not fully regain his health until he had bidden good-bye to Holland, and reached Brabant, when his body in one day felt renewed, and free from a heavy burden."

If we cast a glance backward on what Linnæus effected during his stay in Holland, we cannot fail to be struck with astonishment; no similar case can be furnished in the history of botany, nor in the annals of any other science. During the short time of two and a half years, there were published, besides smaller treatises, no fewer than twelve or fourteen works for the development of botany, the majority being epoch-making. It is true that many of them were already prepared during his student years, but they had to undergo a thorough revision and reworking in the light of greater library help which was available in Holland; and others, especially that gigantic work, "Hortus Cliffortianus," had to be finished and printed in the midst of other and time-consuming duties. All these testify to a power of work and untiring industry, which must awaken unstinted wonder. It is therefore from no boastful self-love, but with justifiable self-consciousness that he gave expression, not for the general public, but for his intimate friends and in his autobiographies, when he truthfully says: "One may judge of the amount of work that I accomplished in Holland: where I wrote more, discovered more, and reformed more in botany than anyone had done before in his whole lifetime"; and in another place—"He who sees what botany was before my time, and what it now is, since I began to write, would hardly recognize it; I have changed all and have been the greatest reformer in that science that ever existed."

But even with his genius power for work and perseverance such as Linnæus displayed, it would not have been possible to accomplish so much without fortunate circumstances. It has been shown how he was helped by the quiet of Hartecamp, with every requisite scientific assistance at hand, and relieved of all economic worries, in order to work exclusively upon that which appealed to his soul's most stirring dictates. One must also ascribe his great success to gaining powerful and devoted friends, who willingly took upon themselves everything which would have crippled his activity, and who by advice and deed, contributed to those departments in which he was not versed. The man to whom Linnæus was deeply indebted, was J. F. Gronovius, who " laboured for him night and day, year in, year out, on the correction of his work, and made it what it was." He who reads his many letters to Linnæus while in Holland can only marvel at the perseverance and self-sacrifice with which he devoted himself to the duty of proof-reading, seeing the work through the press, enduring the many disagreeables which printer, engraver, and publisher caused him. One instance may be cited of his care and industry: for the correction of only one plate in " Flora lapponica," he spent six days, and at last gladly reported that it was beautifully printed. Neither did he confine himself to mere mechanical details, for he subjected the manuscripts to a close and intelligent scrutiny. This led to many and long letters, in which small incongruities were pointed out, suggestions for changes in expression, additions hinted at, errors corrected, and so on. It must undoubtedly be maintained, that without the aid of Gronovius the nervously and abruptly drafted works of Linnæus would no doubt have exhibited many blemishes from which they are free. Similar help, if not so extensive, was derived from Bartsch, especially on the " Flora lapponica," Lawson, Burman and van Roijen. It happened as though a young foreigner sat in the Clifford library as a king, and had a whole staff of diligent and experienced helpers, who took charge of the fulfilment of his wishes and orders.

If Linnæus in this aspect can be reckoned fortunate, it was certainly in a high degree owing to his lucky appearance at the right moment. The work to which naturalists now applied themselves, after the long sleep of the Middle Ages, was eagerly prosecuted, and had produced a very important literature. Unhappily it was credulous, uncritical, and without guiding principles, the works producing the impression of being written in a rude and rough way. Science had sunk to the greatest barbarism, in which confusion and arbitrary opinion reigned together. The time for a thorough reformation was ripe. The reformer must enter with undaunted courage, putting aside what was false in the teaching of the predecessors, and substituting for the discarded statements, such as were supported by accurate observation and irrefutable principles. It was order, clarity, and easy comprehension for which people longed, and this not least in the matter of system, when many, troubled and confused by the numerous unsuccessful and unworkable attempts at systems, declared that alphabetic arrangement of natural objects was the only satisfactory one. A reformer now appeared in Linnæus. He came, well versed in the subject, suave in method, as a deliverer from the universal confusion. He furnished simple, well-founded laws to classify the differences of Nature with such clearness and simplicity, that this revolution, due to his writings, took place without violent rupture or bitter disputes. Naturally, many older men thoughtfully shook their heads at such novelties, having special objections to both principles and details, but these objections led to no open polemics. With earnest or joking utterances, a correspondence, without bitterness or heat, was entered upon with Linnæus, usually ending with recognition of the value of the new views, except on a few minor points. It is not to be wondered at that the younger men (practically as a body), crowded round Linnæus, and soon he was universally greeted with praise in contemporary publications from such leaders as Boerhaave, Haller, Burman, Dillenius, Gronovius, Sauvages, Gesner, Ludwig and others, who, generally speaking, recognized Linnæus as a great and rising light, and as a prince amongst the botanists of his age. A public acknowledgment of Linnæus's services was made by the Academia Imperialis Leopoldino-Carolina Naturæ Curiosorum, which on the 3rd October, 1736, elected him as a member, and according to the regulations, bestowed upon him the name of "Dioscorides secundus." As another testimony to the great regard in which he was held, was Haller's suggestion that he should succeed him as professor at Göttingen, as he himself intended to return soon to his native Switzerland.

Amongst the praises which the young Swede received from various countries, there was one unfavourable criticism. It was from the demonstrator of botany at St. Petersburg, J. G. Siegesbeck, who entirely unexpectedly came forward as his opponent, basing his claims on his work published in 1737, "Botanosophiæ verioris brevis sciagraphia" [Short outline of true botanic wisdom]. In this he attempted to demolish Linnæus's published views on the sexuality of plants, with the system founded upon it, and among other arguments, he put forth the plea that God would never, in the vegetable kingdom, have allowed such odious vice as that several males (anthers) should possess one wife (pistil) in common. or that a true husband should, in certain composite flowers, besides its legitimate partner, have near it illegitimate mistresses; and he complained that so unchaste a system should be taught to studious youth. That Linnæus felt himself unpleasantly astonished at this unfavourable criticism is the less surprising, as the writer, a short time before, was in friendly correspondence with him. Linnæus had named a genus after him and intended to visit St. Petersburg, where he had been invited to stay with Siegesbeck as his honoured guest. Meanwhile, he did not consider himself obliged to reply to this stupid and lying volume (and also he was hindered by illness). In this resolve he was confirmed by Haller, who wrote, "that he could just laugh at Siegesbeck and his like"; an opinion shared by Boerhaave, Gleditsch, and many others. A short history of the quarrels belongs to the next period of Linnæus's life.

It was in the month of May 1738, that Linnæus left Hartecamp for the second time, and at Leyden took leave of his friends, amongst whom were professors of various faculties. The journey was by Antwerp, Trefontain, Mecheln, Brussels, Bergen, Valenciennes, Cambray, Péronne, Roye and Pont St. Maxence to Paris, calling for nothing worth notice. "As soon as he came into Brabant, he saw that he had come out of a garden into a scanty pasture, where the people were ill-favoured, and the houses wretched. In Brussels, where the Kaiser's sister was living, he saw fine fountains in the streets, the valuable Arsenal, and observed the Papist religion in its highest ceremonies. At Bergen there was a strict examination, for no one was permitted to pass with more than 50 livres [about £2], but Linnæus happily was admitted, possessing a couple of hundred ducats [about £94]. In this town, though not large, there were eleven apothecaries. At Valenciennes, Linnæus's trunk was sealed, as he had with him a number of new books, carrying with him a copy of each one which he had published in Holland.

In Paris, Linnæus hastened to visit Antoine de Jussieu, the Professor of Botany, who showed himself very obliging, and constantly offered him hospitality. As he was fully occupied in the practice of medicine, he passed on his guest to his brother, the demonstrator of plants, Bernard de Jussieu. This one also showed Linnæus the greatest kindness and liberality each day. Through him he became acquainted with the botanist d'Isnard, the entomologist Réaumur, the late Tournefort's fellow-traveller, Aubriet, with whom he discussed butterflies, the widow of Vaillant the botanist, and the mathematician Clairaut, who astonished him by conversing in Swedish. An intimate friendship was formed with V La Serre, a physician and naturalist, and a close friend of the Jussieus. Special value was attached to his introduction to a lady, the clever Mlle. Madelaine Françoise Basseporte, who had a situation in the botanic garden as botanic painter to the King, although conversation must have been difficult, as Linnæus spoke no French, and she neither Latin nor Swedish.

Naturally Linnæus did not neglect to see the most notable things in Paris, such as Versailles and the neighbourhood, but most of the time was given to scientific employments. Linnæus went over the fine botanic garden to see the herbaria of the Jussieus, Tournefort, Vaillant, Surian, and others, with d'Isnard's great collection of botanic books, where he found so many unknown to him, that he saw he could bring out a new edition of "Bibliotheca botanica," with twice as many titles as his former one. He also took part in B. de Jussieu's excursions with students, and tradition relates an episode during one of them. The students played the joke (though it seldom succeeded) of asking Jussieu to name plants from mutilated or artificially made-up specimens from bits of different plants—applied to the foreign botanist to name a strange plant, but he, taking the opportunity to pay a compliment to his French friend, referred the students to him by saying it must either be God or Jussieu who could name it. Special enjoyment came from a trip to Fontainbleau at Jussieu's expense, when La Serre was of the party. During this trip, which lasted several days, he saw the rarest plants in France, and amongst them nearly all Vaillant's figured orchids in full flower.

A very pleasant and flattering surprise was prepared for him during his Paris visit. On the 24th June Linnæus was the guest of President Du Fay at the Académie des Sciences. After the meeting, he was requested to wait a little, and then he was informed, that the Académie received him as a Foreign Correspondent. More, the President intimated that if he chose to become French, the Académie would appoint him a member with an annual pension, but he remained faithful to his fatherland.

After Linnæus had achieved all that he wanted from a scientific point of view, he began to think about his return journey, for his object was not to learn French habits or foreign languages, as time is never more dearly bought than when one travels abroad merely to study languages. It was known that Linnæus's habits did not give him time to study these, but nevertheless he did well in conversation everywhere. His stay in Paris, which was intended to last a fortnight, was extended to a whole month, and his purse had so diminished, that he found himself obliged to forego a journey to Germany, and pass direct, by the cheapest way, to Sweden. After receiving on the 18th June from P A. Fleming, the Swedish Minister, an open recommendation, he travelled to Rouen. Thence he sailed with a fair wind to the Cattegat, when the wind suddenly turning, he landed at Helsingborg to visit his old father once again. He was received with extreme joy at Stenbrohult, where he put into the old man's hands the many books he had published on the subject in which his father had so much pleasure. After a couple of weeks resting at home, he continued his journey direct to Falun, to greet his fiancée, who had waited for him nearly four years. Then followed a formal betrothal, and a month later, he journeyed to Stockholm, there meaning to spend his life.

CHAPTER VII

LIFE IN STOCKHOLM, SEPTEMBER 1738—OCTOBER 1741; APPOINTMENT AS PROFESSOR AT UPPSALA

It was in September 1738 when Linnæus, by the advice of his future father-in-law, settled in Stockholm as a practising physician. As he was quite unknown, nobody ventured to entrust his life to an inexperienced doctor, so that he often doubted as to his future. He who had everywhere abroad been honoured as "Princeps botanicorum," was at home a Klimius, come from the underground regions (a reference to Holberg's well-known "Niels Klim's underground journey,") that had he not been in love, he would infallibly have left Sweden.

It was only the want of opportunity that stood in the way of the realization of his plans. At that time Linnæus was intimate with A. von Haller, always changeable and restless, who was seriously thinking of quitting his professorial chair at Göttingen, and returning to his beloved native country, Switzerland, a project he ultimately carried out. In November of the same year he advised Linnæus, "of whom Flora hopes more than of any other botanist," to come to a milder climate, and promised when he himself was recalled to Switzerland, which he hoped would be soon, he would take care that Linnæus should be his successor at Göttingen. "I have," he said, "already consulted those who have the entire disposal of the post." This letter, sent by the medium of a travelling German priest, only reached Linnæus on the 12th August, 1739, when circumstances had substantially improved, so that his answer was really a refusal, but full of gratitude to the renowned man, who showed such proofs of regard and kindness. "It is impossible to express in words my gratitude, but so long as I live, your name will always be cherished."

Several months thus passed, without any real improvement in his prospects. Only one encouragement occurred, on the 27th September he was chosen a member of the Scientific Society of Uppsala. "Laudatur et alget"—he is praised but starved—being an utterance he frequently employed at that time.

As patients did not seek him, he determined to seek them. He began to frequent the quarters of the city where he saw young fellows suffering from chest complaints and indulgence in fast living, sad and depressed. He exhorted them to be of good courage and drink a measure of Rhine wine, assuring them that he could cure them in a fortnight. When two of them, who had consulted physicians without success, ventured to entrust their case in his hands, he cured them at once. To their comrades' amazement they began again to enjoy their wine, declaring that Linnæus was an eminently skilful practitioner; consequently in a month's time he had most of them under his care. His credit then rose in cases of epidemic, small-pox and agues, the result being that he gained such an extensive practice, that he was busy from seven in the morning to eight at night, with hardly time to eat. "This augments my purse, but takes up all my time, so that I have not an hour for my best friends"; and in the new year he recorded that he had each day from forty to sixty patients.

This fame as a skilful physician, particularly in chest disorders, was not without an important influence on Linnæus's future. "Among his patients was a court lady who suffered from an irritating and obstinate cough, and for its relief was ordered pills of tragacanth which she was to have at hand to use when necessary. This lady was playing at cards with the Queen Ulrika Eleonora, when she was obliged to take a pill from her box. On the Queen asking what it was, as she herself had a cough, she was given her excellent physician's name; Linnæus was called in, and his prescription having the desired effect, he became known and consulted among the highest ranks of society."

This had the further consequence, that Linnæus in Stockholm, as in other places where he stayed, found an influential and zealous protector, who in every way smoothed his path. This was the well-known Count Carl Gustaf Tessin, the head of the "Hats" party and Speaker of the Nobility in the Riksdag. He took the liveliest interest in art and science, and making acquaintance with the young physician, heard of the great celebrity which he had gained in other countries by his many writings. This was enough for Tessin to seek for State help to secure support for him till he could earn sufficient remuneration. He called Linnæus to him, to ask if he had any request to make to the Riksdag, being sure that the authorities would regard it as a pleasure to favour a Swede who had so greatly distinguished himself in foreign parts. When Linnæus asserted that he had nothing to ask, he told him to think it over till the next day and to come again. Encouraged by this, Linnæus, on the 24th of November, presented a request addressed to the Secret Committee of the Riksdag in which he related his attainments in Natural History. He further stated that, while abroad, he had published fourteen works on botany, which were used by the Leyden University, so that now he applied for a grant for his support, and until this was supplied, asked for some public appointment. No specified sum was named nor the source from which it might come. A new friend, Captain Mårten Triewald, urged that as Linnæus understood mineralogy, he should apply for 100 ducats [about

£47] yearly from the Mining College, as that award was then vacant. Tessin approved of this, providing the paper for a formal "Pro Memoria" in that day's sitting, 7th December; it was put forward by the Landmarschalk, Tessin, and sent up with his approval. In the end Linnæus was awarded a grant of 600 silver dalers [£45] such as Captain Triewald had in 1726, on condition that the students of the Mining College should receive lectures from him on mineralogy in the winter, and in the summer on botany. The application was supported by many influential members, and finally in the following March, it was granted without opposition. In consequence Linnæus was often styled Royal Botanist, though not officially so named.

This economic advantage was not the only thing for which Linnæus had cause to thank Tessin. He offered him a room in his palace which he himself had occupied as a bachelor, with board, so that he was no longer solitary. Linnæus gratefully accepted this offer, taking up his abode there until he married. Here, amongst the crowd of notabilities, he was playfully named "The Hats' Archiater," and his medical practice became so increased, that "it was as large as that of all the other doctors put together," to use his own expression probably an exaggeration.

Tessin delighted in Linnæus's gratitude, and was unceasing in doing his utmost for him. A short time later, Vice-Admiral Ankarcrona sent for him to enquire if he were desirous of the post of physician to the Admiralty in Stockholm, which was vacant by the death of Dr. N. Boij, which occurred on the 1st January 1739. He further promised that if he agreed, his name should be presented alone. Finally he was appointed by the King, with a salary of 2,700 copper dalers [£67 10s.] annually, this appointment not being questioned by the medical college.

Among the many acquaintances Linnæus made at this time, was Captain Triewald, already mentioned, who was now set upon establishing a Scientific Academy publishing in Swedish, and discussed his project with Baron Höpken, Jonas Alström (afterwards Alströmer) and Baron S. C. Bjelke. As a result of these deliberations an invitation was sent out for a meeting at nine a.m. in the lecture room at the Riddarhus [Knights' or Nobles' House] to start the said Royal Academy of Science for the investigation of Mathematics, Natural History, Economics, Trade, Useful Arts, and Manufactures.

Such was the modest beginning of this famous institution, which has now so powerfully developed, and contributed so much to the progress of science in Sweden.

One of the duties of the five persons above mentioned in founding the Academy, was to decide by lot who should be the first President, and this fell upon Linnæus. Thus it happened that within a month he became a public lecturer at the Riddarhus, physician to the Admiralty with a salary, and first President of the Academy.

Thus everything had gone well with him in Stockholm, surpassing his boldest hopes. He had won both respect and income, amounting annually to 9,000 copper dalers [£225] so now he trusted that the time had come for his marriage. On Trinity Sunday, 17th June, he started from Stockholm for his future father-in-law's house ("Sweden") at Falun, where the wedding was celebrated on the 26th June, according to the old time customs and rejoicings, the bridegroom being greeted by verses upon his system as having obtained in Dalecarlia "a monandrian lily"; an allusion to the Linnean system, as of a single-stamened, or "one-man" flower. This marriage was followed in due course by the birth of a son on the 20th January 1741, who was baptized with the name of Carl. He was able to say, that with the good fortune he now enjoyed, thanks to God and Count Tessin, he lived in great enjoyment and comfort.

Hard work was his portion, as his numerous private patients, and his naval duties, caused his position to be no sinecure. The naval hospital contained from one hundred to two hundred sick, demanding his care with only the help of two assistants. Linnæus not only devoted himself to consideration for the welfare of those committed to his care, but earnestly sought for simple methods of care, beginning to form a garden, chiefly for raising medicinal plants. At the same time he had opportunities for autopsy, and he became one of the first pathologic anatomists in his country, which led during his time, to great development in the medical faculty at Uppsala.

Then too he had his lectures to deliver, in 1739 and 1740, on botany and mineralogy according to the season. He not only spoke from his chair, but invited his hearers to excursions in field and meadow, and rich flower gardens round about. As to mineralogy, he remarked in early spring, that zoological subjects were not yet available, being still in their winter sleep, some in southern lands, some in deep waters, and some in holes and corners of forests. All the flowers were in their winter-quarters and "were sleeping with the bears." The subject, however, drew so large an assembly when he lectured on the rock specimens of the Mining College, that Triewald's room could hardly hold them all, to Linnæus's great surprise.

The young Academy demanded no little time, especially that being the case during the first four months when Linnæus was President, thus compelling him to do his very best in many different directions. Difficulties were so much the greater, as Sweden had not hitherto possessed such an institution, consequently it had more or less to serve as a model. It is true that at Uppsala since 1710, there existed Sweden's first scientific society, named the Royal Scientific Society, and the Academy had to form, as it were, a complement to it. The former acted as a link between home and foreign naturalists, Latin being the medium of language, and the substance being pure science; the latter, however, was imparted to the natives in Swedish reports, consisting of those that were valuable from the economic point of view. The Academy consequently in its early days had almost the same objects as the present Agricultural Academy. The original name proposed was " Economic Scientific Academy," but on the advice of Anders Celsius, it was changed to that now in use.

The regulations entailed but little trouble, as they were drafted by Höpken. It was necessary to obtain a large number of suitable members, and under the presidency of Linnæus, there were thirty such elected, among these being many of his supporters, such as Governor Reuterholm, Dean Olof Celsius, Professors Anders Celsius, Roberg and Klingenstierna, Assessor Moræus, and others. That the weight of Linnæus's influence was employed in these early selections is shown by a minute recording that he was shortly obliged to leave the capital (to be married) and that during his absence none should be elected, to which the members agreed.

More difficult than obtaining members was to provide for expenses, particularly for the printing of the Transactions, which had to be published. The wealthier gave liberal contributions, Linnæus and the less well-to-do offering each his ducat [9s. 2d.], while others, the President amongst them, after a part of the Transactions had appeared, gave some copies of their works for sale. Even presents in kind were made, as for instance furniture, and on the 20th June, Linnæus presented a copy of his " Hortus Cliffortianus " and two Chinese books on Rice and Silkworms, which had been given to him by a supercargo of the East India Company, thus making a beginning towards the present rich library. Others made gifts to the Riksmuseum as at present constituted.

Still further activity was displayed by Linnæus to

obtain contributions to the Transactions. He set an example in his "Report on planting, founded on Nature." In his speech on relinquishing the presidency, he counted up eighteen observations on diverse subjects by fifteen members.

Linnæus had a strong influence in this direction. Pure science was less frequently in evidence, but much discussion took place on such subjects as to how stones became loose in the earth, or on mountains, and if they came from the Flood. The diminution of the water-level was also discussed.

Much time was taken up by matters of organization, details of printing, ceremonies at the meetings, and sending an application to the King to prevent any reprinting from the Transactions, also making a device for the title-page, which by Linnæus's suggestion was the representation of an old man planting a palm-tree, with the inscription: "För efterkommande" [For those who come after]. For style, either in Latin or Swedish, Major Pihlgren was chosen, as being proficient in Swedish, to read through the papers passed for printing. He himself felt weak in Swedish, therefore he strongly urged that the mother-tongue should be purified and trimmed before publication.

At the end of September, 1739, the time came for Linnæus to lay down his office, and according to the rules, to deliver a short discourse on doing so. But Linnæus, instead, delivered a formal oration, "On curiosities in insects," which set the example for all succeeding presidents. At the request of the members this speech was printed at the expense of the Academy. The satisfaction given by Linnæus as President was shown by the fact, that when a new one for the last quarter of 1740 had to be chosen, Linnæus was one of the four named, though the lot fell upon another.

After Linnæus had ceased to be the official leader of the Academy's labours, during his stay in Stockholm he continued to work hard for its success. He rarely missed a meeting, until his contributions numbered ten, in addition to small paragraphs. Amongst these may be mentioned an observation from Dr. Wallerius on Gadflies on cattle, and another on Meadow-Sweet, *Spiræa filipendula*, which he suggested should be used as food in place of almonds. Definite decision as to the latter was postponed until Linnæus in the summer of 1740 made experiments on it, but the paper was declined on the ground of gross blunders and mistakes. It may be taken that the savage attack which Wallerius soon afterwards made upon Linnæus, to be related later, had its origin in this rejection.

The quick success of the Academy naturally gladdened Linnæus as well as the other founders. Two years after its foundation, it was able to add the epithet "Royal" to its title. It was also recognized by the "Ständerna," or Estates of the Realm, as an authority, and questions were referred to it for its opinion. Thus the Commercial Committee wished for its verdict upon the native plants which might be serviceable as drugs, and a catalogue of such was drawn up by Linnæus, for which he was thanked. The Manufacturers' Committee put the question regarding the freedom from customs which the apothecaries enjoyed and the importation of foreign drugs, asking that they should be limited to a specified amount, but it was decided not to alter the regulations until equally good drugs should be raised in the country. For this reason Linnæus's fear of ill-will on the part of the apothecaries passed away, and the said catalogue was soon afterwards published in the Transactions.

If we now put together what has previously been told of Linnæus's activity during his stay in Stockholm, one cannot refrain from recognizing that it was astonishing. Yet there still remains something for his authorship besides that already enumerated. On his arrival in the capital he announced his intention to refrain from writing, and especially to let his botany rest. "They laugh," he says in a letter to Haller, "in Stockholm at my botany. How many sleepless nights and weary hours have I spent on it, but with one accord they say, that I have been overcome by Siegesbeck." He found himself challenged not to let this attack pass without an answer, but he would not, without the counsel of his foreign friends, appear against the least worthy of his opponents. In his friend Mennander he found a person who was willing to appear on the title-page as author, if only Linnæus himself wrote the refutation. In the meantime nothing came of this, as subsequently the intimate bond of friendship was again formed between him and Browallius (against whom, rightly or wrongly, he had formerly harboured distrust), for he came forward as a champion on the scientific field. In the year 1739 appeared at Åbo the work "Examen epicriseos in systema sexuale Linnæi auct. Siegesbeckio" (Examination of the determination in Linnæus's sexual system of Siegesbeck) in which the statements in the named "Epicrisis" were reduced to their true value. Much of it, and that the most important, was undoubtedly from the pen of Linnæus. Similar in some respects was the condition under which J. G. Gleditsch published his "Consideratio epicriseos Siegesbeckianæ" (Consideration of the Siegesbeckian determination) in 1740 in Berlin. This author, even before he began corresponding with Linnæus, felt himself induced to write his answer, and to impart the views of other naturalists by his criticism of Siegesbeck's polemic.

About the same time that Browallius's pamphlet was printed at Åbo, there came out another in Stockholm, which indeed did not bear Linnæus's name, but which in reality came from his pen, and was seen through the press by him. At that time there was living in Karlscrona, Assessor J. E. Ferber, who in 1711, laid out a large garden on his estate at Agerum, which Linnæus himself never saw, but whose riches had awakened the admiration of both Swedes and foreigners. An account of the plants there cultivated seemed desirable, and it would be the first book published in Sweden in which the Linnean system was followed. But although Linnæus himself gave the proper shape and scientific value to it, he considered it would be improper to usurp the credit, and so much the more, as his old friend and benefactor, Rothman, had drafted the work, as well as written a preface of fourteen pages, in which he declared his judgment on the method of arrangement used in this catalogue.

In the preparation of this little treatise there were three concerned: Ferber, Linnæus and Rothman. To these may be added a fourth. In a letter to Olof Rudbeck, dated 15th March, 1739, Linnæus begged him to read through and amend it, as the censor of books in Stockholm could hardly be induced to pass a botanic work, but all blame would be avoided if this were done. Submitting it to Rudbeck's approval as to how far Rothman's utterance should be supported, as it seemed quite too flattering. But Rudbeck shared Rothman's views, and his judgment therefore stood.

During the time which followed this publication, Linnæus was so immersed in medical practice, that he had no time to think about plants; and he was not entirely dissatisfied in that he found that " Æsculapius bestowed good fortune, but Flora only Siegesbecks." More than once he had had thoughts of selling his collections, but better economic conditions awakened new interest in botany, the result being the issue of a new and enlarged edition of his " Systema naturæ " in 1740, dedicated to Tessin, " which shall praise the name of my great Mæcenas's name when we are silent." Also he printed two new editions of his " Fundamenta," in 1740 in Stockholm and 1741 in Amsterdam.

To this stage of Linnæus's writings may be added two papers which he contributed to the Scientific Society at Uppsala for its Transactions: one on "Animals observed in Sweden," and another on "Known orchids." According to a letter to Tessin in 1740, he had besides, a tract ready to be printed that summer in Holland, probably Gronovius's ' Index supellectilis lapidea" (Lugd. Bat. 1740; ed. II. 16. 1750), said by Linnæus to be almost wholly his own. Such was his literary activity at a period when he was nearly overwhelmed with a multitude of other time-consuming duties.

It was not long, however, before he began to find his extensive medical practice as an oppressive burden, and longed for a return to the quiet world of botany. The success he had met with as a lecturer awakened his old desire to work as a university teacher, his future plans concentrating more and more upon obtaining the professorial chair of the aged Rudbeck. If he could not obtain that post, he was ready, whenever Haller called him, to move to Göttingen, if he could take his little wife with him. This was only in case of necessity, for he desired most of all to labour in the medical faculty at Uppsala.

Officially, conditions were similar to those of his student time. Olof Rudbeck the younger was still professor, but enjoyed continuous vacation to work at his great philological work, "Lexicon harmonicum," while Adjunct Rosén, now physician in ordinary to the King, and Assessor in the Medical College, was appointed his deputy. It was printed in the list of lectures year after year unaltered, that he would lecture on anatomy and botany, also that Professor Roberg would publicly lecture on Human Physiology and Characters of Diseases, sometimes on Chemistry. Actually the official promises were not fulfilled. Old, and with weakened powers, Roberg seldom lectured in public, and never in private during later years, while Rosén, though appointed to discharge Rudbeck's duties, had been obliged to take up some of Roberg's. The result was that he neglected some of the topics which properly belonged to Rudbeck's part, so that Chancellor Gustaf Bonde, when in February, 1738, made himself acquainted with University conditions, desired to know, if any botany was taught. Rudbeck answered that Assessor Rosén was responsible for that duty, but he did not know if any lectures had been delivered during the past two years or not. To the question, if anatomy was often taken up, the information was given that Assessor Rosén was specially diligent in this particular. Rosén also taught physiology, pharmaceutic chemistry and physics, and thought this perfectly right, these sciences being the right arm for a surgeon, "whilst botany was a velvet cuff which adorned it."

Rudbeck's perfectly truthful answer to the Chancellor's question concerning botanic teaching, gave rise to a long and very bitter quarrel, that laid bare the slack control reigning in many directions, to the detriment of important academic duties. The signal for this was given by Rosén, when, in March 1738, he tendered an application to the King that he should be appointed as a third ordinary professor with Adjunct's salary. He made this appeal on the ground of many years' valuable service in the University, and might have attained his object without trouble, although some (Rudbeck and both O. and A. Celsius) favoured Linnæus's appointment, because in his application he had not referred to Roberg's slack method of discharging his duties. The medical faculty, to whom this application had been remitted, referred it to the Consistory. A whole year was taken up in consideration of various details; but at length Rosén's desire was approved by a majority. Rudbeck declared he could not see why the Consistory had made their decision so promptly, nor could he understand what duties were to be assigned to Rosén, should he teach botany; on oath and conscience he could testify that Dr. Linnæus, who had lived in his (Rudbeck's) house as tutor, was more eminent and more fitted for the post. Roberg also interposed difficulties, and the result was, that the third professorship was not instituted. Probably it was seen that both the chairs would soon be vacant.

At the request of the University Chancellor, the Estates declared that in case of old age or other reason, a professor might resign, but enjoy his salary during life. The Chancellor, then Count Carl Gyllenborg, requested the University to state if the students were likely to suffer by the age and weakness of professors. It was easy to see that the Chancellor earnestly wished that both Rosén and Linnæus should become professors at Uppsala, presumably by the resignation of the veteran professors.

This affair was not remarkable for any especial promptitude in the Consistory; after many delays, Rudbeck declared that he did not think he should be included amongst those who could not discharge their duties, as although he was now eighty years of age, he was still in full vigour, and was daily at work upon his "Thesaurus harmonicus." He still could teach, if a deputy were appointed in order that the other important matters he had in hand might be completed and printed. Roberg also demurred to the Chancellor's suggestion, and two other professors, Grönwall and O. Celsius, aged respectively sixty-eight and sixty-nine, protested against being superseded.

Thus no one admitted the weakness of old age, and the Consistory had to declare its belief, that both Rudbeck and Roberg were able to discharge their duties, in spite of their great age. This statement was combatted by Rosén, who hastened to inform Linnæus, that the two professors in question were regarded by the Consistory as coming under the rule concerning resignation. They protested vehemently and so furiously, that he (Rosén) had never seen the like.

So this "consilium abeundi" had no result, either with Rudbeck or Roberg. In the end, the question resolved itself into a complaint against Rosén, who did not deny the neglect of lecturing in the botanic garden, but asserted that there was no material there for his lectures. Upon this Rudbeck remarked that at first there was plenty to lecture on, as shown by the fact that Martin and Linnæus, shortly before, had been able to teach successfully; further, that so recently as 24th October, 1739, there were actually two hundred and eighty flowers in the garden; although he admitted a decline since that time. He also declared that he (Rudbeck) was not to blame, as he had been specially allowed to work on his philological books, while Rosén had been appointed his deputy; further that an incompetent gardener had been put in charge. It is plain that in 1730, there was no prefect of the garden, but it was left to the four gardeners who quickly succeeded each other to the office.

In this unedifying quarrel Linnæus was so far involved, that each of the disputants in turn solicited his help. Evidently uneasy about the bad state of the garden, Rosén hastened to send a letter to Linnæus asking him for good advice. On his side, Rudbeck travelled to Stockholm to beg the loan of Linnæus's "Adonis Uplandicus," which showed by plans and descriptions the former poor condition of the garden, and the many plants at present in it to be used for demonstration.

This volume had but little effect on the quarrel, but Rudbeck, wearied with wrangling, withdrew his statement concerning Rosén. An improvement, however, in the management took place at this time. At the beginning of 1739 the gardener, Samzelius, died, and at the Consistory, Rudbeck produced a letter from Linnæus, urging the appointment to the vacant post of the skilful Dietrich Nietzel, whom he had known at Hartecamp and had greatly valued his work. The Consistory at once replied to Linnæus with the request that he should offer the appointment to Nietzel on advantageous terms. The offer was accepted, and Sweden secured the services of this extremely skilful and diligent man, who afterwards contributed to the world-wide reputation of the Uppsala garden. He arrived at the end of June, with many rare trees and plants, and threw himself energetically into his work. Soon after the garden had taken on an improved appearance, Rudbeck died, after ten days' illness, on the 23rd March, 1740.

Thus the chair of botany, anatomy and other departments at length became vacant. On the 28th April, Linnæus sent in his application, recounting his scientific merit and experience. Rosén and Wallerius did not delay to make similar applications.

According to academic constitutions, these applications were referred to the medical faculty, that is to say, to the surviving member, Roberg. Because of his unquestionable merit, he assigned the first place to Rosén, the second to Wallerius, also on the ground of his long service at the University, and the third place to Linnæus, for his European reputation in botany.

In the Consistory which discussed the question two months after Rudbeck's death, opinions were divided; most members put Rosén first, only Professor Roberg, A. Celsius and Dean O. Celsius preferring Linnæus. The votes appeared thus: Rosén, 12 first, 2 second and 1 third; Linnæus, 3, 6 and 6; and Wallerius, 6 in the second and 8 in the third place, with one vote for Professor Spöring at Åbo, although he was not a competitor.

It cannot be denied that both those who supported Rosén, and those who preferred Linnæus, had good reasons for their choice. For the vacant professorship, the chief objects were botany and anatomy; Linnæus's overwhelming superiority in the first was balanced by Rosén's in the second. While botany at that time was a main subject in the medical curriculum, it would have been unjust to overlook the nine years Rosén had acted as deputy.

Following the practice prevalent at the time, to bespeak a patron's recommendation, Linnæus applied to Tessin, then in Paris; for a single word to Count Gyllenborg would have a powerful effect. Tessin hastened to carry out this wish, the result being that the Chancellor so arranged between the competitors, that Rosén should fill the vacancy and that Professor Roberg, when he resigned on account of his age, should be succeeded by Linnæus, and that both should exchange functions. These views were submitted to the King for his approval.

Before this appointment took place, on the 10th July, 1740, the other medical chair in Uppsala in consequence of the Chancellor's powerful action became vacant by the resignation of Professor Roberg, he stipulating that he might still continue to live in the little hospital house, which he had built at his own expense. The Chancellor at once wrote agreeing in the main, and on the 3rd of May Roberg sent in his application to the Consistory, which was approved finally by the King.

To this chair, now at last vacant, Linnæus put in his application on the 13th of July, shortly specifying what he had recently set out on his own behalf.

It might be thought that this would readily be arranged, as Linnæus was the sole applicant for the place, but this did not prove to be the case. When the matter came before the Consistory, Professor Asp brought up a request from the Vice-Chancellor that Linnæus and any other applicant, should, according to the Constitution, produce proof of his competence, either by disputing or any other suitable means. The majority of the Consistory agreed to this, but would not be content with any theses being printed or written, evidently showing that these professors wished to satisfy themselves as to Linnæus's power to speak Latin.

To this the Chancellor without delay replied that Linnæus had been a teacher for years in Uppsala, and that according to the testimony of many learned men in Europe, hardly any Swede had attained so great renown, and hoping that matters would be taken in hand without delay. The Consistory briefly discussed this, but considered the presence of the Vice-Chancellor was necessary; on his arrival he stated that he did not willingly depart from his former opinion, as based on good and safe grounds, but only sought the best method of preventing any incompetent person being appointed to the University, as he must preside and carry on the duties which belong to the chair, etc. Then it was decided to send a new letter to the Chancellor, asking why the previous decision was maintained, pointing out that Linnæus had gained abroad the highest esteem in botanic matters, which did not properly belong to this chair.

The opposition which the Consistory showed by being disposed to negative the Chancellor's order, began to fail. Two meetings were held to justify the decision arrived at, and at the last one, Rector Frondin admitted that he was somewhat uneasy, feared his Excellency's displeasure, and wished to escape signing the letter. Professor A. Celsius thought it advisable to agree with the Chancellor's reply. Dean Celsius and Professor Roberg held the same view, while Professor Beronius considered it inadvisable to oppose a man, a Swede, whose name was regarded throughout Europe as illustrious, pointing out that in a recent instance, Linnæus, without being asked to show credentials, received the votes of all. Finally it was resolved to send the letter to the Chancellor, the Rector signing it, though very unwillingly.

This refractory conduct of the Consistory towards the Chancellor had this result: that a new aspirant for the post appeared, Adjunct Dr. J. G. Wallerius, who in his application declared his readiness to submit a specimen of his capacity; this application stood over awaiting the Chancellor's intervention.

That the latter would be satisfied with the action of the Consistory was inconceivable. He hastened to inform the King of what had taken place, and to beg his Majesty's most gracious warrant and command. The matter was referred to the Chancellery, and this body demanding an explanation from the Consistory, pointed out that this question had been grievously mishandled. The Chancellery demanded a reason for their disregarding for the first time a paragraph in the academic constitution of the Consistory by unanimously supporting Linnæus in his claim for the then vacant chair to which Rosén had been lately appointed. The answer was simple, that the Consistory deemed it superfluous to require a proof by disputation, for his competence could be inferred from his published papers. For the new rule that Linnæus should defend in Latin some printed or written thesis, the Consistory had only this weak excuse to offer, that it did not understand that he was proficient or experienced in those studies which belonged to Roberg's professorship, and they also pleaded the need of keeping a strict hand over those who wished to escape delivering proofs. To this Professors A. Celsius and Beronius added their opinion, sharply criticizing the procedure adopted. In the paragraph above referred to, it is clearly stated, that the applicant for a vacant professorship shall produce proof by disputation or by some other satisfactory method, which latter clause had been excluded by the Consistory. What was specially requisite for a University teacher was erudition, particularly in the Latin tongue. Dr. Linnæus was invited only to dispute, to know if he understood Latin; but they were satisfied that he could write it, as all his books had been published in that language, and that he was accustomed to speak it, was known from the fact that he had passed through Växjö Gymnasium and had been heard to "oppose" at Uppsala. The Consistory admitted their forgetfulness of a clause in the Constitution, and finally consented that proof should not be asked of Dr. Linnæus.

As a result of this correspondence the King declared that Linnæus should be freed from what the Consistory demanded of him, but other applicants must submit theses. Consequently, not only was Wallerius to furnish a thesis, but also the recently promoted Dr. Abraham Bäck, who now, at the eleventh hour, announced himself as an applicant.

What the Consistory really thought about their shifty procedure, cannot now be determined, but it soon became manifest that an attempt was being made to shut out Linnæus from the coveted post. To obviate this, at the end of 1740 or beginning of 1741, he printed a small tract entitled "Orbis eruditi judicium de Car. Linnæi, M.D., scriptis" (The judgment of the learned world on the writings of C. L.). In this he collected the judgments of many of the most eminent naturalists, prefixing a short history of his life, and a list (twenty-one in all) of the writings he had published. Amongst the witnesses quoted were Boerhaave, Sloane, Albert von Haller, and Ant. de Jussieu; altogether there were five Germans, five Netherlanders, two Swiss, four French, and as many English, their praise not being scanty, nor stinted. This small tract became a sharp weapon, of which Linnæus and his friends made good use; without it, possibly Uppsala University might have lost the honour of numbering him among its teachers.

In February, 1741, Wallerius gave in his material for his thesis, but rumours began to circulate as to the contents, that they were simply an attack upon Linnæus. Rosén, as the only member of the medical faculty, had passed the thesis, and remarked that although it upset Dr. Linnæus's works, he did not for his part find anything therein against the work of God, the general state, or good manners, and therefore could not see what should hinder the disputation from being held. This was the prologue to a tragicomic play, which created general scandal, not only in Uppsala, but also in wider circles.

The 27th February, 1741, was the day appointed for the disputation on "Decades binae thesium medicarum" (Two decades of medical theses). Only a preliminary glance is needed to find that its contents were nothing but a detailed attack directed against Linnæus's "Systema Naturæ," "Flora lapponica," and before all, his graduate disputation. That in one or two small things Wallerius was right cannot be gainsaid; but on the other hand, on other points it was obviously incorrect, petulant, and mean. Moreover, one notes a tone of spite with a desire to distort statements and to slander rather than to honour an author, thus betraying his lack of learning and experience. The writing aroused great displeasure, even the title, "Jussu Max. Ven. Senatus Acad.," giving offence, as it was stated to have been used by order of the Consistory. This title had never before appeared on a disputation. In thirty-eight pages, Linnæus's name occurred between sixty and seventy times.

Under such circumstances, and apprehending that the act would be of an unusual character, the students filled the largest auditorium of the University in great numbers. At first all was quiet and normal, but when the first opponent, after an hour's speech, gave no hint of ending, the students, who were prepared for something different, showed their impatience by stamping. Rosén, who for the first time as Dean, should have kept order, seems to have been irritated, and as the disputation went on, his bad temper increased, so that time after time he broke out against both audience and the other opponents. First among these Professor Beronius (if he might be termed an opponent) rose and soon showed himself a strong disputant, who took the author to task for not observing the rules which Christian love and good manners prescribed. Rosén vainly interrupted more and more heatedly, declaring that all the arguments that Beronius brought forward were "scommata" (derisive and wounding expressions), inept, childish, unbecoming in a theologian, so far forgetting himself as to apply the epithet of "opponens impudissimus" to Beronius, at which some of the students showed their displeasure by shouting. Even Wallerius, calm at first, afterwards gave vent to his feelings by stamping and uttering derisive cries. The students took sides, and shouted approval or the reverse. Beronius, who seems to have preserved an unruffled temper, went on ignoring the Dean's order to be silent, and turning to the students, quoted with unequalled effect from the "Orbis eruditi," begging them to remember that Linnæus was an honour to their country. He related his struggles against poverty in his student days, and his constant trust in God's wonderful providence, to which speech the students listened in silence with bowed heads.

When Beronius ceased, Mag. Carl Clingenberg continued the discussion, but declared himself only a friend of Linnæus and not an expert. The contest between the speaker and Rosén increased in heat, and the students then began to stand on the benches, jumped, stamped, laughed and shrieked, whilst the Latin abuse flowed on. Said one who was present, "Never did I hear such an amount of abuse in Latin, nor so coarse." It went so far that many in the audience tore up their copies of the thesis, and after the end of the Act, the floor was covered with fragments. It was a complete academic scandal, in which all those concerned were more or less to blame.

Naturally after the storm were heavings as of the sea. Rosén, certainly, during the disputation had gone so far, that the audience suspected him of having encouraged Wallerius in this attack on Linnæus, because anticipating disgrace in his office as Dean, he found himself compelled to seek succour from the Consistory. First of all complaint was made against Clingenberg, as having himself transgressed the rules of the academic constitution and of good order, while Beronius being accused of much the same, he denied the charge, the Consistory refusing an enquiry as tending to show partiality. Rosén then declared that as this decision diminished his authority as Dean, he could not permit Dr. Bäck to put forward his disputation, lest another disturbance should arise. The Consistory decided that an intimation should be given to the students exhorting them to behave well during future dispensations, and a similar proclamation was issued to the "Florentissima juventus academica."

Evidently the Consistory thought that the incident should now close. It was therefore an unpleasant surprise to receive a letter from the Chancellor intimating his displeasure at not receiving a report on recent events, which he had heard of by private letters. Ultimately, it was decided to send a reply in the most humble terms, stating that the Consistory regarded these occurrences with extreme regret, but that they could not furnish a detailed account, owing to the confusion which prevailed at the time, but they had requested the Dean and opponents to furnish reports which should be forwarded as soon as possible.

The Chancellor did not wait long; only a fortnight later, he reminded them that the reports were still delayed, and that the regulations required all vacant posts to be filled in a certain time. He therefore called upon them to make a suggestion as to who should fill Roberg's vacancy, and to forward it to him. After further discussion the affair was closed, Linnæus soon after being appointed professor and Wallerius filling the post of adjunct, with a caution.

While these things were happening in Uppsala, another question arose in the capital, which might have prevented Linnæus being ranked as one of the professors at Uppsala. The Riksdag [Parliament] decided to send travellers, at the State expense, to certain parts of southern Sweden, and this arrangement coming before the House of Nobles, fear was expressed that as Linnæus was applying for a chair at Uppsala, the public service might suffer by his being unable to undertake any commission. Count Gyllenborg considered that so thoroughly equipped and learned a man would be more useful to the nation on such an expedition than if he gained the desired post; but events proved otherwise.

As previously mentioned, the Chancellor had urged the Consistory to make a speedy decision concerning the vacant professorship. It finally resolved to put forward three names, Linnæus, Spöring and Wallerius, the Chancellor naturally approving the first-named. On the 7th May, he was able to inform the Consistory that His Majesty two days before had appointed Linnæus as Professor of Practical Medicine.

On the 23rd May a letter from Linnæus was read enclosing (1) a copy of the King's warrant to succeed to Roberg's place; and (2) a copy of the order of the Estates of the Realm authorizing him to travel through Öland and Gotland to investigate their natural productions. Linnæus remained at Stockholm until the 15th May, when he began his journey, returning on the 29th August. For a month longer he stayed in the capital, and on the 6th October removed to Uppsala, there to remain for the rest of his days.

In a letter to Sauvages at this time Linnæus wrote: "Through the grace of God, I am being now freed from the wretched practice in Stockholm. I have obtained the post which I have so long desired: the King has appointed me Professor of Medicine and Botany at the University of Uppsala, and thereby again given me to botany, from which I have been exiled for three years, spending that time among the sick in Stockholm. Should life and health be vouchsafed to me, you will, I hope, now see me perform something noteworthy in botany."

He kept his word; during more than a third of a century, Uppsala became, through Linnæus's activity, the central point for the study of natural history, especially botany.

CHAPTER VIII

TRAVELS UNDERTAKEN ON PUBLIC COMMISSION: ÖLAND AND GOTLAND, 1741; WEST GOTHLAND AND BOHUS COAST, 1746; SKÅNE, 1749.

THE appointment as Professor in Uppsala is a noteworthy boundary line in Linnæus's career, which is thereby divided into two parts of nearly the same length, but of differing qualities. The former, which has hitherto been described, may be taken as his "Sturm und Drang" period, during which, often under untoward conditions, he laboured with youthful energy, rising from an undistinguished position to one of world-wide reputation, and in various countries, helped by their sympathy, acquired devoted friends and warm-hearted benefactors. Over the whole of this period rests therefore a poetic glamour which entrances both biographer and the general public. The later period, on the other hand, though most important and not wanting in episodes which arouse admiration, are to some extent somewhat prosaic.

Here we find Linnæus living a quiet, little-changing life, in an honourable and secure position, engaged in teaching and strenuous research work. He was no longer the young enthusiast, obliged to fight undauntedly for recognition of his new ideas, but the generally received master, whose word was law, and round whose chair a crowd of eager enquirers gathered to find nature interpreted. That these changed circumstances, in a certain sense stamped their mark on his person and life, is quite natural, though he remained substantially the same, as youthfully ardent and gaining the love of all those with whom he came in contact.

A connecting link between the two stages of Linnæus's life is found in the three journeys he made in his native land by public commission. The first of these took place at the junction of the two periods of his life, the two latter at a later time were shorter, but all three display the same qualities of rich results, so that they may be regarded as direct continuations; together forming a whole, and as such, a short account of them must be given.

Amongst the many persons Linnæus met in his Stockholm life was the bank public prosecutor, G. A. Rutensköld, a member of the Manufacture and Trade deputations, who succeeded in getting Linnæus appointed to visit Öland, Gotland and other places to report upon their natural productions, likely to prove of use to the State. The journey was estimated to take three months. As soon as the Estates had sanctioned the project, Linnæus began his preparations, similar to those for Dalecarlia, and to find some young men ready at their own expense to accompany him and help him in his task. Those selected were: P. Adlerheim, notary of the Royal Mining College; J. Moræus, "auscultant" in the same; H. J. Gahn, mine-owner; G. Dubois, student of medicine; F Ziervogel, royal apothecary; and S. Wendt, student in botany and medicine. All showed themselves diligent, especially Moræus and Gahn, signing and promising on their honour not to draw back from the rules which Linnæus had made.

On the 15th May Linnæus and his companions left Stockholm to ride to Kalmar. The spring was well advanced, most trees being in leaf; the cuckoo was heard and swallows were seen. They rode quickly and paused for the night at Svalbro inn, where they suffered from want of horses, fires, and good beds. The next day they passed Nyköping, and the day after reached Norrköping, there visiting a sugar factory and a tobacco spinning shop, operated by children. Though dust troubled them afterwards on the roads, the boundaries of Småland were reached, where the peasants spoke the dialect familiar to Linnæus in his boyhood. Växjö was visited, and Linnæus hastened to call on his old teacher Rothman, and others. Hence the journey was rapid to Kalmar, but food was hard to get, and only by entreaties at the apothecaries could they procure any. They had been able to gather morels in the forests.

Their intention was to cross over the Sound on the next day, but a succession of cloudy, cold and dreary days, with strong winds following, the travellers hardly ventured out of doors, much less crossed the sea. Nevertheless on their short excursions they were successful in finding near the castle, a plant locally called "Mannablod"—man's blood—which proved to be *Sambucus Ebulus*, our own "Danewort."

The bad weather continued, but tired of Kalmar, they resolved to pass over to Färgan, a small town in Öland, which they soon reached, aided by a strong south-west wind, and stayed there for three weeks. They at once saw that the island was different from the mainland, and few naturalists had previously collected there save O. Rudbeck, senior, gathering *Helianthemum œlandicum* and *Euphorbia palustris*, and the physician J. Linder hurriedly taking *Adonis vernalis* and *Viola odorata.*

In splendid weather, the party passed on to Borgholm, where Linnæus was delighted to see a multitude of orchids; here too, a quarry which yielded pavement stones (flags) and many fossils was inspected. Next, the west coast southward was searched, where they found fine woods and abundance of flowers. One tired with the world's shifting state and desirous of enjoying a quiet retreat, could not find a more agreeable one.

On the 6th June they reached Kastlösa. Here

they wished to investigate the place where coal was said to have been found, but the people tried all they could to prevent them, though with politeness and fair speech. The journey was therefore continued to Möckleby, and its alum quarry with its twelve strata, where, during the past two years, a fire had been burning among the store of alum-shale. Linnæus thought that Öland would have the curious appearance of a fiery mountain in Sweden, should the flames spread and kindle the whole of the unworked shale.

During the work of investigation, Linnæus had the ill-luck to be struck by a bit of broken rock from the cliff, which bruised him on the left ankle, and if he had not been on the alert, the bone of his leg might have been hurt, and his foot crushed; he was carried back to the inn at mid-day to rest his contusion.

But Linnæus not being able to rest, caused a giant's grave to be dug up, in which many bones were found, to try and discover if the people in former times were taller than now. As the result of several measurements, he concluded that the men whose bones these were, were evidently four ells long—six feet tall. They resumed their journey, but had to go farther than they intended, as the inhabitants, though forewarned, were afraid of the party. Ultimately the Comminister at Åhs took them in, but the pain Linnæus suffered prevented his having a single minute's sleep, no remedies being procurable.

The next day was Sunday, and a resting day. In spite of all the minister could say, the people believed the party to be spies, and to guard against ill-treatment, they engaged a guide.

After nursing his foot for two nights and a day, Linnæus went forward, to the southernmost point of Öland's coast, and then by extensive sheep pastures, northward by the eastern coast, to where the road ended. Many fine botanic and zoologic discoveries were made, and Linnæus was gladdened to come upon the notable bird, the avocet. A cave of some interest was visited, and the customs of the place as to fishing, quarrying, lime- and tar-burning, agriculture, etc., noted. They saw with interest the women knitting stockings as they walked. Oppressive laws forbidding the natives to sell the Öland ponies, the breeding of them had almost ceased, and the bearing of firearms was also forbidden in order that the peasants should not shoot the deer. Both red deer and fallow deer had been introduced and had spread themselves over the whole island, doing much mischief, wild pigs also being a pest.

Gaxa, where they arrived on the 13th June, was the starting point for Gotland, but as the post yacht would not start for some days, Linnæus made an excursion to the little island of Blåkulla or Jungfrun (the Maiden). Strong wind hindered the passage at first, but when it had moderated, the rowers told a tale of the witches who visited the island and raised storms. After hard work by everybody, they reached the island, narrowly escaping disaster against the rocks.

The impression the island gave was not favourable, for it was very steep and the bushes so grown together that it was hard to climb, but from the top a wide prospect was obtained. After noting the plants they returned to Gaxa late at night.

To get to Gotland was not easy, the only craft for the transit being so wretched, as to be absolutely dangerous. The party therefore remained for several days on Öland, and going to the northernmost point, observed the scanty growth on the sand dunes. They found plants that were new for Sweden, with insects and fossils, also an abundance of the ant-lion. At last they were able to hire a sailing boat to pass over to Gotland.

The party embarked on the 21st June, the sail was hoisted at nine in the evening, when the sun sank below the horizon. The next morning at two o'clock, having been well carried forward by a south-west wind, they cast anchor at Visby, where they were welcomed.

Devoting a couple of days to seeing Visby, they were shown some "giants' bones" which proved to be whales' bones. Thence the party journeyed by the west coast to Bunge with success in botanic results, including *Cladium Mariscus*. Many fossils were noticed in the cliffs, but the roads were bad, even dangerous for the horses. Seal flesh was eaten fresh, salted or dried, seal fat was used instead of butter, and sea birds abounded. The use of *Elymus arenarius* was noted for binding sand-dunes. The east coast was then followed to Hoburg, with its weathered rocks, and they discovered *Coronilla Emerus* for the first time in Sweden. Turning northward, on the 25th July, the party visited Karlöarne [Charles Islands] with good results in animals and plants. Intending to pass over to Öland, they declined to entrust themselves to the post-yacht as being too dangerous, but after waiting a week, they hired a boat, and in a strong wind with some risk they reached Öland, afterwards crossing to Kalmar, where the party broke up. Linnæus was tired with his two months of travel, but went on to Växjö and then to Stenbrohult, arriving on the 9th August, greeting his old father, his sisters and brother, for the last time as he expected. With Moræus, Gahn and Dubois, who now joined him, they travelled homeward, by Jönköping, Vadstena, and Medevi with its celebrated medicinal waters, there meeting Carl De Geer, the entomologist; next to Örebro, where Linnæus called upon his benefactor, Governor Reuterholm, and through Arboga to Uppsala. After twenty-four hours, Linnæus went to Stockholm, the entire journey having taken fifteen weeks, his expenses being returned at 536 silver dalers, 221 for posting and 315 for food [respectively £40 4s., £16 11s. 6d., and £23 12s. 6d.].

As soon as he completed this journey, he had to

take up his duties as professor, and the numerous duties which then devolved upon him obliged him to postpone the other journeys. It was not until the close of 1745 that he found time to print his account of his travels in 1741. The West Gothland journey, in 1746, was entered upon with only one companion, E. G. Liidbeck, who acted as secretary throughout. Two other friends went with them during the first fortnight, for their health's sake only.

At Vesterås he noted with pleasure the methods employed by Bishop Kalsenius to interest the school children in astronomy and natural history. At Örebro he visited Governor Reuterholm. Kinnekulla was reached on the 19th June, and investigated during four days; Lidköping, Skara, and Höjentorp, where the establishment of Jonas Alström and his sheep farm were inspected. The rainy and boisterous weather, however, hindering observation, they journeyed on to Falköping, where they rested on Sunday, 29th June. Three days later they visited a peasant named Sven, reputed as being famous for healing diseases, but found that he had no knowledge of medical practice, only using certain drugs, which amazed Linnæus. At Ållestad there was an Englishman, Dr. Blackwell, who had come to Sweden on a visit at an opportune moment when national economy was the rage and to teach it. (In 1747 he was executed in Stockholm for plotting high treason, but the charge was based on unsatisfactory grounds. His wife is known for her illustrated work on plants, "A curious Herbal," London, 1737-39, 2 vols., folio.) They visited Borås, rebuilt after a disastrous fire in 1727, which was found to be full of industries, then Alingsås, in the West Gothland fells, well known for its wool manufacture, energetically pursued by Jonas Alström, thence to Göteborg [Gothenburg] which Linnæus described at length, mentioning many inhabitants whom he visited. A week later the journey was directed to Bohuslän, with Marstrand

and Uddevalla, where many interesting objects were found. Next came Trollhättan, but its waterfall did not much impress Linnæus, who had seen the Lapland rivers; then by Hunneberg to Vänersborg, turning homeward by Wermland, Karlstad and Filipstad, where there was not much to record; then through Nora and the ironworks of Wedevåg, where the travellers were hospitably entertained by Count Jacob Cronstedt at his estate of Fullerö. On the 11th August the travellers reached the boundary line of Uppland, and in the evening they came to Uppsala.

The account of this journey appeared in the spring of 1747, but before that, Linnæus was commissioned to undertake yet another journey, at the public expense, this time to the southern province of Skåne (Scania). This time Linnæus stipulated to be accompanied by a paid amanuensis, and eventually they set out on the 29th April, O.S. (10th May, N.S.) They were joined by the student Olof Söderberg, and during a portion of the time by the Lund student, Lars Aretin, as a companion of Söderberg. This time Linnæus travelled with greater comfort than previously, as he bought a coach to journey in, rather than to ride on horseback. Örebro and Växjö were again passed through; and Linnæus spent three days at Wirestad with his sister Anna Maria, her husband being Linnæus's old tutor, Gabriel Höök; thence he came to Stenbrohult, on the 15th and 16th May, with a feeling of melancholy. "Here," he says, "I found the birds vanished, the nest burnt, and the young scattered, so that I could hardly recognize the place where I was brought up, and where my late father laid out his garden. I, who twenty years before knew every inhabitant, now found hardly twenty left, who were youngsters during my childhood, and they were now with grey hair and white beards," but he had the joy of seeing his only brother occupying his father's place, as rector of the parish.

Since the date of his eldest son's last visit to the

old home, Nils Ingemarsson Linnæus had died at the age of seventy-four. Concerning the latest years of his life and his departure, the following account is given by his son Samuel, who succeeded his father in November, 1749.

"When he heard in 1741 that Carl, his elder son, had made so good a marriage and had been appointed Professor, that a medal was struck in 1746, succeeded by the title of Archiater in 1747, the old man was overjoyed and often said, 'I have had so many tokens of God's grace and goodness towards my children, one after another, to gladden me, that I cannot die.' In 1748 he became very ill, and as he lay sick in bed, he ordered all his children who were at home to place themselves the day before he died by his bedside. All four of us had to place ourselves in order of age. Carl's place was vacant, but there were present Anna Maria, the pastor's wife in Wirestad; Sophia Juliana, the pastor's wife in Ryssby; Samuel, adjunct to his father; and lastly, Emerentia, wife of C. A. Branting. Our late father, looking long at his children, said: 'Carl is absent. He has caused me great joy. God has blessed my five children, all have gladdened me, and none have caused me sorrow. Now I desire to bestow my fatherly benediction before I go from you. Carl is absent.' Afterwards he left his blessing for him, and for his whole house and family, and placing his hands upon each of them, hoped for a happy meeting in heaven."

On the 17th May, 1749, the journey began in Skåne, towards Kristianstad, where Linnæus stayed four days, noting the sand-dunes and their flora. Then to Råbelöf on an excursion to Balsberg in company with N. Retzius, observing the fossil oyster shells. At Tunbyholm Linnæus had a little adventure, the only one in his journey, with the testing of a divining rod. When the Secretary took a forked branch of hazel, one of the company hid his silver tobacco-box, another his watch, and Linnæus cut up a turf and hid his purse under it, a mark of the place being a tall buttercup close by. The Secretary searched with his rod for nearly an hour, but when at the end of the vain experiment, Linnæus went to take up his purse, the diviner had so trampled the grass, that the spot could not be recognized. Finally one of the company put his finger on the spot and drew out the purse, which had 100 ducats [nearly £48] in it.

At Lund he stayed for two days, 10th and 11th June, then on to Malmö, and spent a week there, observing the rare plants and trees in the gardens in that mild climate. In heavy rain and strong wind, the journey was continued to Trelleborg, then to Skanör, Falsterbö and Ystad, back through Lund to Helsingborg, and by Kristianstad again, with six days' sojourn at Stenbrohult and at Ryssby with Linnæus's sister and her husband, Johan Collin. Here he bade farewell to all the well-remembered places and plants. In this last visit, Linnæus had the good fortune to find *Isoëtes lacustris* for the first time in Sweden, which had escaped his notice in his boyhood. On the 7th August he started on his return to Uppsala, which he reached six days later.

Thus was concluded this journey, during which he had enjoyed much fine weather in contrast to his West Gothland trip, and he ended his account of the fair province, by recounting its favourable climate and products compared with the more northern parts.

In the course of these three journeys, he made many observations on the products of agriculture, trade, and customs he had noticed, bearing in mind the commission he was fulfilling. Economic and medicinal plants also were laid before the public in reports or tracts previous to the publication of his three volumes. He also touched upon the geology of the places visited, all this being written in a fresh and naïve style, even if a little weak at times. These statements are a gold mine for the present-day student in acquiring a knowledge of Sweden gleaned from the country folk in the middle of the eighteenth century. The complete relation of his travels was received with warm approval. But a slight misunderstanding arose between him and his patron, Baron Hårleman, as to the account given of paring and burning the turf, which led to the cancelling of one of the printed leaves, and the substitution of a modified statement, an occurrence which vexed Linnæus.

Fresh work and new progress soon dissipated this vexation, but although the Riksdag desired him to investigate other parts, he did not undertake any similar task again. "Often have I ventured on the sea to fetch gold from Ophir; I have come back with broken powers, wrecked ship, and torn sails, and were I to venture out again, I might easily be lost" was his expression in a letter to Wargentin, the secretary of the Academy of Science.

CHAPTER IX

LINNÉ AS A TEACHER—HIS PUPILS AND HIS RELATIONS WITH THEM

On the 25th October, 1741, the Rector of Uppsala University issued a Latin invitation addressed to all and sundry to attend in the Caroline audience hall on the 27th at nine a.m. to hear the Latin lecture, with which the newly appointed Professor, Carl Linnæus, would enter upon his duties. The lecture was on the discoveries which could be made, and the benefit which might result from natural history travels in Sweden, and he took the opportunity of imparting the most important details which he had himself observed in similar journeys. After this, with the accustomed ceremonies, the new Professor took the oath and placed himself at the table amongst his colleagues.

With this, his professorial activity began. There was, however, this abnormality, that he and his fellow professor, N. Rosén, were appointed to teach science in those departments in which each was weak, whilst the other was eminent. They therefore presented a joint appeal to the Chancellor, begging that their lectures should be so ordered that Rosén should undertake Practical Medicine, Anatomy, Physiology, Pathology, and Pharmaceutic Chemistry, whilst Linnæus should teach Botany, Metallurgic Chemistry, Semiotics (Pathology combined with symptoms), Dietetics, and Materia medica, also to superintend the botanic garden. As the Consistory approved this exchange of duties, the Chancellor confirmed it on the 21st January, 1742, on which day Linné attained the place he had so ardently desired.

His first discourse after installation was held on the 2nd November, 1741, and he continued thus to the end of 1776 with never slackening industry, so that except during the time that he was absent from the town, ill, or hindered by other causes, he did not neglect a single lecture. On the contrary, he gave to his pupils' instruction more time than he was obliged to by law. Without fear of an unfavourable judgment therefore, he was able to say: "With what energy I have prosecuted my professorship, I leave others to judge." His eagerness or rather delight in teaching was so great, that when at the close of the spring term, shortly before Midsummer, he felt himself overstrained by the many labours, the feeling lasted only a couple of weeks, when he complained that he was as weary of the holidays as he was formerly tired with work.

The statutory lectures were delivered at ten a.m. in the Gustavian building, usually in its largest hall, but sometimes in the botanic garden, so as to have abundance of plants to show, without having to carry them backwards and forwards. He never omitted, when he had opportunity, to show the *Musa*, which was specially dear to him since his days at Hartecamp.

At the same time as these public statutory instructions, were given his private coaching lectures, which took place in his own house. He gave these partly because of his own zeal and his pupils' expressed wishes, and partly for the pecuniary gain, which gave a welcome increase to his scanty stipend.

With both these lectures and teachings, Linnæus, especially at the beginning of his professorial career, had occasion to rejoice at the large audience, so large as to awaken astonishment. Although he was Professor in the medical faculty and at first had very few pupils to examine, he had among his numerous audience, so many belonging to other faculties, that few of his colleagues were so successful. In some subjects, such as the medicaments derived from animals, the attendance fell to twenty, yet when the lectures dealt with the system of diseases, or the philosophy of botany, the auditors rose to fifty or sixty. Still greater numbers came when natural history, especially zoology, was discussed; but without contradiction, the most attractive subject was dietary. For instance, in the spring terms of 1748 and 1756, the numbers were respectively 144 and 101, but Linnæus declared that he had double the audience than the stated figures, and that at a time when the total number of students at Uppsala did not exceed 600. The highest figures seem to have been reached in 1760, with no fewer than 239 students on the list, probably due to the Pomeranian war (the students being immune from conscription), and at its height, reached 1,500.

It is not so easy to determine the number of those who took part in the coaching lectures; only for the spring terms in 1748 and 1754 can accurate figures be supplied, respectively 165 and 88, but one of his pupils relates that late comers had to stand in the lobby, because of the crowd. Although during the whole of his career, the entire number of matriculations in the medical faculty amounted only to 344, altogether his pupils must have reached many thousands.

This flourishing state of things was due to many coincident causes. The long period when this subject was treated by aged professors, created an impulse at this time when it was handled by Linnæus, with his lively and pleasant teaching. A brother professor, D. Melanderhjelm, relates that botany was presented as a new and unknown thing, and the fashion was to run after a new subject. To see a flower from the Cape or from Asia, monkeys and snakes from Africa, and parrots from South America, in Sweden, was to see a miracle, which no one but Linnæus could show. The book of Nature had till now been closed to the students who came up for divinity or classical languages; but through Linnæus's teaching, obscurity was cleared away, and the rich fields of Nature were illuminated before their eyes. He was a skilful guide who threw a new light over natural objects. Further, to have been a student " under Linné " was a memory for life, and each one wished to take with him from the University, the proud title of having been " Linné's pupil."

Another weighty cause of his success was that during the Era of Liberty, began a reigning utilitarianism, or a desire to make the most of the country's productiveness. Those who were intended for the priesthood, saw that they would succeed better by acquiring some knowledge of natural history and medical science. Their note-books were as text-books, and in after life became valuable.

For his own part, Linné (to use the Swedish form of his name), though belonging to the " Hats " party, quite sympathized with his students' views in their desire to attain to a useful knowledge for their country's welfare.

There is still another reason to be assigned for his popularity, namely his sympathetic personality. He had an almost magic power of attraction while possessing a charming personality, and as he was fresh and lively in his teaching, it formed a contrast to the dry reading of a written lecture, which then, and for long after, most of the professors inflicted upon their hearers. The best representations of his lectures are obtained in the sketches which some of his pupils gave. Thus J G. Acrel remarks: " In the professorial chair he had a special eloquence peculiar to himself, and although he was helped neither by a powerful nor melodious voice, nor by particularly winning utterance (for he spoke the Småland dialect) he never failed to interest his hearers in the highest degree. He seemed to give expression in his short sentences to the weight by emphasis he gave his meaning, so that it was impossible for anyone to fail to be convinced of what he argued. He who heard him discourse on the introduction to his 'Systema Naturæ,' on God, Man, Creation, Nature, etc., was more moved than by the most eloquent sermon. With this power of convincing, he had the advantage of an incomparable memory and a clearness of thought, so that he could deliver from a few notes on a scrap of paper, a long oration or a lecture. His lectures were rarely written on a larger sheet than a strip of paper, which he held folded up between his fingers, and with his thumb marked the latest place where he stopped." This completely agrees with S. Hedin's report, from which the following is extracted: "He mingled quickness and thoroughness so marvellously, that both curiosity and understanding were satisfied. If Linné spoke of the Creator's power and majesty, reverence and wonder were depicted on all faces; if he talked about dietary, he permitted his hearers to laugh unrestrainedly, when depicting the then fashionable whims, and with easy and delightful humour, he imparted the most useful wisdom concerning the care of health and its preservation."

There was something more yet which attracted pupils to him, namely, the botanic excursions which he conducted with his pupils in the neighbourhood of Uppsala. These celebrated "Herbationes Upsalienses" (in which not only plants, but animals, minerals, and all that they could capture, were explained by him) are thus recounted by a participant therein, J G. Acrel, thus: "The botanic excursions which he instituted each summer, were not less enlightening and amusing for youth, than useful in kindling a desire for Natural History. They took place according to a certain order, as defined in his 'Herbationes Upsalienses,' to eight places round the town. At this time he had no fewer than 200 to 300 members who accompanied him afield, all clad in an easy suit of linen and provided with everything necessary for collecting plants and insects. From his auditors he himself chose certain recognized

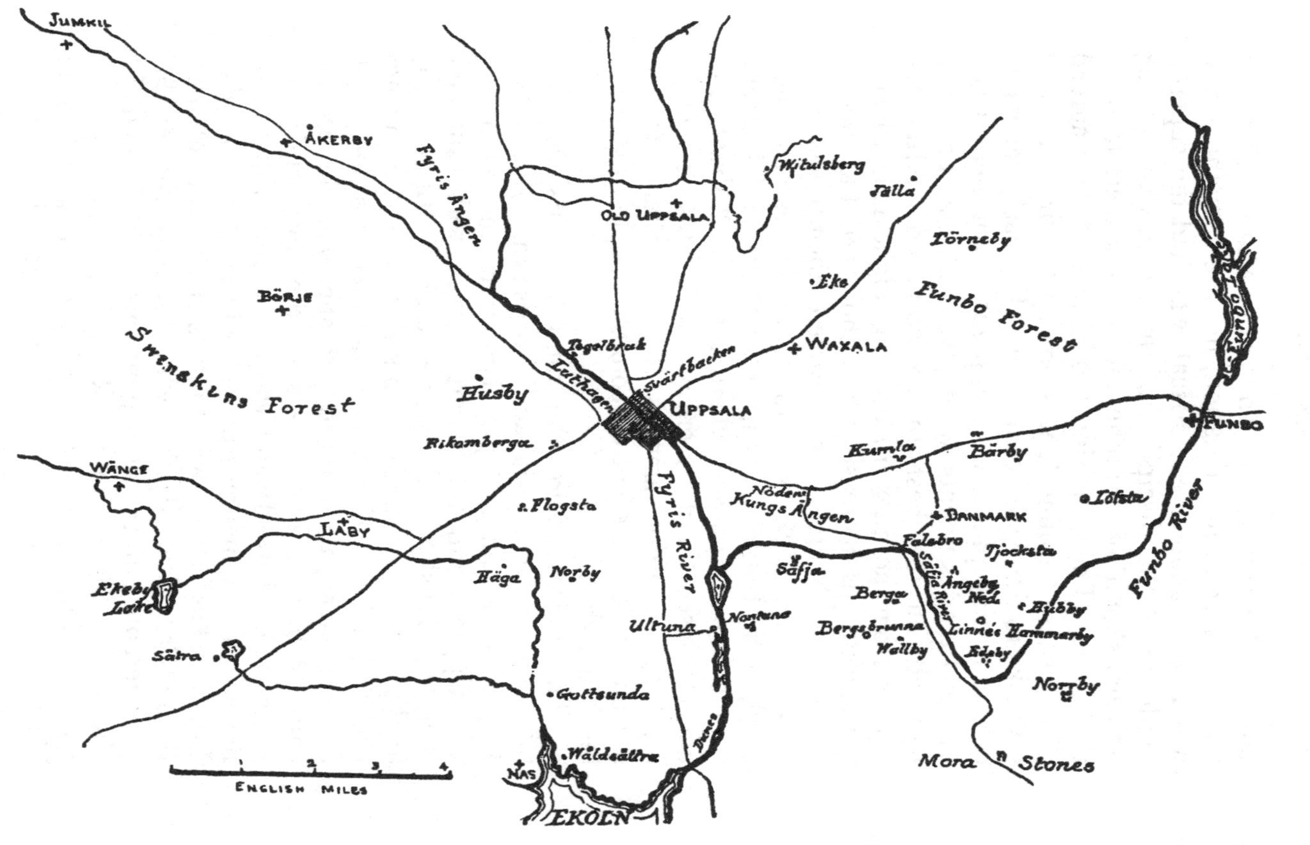
JUMKIL
ÅKERBY
Fyris Ången
OLD UPPSALA
Wittulsberg
Jälla
Törneby
Eke
Funbo Forest
Funbo Lake
BÖRJE
Swenskuns Forest
Tegelbruk
Svartbacken
WAXALA
Husby
Luthagen
UPPSALA
FUNBO
Rikamberga
Kumla
Bärby
WÄNGE
Flogsta
Nöden
Kungs Ången
Löfsta
DANMARK
Funbo River
LÅBY
Falebro
Tjocksta
Säfja
Säfja River
Ångeby Ned.
Häga
Norby
Ekeby Lake
Berga
Hubby
Ultuna
Nontuna
Bergsbrunna
Linnés Hammarby
Sätra
Wallby
Edeby
Norrby
Gottsunda
Danes
Wäldeälfre
Mora
Stones
NÄS
1
2
3
4
ENGLISH MILES
EKOLN

officials; for instance an Annotator, whose task was to take down from his dictation, in case something new was found; another was Fiscal, who had the superintendence of the discipline of the troop, that nothing unusual should occur; others were marksmen, to shoot birds, etc. The gathering was always at some agreed place, where he himself was among the first, ready to set tasks to those who came late. For each excursion, certain resting places were appointed, and here, when the scattered students were gathered, the Professor lectured on the best things collected. After the youths from morning till evening, eight a.m. till nine p.m. (until his increasing age shortened the time to five—seven hours), had enjoyed themselves thus, they marched back to the town, the Professor at their head, with French horns, kettledrums and banners, to the botanic garden where repeated "Vivat Linnæus" closed the day's enjoyment. This cheeriness, rejoicing and ardour amongst the young men, attracted not only foreigners but up-country people to share in these delights.

The places where the annual excursions were taken, were Gottsunda and Vårdsätra, Ultuna and Liljekonvaljeholm, Håga, Danmark and Nontuna, Old Uppsala and Vitulfsberg, Vaxala with Jälla and Törnby, Husby with Börje and Kättinge also; finally, the far distant Jumkil, where "Sceptrum Carolinum" [*Pedicularis Sceptrum*] was the chief floral treasure; these places are shown in the accompanying map. The streets of the town and open places offered weeds and wayside plants which there flourished. Sometimes the expedition was to Funbo-Löfsta, whose owner was Linné's friend and admirer, Baron Carl Sten Bjëlke, who not only showed his considerable cultures, but generously entertained the glad and hungry crowd. Reports still in existence testify to the fire which Linné knew so well how to kindle in his pupils. Even the most insignificant plant or animal had something worth speaking about, its properties or life-history or its use in practical life, or its application as interpreting other obscure factors in nature; and he did not neglect to spice his remarks with humorous episodes, which contributed to the interest. This sketch of Linné's method of teaching applies to the first two decades of his professorship, concerning which he himself admitted that science had now reached the summit. After this a decline became noticeable. After Linné had botanized for so many years with his pupils, both his bodily and mental powers began to fail. The result was that though he continued to be surrounded with devoted pupils, they latterly consisted of such as made natural history their main study, or else belonged to the medical faculty and obliged to undergo examination in Linné's departments. Naturally he saw this with some melancholy; he said, "It is with science as with *Cynosurus cœruleus;* one marks its beginning, but it spreads all round."

Through Linné's teaching and writings, natural history both at home and abroad had gained many friends and earnest workers, but at the same time the throng, formerly so thick round his chair, lessened. But quality had superseded quantity. Those belonging to other faculties fell off, but there were not wanting others who gave themselves heart and soul to biology. These were increased by many foreigners, and in spite of the difficulties of travelling at that period and of speaking foreign languages, they came over to receive from Linné's own lips, the solution of Nature's riddles. The fame of his extraordinary power of teaching had spread widely, so that from far distant lands, even North Africa, Siberia and America, people came to the little, unpretentious town of Uppsala. Nothing like it had been seen before; it aroused notice, gladness and pride in the whole country, his brother professors glorying in the reputation gained for the old seat of learning, even though feelings of envy could not be suppressed.

The town's inhabitants were satisfied with their gains, as the visitors were not merely hasty trippers, but often remained for several years.

Linné saw this recognition of his scientific merits with gladness and pride, and it is certain that he bestowed upon these foreigners endless trouble and even involved himself in pecuniary sacrifices. Many of them were entertained at his table, and payment for lectures was left for their discretion. Thus when the German Giseke, on taking leave, gave him a Swedish banknote, Linné roundly refused to accept it, till after Giseke's repeated requests he said: "Now tell me truly, can you afford it? Do you require this money for your journey home? If the former, give it to my wife, but if you are straitened, so help me God (*ita me Deus*, Linné's usually confirmative expression), I will not take a single farthing from you." In the same manner he would not take the smallest sum from Ehrhart, whose teacher he had been for years. "You are a Swiss and the only Swiss who has come to me. I will not take anything from you, but you have given me the pleasure of teaching you what I know gratis." The same tale is told of the Danes, Fabricius and Zoega, and it seems to have been his rule with all foreign pupils, unless they evidently had plenty of money; further he helped diligent students and provided them with stipends or occasional gifts from the University funds. He looked upon the foreigners as his beloved children and was to them a tender father. For him it was sufficient and at the same time a pleasant enjoyment, to exchange views with those who really loved his science, and who had reached a certain measure of insight, as for such conversation he had otherwise but little opportunity in Uppsala. It was a pleasure for him to show his countrymen how highly he was esteemed abroad.

But it cannot be said that he had no advantage from his coaching; on the contrary, wealthy pupils, it is true, according to their discretion, often gave generously, sometimes in princely fashion. Thus he received for a course of lectures which he conducted for the Russian Barons Demidoff not less than 3,500 dalers [£262 10s. if in silver or £87 10s. if in copper], but these teachings were specially intensive, as in September, 1760, he devoted three hours daily to them : at ten o'clock, botany, at eleven, zoology, and at twelve, mineralogy.

Their lodgings in Uppsala during term, were in the neighbourhood of the botanic garden, that is, in the least pretentious part of the town, Svartbäcken, where were mostly low timber houses, on whose thatched roofs such plants as Linné gave the name *tectorum*, as *Crepis*, *Bromus*, etc., found a favourable site. During the summer, on the other hand, they obtained quarters in the humble peasant cottages in the vicinity of Hammarby, where convenience was reduced to a minimum. At stated times daily lectures were held, and in consequence of Linné's ignorance of current foreign languages, exclusively in Latin, in which he easily expressed himself, though not always in classic diction or construction. The superlearned laughed compassionately at his "Svartbäck's Latin," as in his eagerness, he did not give proper regard to the niceties of the Latin grammar, correctness of meaning weighing more than words. He owned his weakness in Latin, but also declared that he would rather have three slaps from Priscian, than one from Nature—Malo tres alapas a Prisciano quam unam a Naturâ.

Of the relations between Linné and his pupils, the best account is from one of them, afterwards the celebrated entomologist, Fabricius. "For two whole years, 1763 and 1764—Linné being then in his fifty-sixth and fifty-seventh years—I had the happiness of enjoying his teaching, his guidance, and his intimate intercourse. No day passed that I did not meet him, hear his lectures, often spending several hours with him in friendly talk. In summer we three foreigners,

Kuhn, Zoega and I, accompanied him into the country. In winter we lodged close to him, and he came to see us almost every day in his short, red dressing-gown, green fur cap, and pipe in hand. He generally came for 'half an hour,' but stayed one or even two hours, his conversation being extremely animated and pleasant. Either it consisted of anecdotes of the learned in his science, whom he had met at home or abroad, or he cleared up our doubts and questions in science. He laughed heartily, and his face beamed with gladness and high spirits, which plainly showed how ready his soul was for society and intimacy.

"Still happier was our life in the country. We lived about three-quarters of a mile from his house at Hammarby in a peasant's cottage, where we had established ourselves after our own fashion, and had our own household." (They constantly had their meals with the Linnean family.) "He (Linné) in summer rose early, usually about four a.m. About six o'clock, as his dwelling house was being built, he came to breakfast with us, and lectured on natural orders as long as we liked, generally from ten a.m. onwards. Afterwards we went about noon to the rocks near by, which, under his guidance, provided sufficient occupation and interest. Towards evening we went to his garden, and later on we played at trisett [' three sixes '] with his wife, her favourite card game.

"On Sundays the entire family was with us at our place, and sometimes we let a countryman come with an instrument looking like a violin [hurdy-gurdy], when we danced in the barn to our great contentment. Truly our balls were not particularly brilliant, the company not numerous, the music wretched, but we danced in turn minuets and polkas and enjoyed ourselves not a little. The old man, looking on, smoked his pipe with Zoega, who was delicate, and even he himself, though rarely, danced a Polish dance, in which he excelled all us youngsters. Unless he saw that we were cheerful, and even noisy, he feared that we were not enjoying ourselves. Those days and hours will never be forgotten, their remembrance being delightful to each one. He was also my teacher, and with grateful heart I recall how much I have to thank him, both for his instruction and for his gracious behaviour."

Fabricius was not alone in these feelings of respect, affection and gratitude; but to this witness we owe many valuable expressions of regard, printed, or in letters to Linné, both father and son. A particularly handsome testimony was given by one of the Demidoffs. When the Swedish prisoners of war passed by St. Petersburg on their return home in 1790, he met some of them, giving them help and in other ways showing his kindness, in order to make manifest his gratitude, as he said, for the pleasant time, which he, as Linné's pupil, had spent in Sweden.

As may be seen from the above, he practised his oft-repeated rule of life: "Mingle your joys sometimes with your earnest occupation" (Interpone tuis interdum gaudia curis). Work was for him his principal aim, finding in it his chief pleasure, and even after lectures, he would spend some hours in steadfast investigation and authorship. But it is hard to understand how he could go through such bodily and mental toil, when one calculates the time he devoted to teaching. An extract from a letter to his intimate friend, Bäck, in 1761, will show this. "I lecture five hours each day: at eight o'clock with Danes; at ten, publicly; at eleven and twelve, with Russians; at two, privately with Swedes. On Wednesdays and Fridays three hours are spent in proof reading on my 'Fauna.' The rest of the time is hardly enough for writing additions to the same work; I have no time to think about myself, so I write till two in the morning." On another occasion, in 1766, he writes: "My dear fellow, do not talk about a trip to Drottningholm [a palace near Stockholm], I now have my chief lectures, public, private and most private, the last to Danes, Russians, Germans, and others, who come hither from distant parts to listen to me." And during the holidays when his colleagues were enjoying entire quiet at watering places to restore their health, he was writing in 1771, when he had completed his sixty-fourth year: "I am in obscurity at Hammarby, but have to lecture eight hours a day to my foreigners, and more, if it rains."

Just as his pupils took home pleasant memories, so did their teacher retain similar recollections. An exception must be made in the case of the first who came, Henri Missa, from Paris, who, provided with a recommendatory letter from Haller, in August, 1748, found himself listening to Linnæus in botany. This event caused great attention, for never before had a student come from France to Uppsala University. He was received with open arms, and was thus described: "Missa is fairly quick, and has an incredible temper. He is poor, does us no credit, but one must teach him so that others may follow. He is here for a year. I have boarded him, which sum I reckon at 600 dalers [£45], in addition to advancing him money; I shall not spare my instruction."

By the beginning of the next year there was a change in Missa's behaviour. After a visit to Stockholm, he declared that botany should no longer detain him, and he would go back to France. He attacked Linné, who wrote, "He boarded with me for two days and since then I have not seen him. He repaid me the money I lent him, but went away without any thanks for the long period I had maintained him, when I had done more for him than a brother or a Swede, and now he slanders me, as I have heard from several people. So I harboured a snake in my bosom, but did not know it." Thus happened the unfortunate parting between them, to Linné a great trouble; and among his pupils he thenceforth never mentioned Missa, whose after fate is unknown.

The delight and satisfaction to which the many foreign visitors to Uppsala gave expression, was balanced by the trouble experienced from Th. E. Nathorst from Silesia, who was registered in 1755, but did not prosecute his studies.

There was a prospect of trouble, which, however, did not take place, when the rich Russian noblemen, Matthæus Aphonin and Alexander Karamyschew, at the Tsar's expense, came to Uppsala to study in October, 1761, and there throve so well, that the former stayed till 1769 and the latter till 1767. Although they did not show great diligence, they readily joined in every kind of frolic, and Karamyschew especially seems to have been very popular among the other students, notably when in 1766 his disputation was accompanied by eight congratulatory epistles, in French, German, Russian, Latin (prose and verse) and Swedish.

There was much excitement in 1762 when the Russians were personally abused by some townspeople and badly used. The Rector at once enquired into the matter, and punishment of some days' prison fare and fines was imposed upon the offenders, and a reprimand issued warning all that the credit of the University was imperilled by such disorders. The Russians were again in trouble in 1769 during some wedding festivities, and on at least two other occasions their escapades were brought before the Consistory.

In the Appendix will be found a complete list of Linné's foreign pupils, but certain others stood in such intimate relation to their teacher, as to deserve special mention.

First may be named the German, J. C. D. Schreber. In 1758 he was corresponding with Linné, who said of him, that he was a quick fellow, who wrote little, but liked mineralogy and sent many fine insects. In 1760 in spite of the war, Linné bespoke the influence of the Chancellor to enable Schreber to reach Sweden, which was effectual. The newcomer intended to enjoy Linné's teaching during the summer and in the autumn to offer himself for examination for M.D. But he was found so well grounded, so extraordinarily quick in biology, and possessing such ready insight, that no one in the medical faculty had previously met with his equal, so that he passed in June, with special approbation, and in the same month was promoted Doctor. During the summer he heard lectures at Hammarby, after which he went back to Germany; afterwards remaining in permanent communication with Uppsala; he also edited the second editions of "Materia medica," and "Amœnitates Academicæ."

Another in whom Linné felt great interest, was P. D. Giseke, who came to learn direct about natural orders. He sent to Linné a copy of his dissertation "Systemata plantarum recentiora," for which Linné returned a most friendly letter, beginning: "You want to get from me the characters of natural families: I confess that I cannot give such." Giseke spent the summer of 1771 at Hammarby, in constant touch with Linné, spending so much time, that when he took leave of "optimus senex" (the most worthy old man), he was reminded that no pains for him had been grudged. On being asked what would satisfy his teacher for what he had received, he was told: "What you please." Twenty years later he joined with Fabricius in a volume on the subject of natural orders, but that came out after their teacher's death.

The only Englishman among these was John Rotheram. At the end of June, 1773, he arrived at Uppsala, and was registered in the University and the Smålands Nation. Some time after he fell ill, but was carefully nursed by the wife and daughters of Linné, as a warm letter of thanks from his father in 1774 testified. After that he seems to have become so intimate in Linné's home, that he was one of the two at the bedside when the great naturalist passed away During his stay at Uppsala he learned to speak Swedish fluently, and was one of the few who became friends with the younger Linné. On his return to England, he had a sharp exchange of words with Daniel Solander, when the latter called the young Linné "a thoroughly worthless fellow." He wrote later, "At last it was with little better than mutual abuse, when we bade each other good-bye."

How he prosecuted his studies at Uppsala is now hard to say. It is, however, certain that he was a loyal listener to Linné's lectures, passed the several stages of examination, and was promoted Doctor in 1775. He fared so well at Uppsala, that though he had finished his course of studies, he stayed on, and his father wrote complainingly to Linné that his son would not come home. In the spring of 1778, after the death of Linné, he submitted a thesis "De Variolis" [On smallpox] soon after which, he returned to England and became Professor of Physic in the University of St. Andrews in Scotland.

Friedrich Ehrhart takes a special place amongst the rest of the foreign pupils. He had not received any university training, but was dispenser at an apothecary's in Hannover, when, attracted by some Swedish scientific writings, he determined to betake himself to Sweden. First he had occupation for six months at the Court Apothecary's in Stockholm, where he found himself so happily placed, that "if he had not left his dearest maiden in Hannover, he would have probably remained in the beautiful country of Sweden."

A desire to listen to the lectures of Linné and Torbern Bergman, drew him meanwhile to Uppsala, where he first served a year with the university apothecary, but afterwards kept himself for two and a half years, entirely free at his own expense, during which he became intimately acquainted with C. W Scheele, T. Bergman, and many other eminent men,

He was Linné's pupil from the 20th April, 1773, till the 26th September, 1776, much enjoying the tuition, though he lamented that he had not come before, when the late Archiater himself took excursions in the fields of that fine country. "My teacher had already aged, and expected the end and dissolution. If I questioned him about cryptogams, he answered frankly, that he had for thirty years studied such plants, but must now give their investigation and determination over to others. I often went with doubts to Hammarby and turned back to Uppsala with my questions unsolved."

Few pupils perhaps have been so diligent as he. Every day he spent his free hours in excursions to the surrounding districts, but Sundays he usually passed in the botanic garden; especially during the holidays he went, often accompanied by other pupils, from early morning till late at night into the meadows forests, marshes and bogs, seeking plants. His discoveries he imparted to Linné, who showed him special favour, and he did not scruple to remark upon certain points in his teacher's works. Linné was at first astonished, but after a day or two's thought cried out, "You are right." "And when in his house at Hammarby, I took leave of him and lamented that it would be the last time I should see him, he pressed my hand once more and said, 'Write to me; I trust you entirely.'"

Returned to Germany, Ehrhart first settled in Hannover, then in Herrenhausen, and was appointed botanist to the English King, and Prince Braunschweig Luneberg; he published many botanical works, so that he must be reckoned as one of Linné's most distinguished pupils. The remembrance of his residence at Uppsala remained as a brilliant episode. When in 1790 he brought out a list of wild plants round Uppsala, he penned a warm-hearted passage of the innocent delight and quiet enjoyment he experienced there, with Linné, "the divine Linné."

There are, and always have been, teachers who fulfilled their duties implicitly, but never came into intimate relationship with their pupils. Linné was not of that number, for it was for him a true joy when an enquiring young man came to him with questions. "A professor," said he, "can never better distinguish himself in his work than by encouraging a clever pupil, for the true discoverers are among them, as comets amongst the stars." Thus he established round him a close intimate circle of students, for whom he showed a fatherly regard. A sketch of his relations with this select band must be touched upon, but in a brief fashion.

In the first place we must look at the numerous youths, who, incited by Linné's stories of nature's life in foreign lands, betook themselves in high spirits and burning zeal to the quest of natural objects. We must remember the great perils which then were connected with such journeys, and the scanty appliances which could be got together for such enterprises. Many of Linné's "apostles," as he loved to call them, these naturalist pioneers, suffered a martyr's death, but that did not prevent others from offering themselves to similar tasks, to the same hunger, the same struggle, the same death; an everlasting memorial to their memory, *ære perennius*.

The first of these was Christopher Ternström. Although belonging to the divinity faculty, he had for years accompanied O. Celsius and Linné on their botanic excursions, by which he had advanced so far in botany, that "no one in the kingdom could be compared with him except Kalm." His ardour did not diminish, after he had been ordained, and though married, and a father, he begged to be allowed to sail to the East Indies, partly as priest, but also to botanize, which permission he obtained through Admiral Ankarcrona. Linné gave him instructions, and he was especially charged to procure a tea plant in a pot, or at least seeds of it; to take thermometric observations, and obtain living gold-fish for the Queen.

At the beginning of 1746 he went to Gothenburg to embark, but before sailing, a disastrous fire broke out in the town, destroying more than a quarter of it. Letters came from him from Cadiz, but the next intelligence was that he died at Pulo Candor on the 5th December, 1746.

Linné was terribly grieved, not only for the lost hopes of scientific results, but for the widow and orphans; he tried to procure some help for them, and said that the widow accused him of having enticed her husband away, thus making her a widow. He wished to publish Ternström's observations, but finding no opportunity for that, the genus *Ternstrœmia* was founded in memory of his former pupil.

Before Ternström had started, Baron Sten Bjelke had urged that a naturalist should be dispatched to some country in the same latitude as Sweden, with a view of introducing plants for food, medicine or manufacture. The most suitable land seemed to Linné to be North America, where *Morus rubra* if brought over, might supply food for silkworms and so set up a new industry. For this expedition, his pupil Pehr Kalm was selected, he having previously travelled in Sweden, Finland and Russia, with good results. The difficulty was to provide funds for a long trip, but ultimately help came from the universities of Uppsala and Åbo, as well as from a manufacturers' union, chiefly through Linné's exertions.

At last Kalm, accompanied by a gardener, L. Jungström, journeyed to Gothenburg; after sailing, the vessel was driven into a Norwegian port by stress of weather, so that he did not reach England till 17th February, 1747 He was forced to remain here till the 5th August, but spent his time botanizing round London, reaching Philadelphia on the

15th September. After extensive travels, he quitted America on the 13th February, 1751, and came to Stockholm the 3rd June. While frequent letters had passed between Linné and Kalm, an illness which attacked Linné soon after, caused a delay in Kalm's visit to Uppsala with his large collection of dried plants and seeds. Amongst these may be mentioned *Vitis hederacea*, the Virginian creeper.

Kalm had scarcely begun his travels, before plans for another expedition were in progress. Among the then Uppsala students was Fredrik Hasselquist, who had distinguished himself in his studies, and of whom Linné wrote in eulogistic terms to the Academy of Science—that he was modest, polite, cheerful, and intelligent, but very poor. In his lectures, Linné had mentioned Palestine as one of the countries not sufficiently known, and this fired Hasselquist with the desire of travelling thither. He confided his wishes to his teacher, who felt himself obliged to point out the long distance, the many toils and dangers, the great expense which stood in the way, as well as his weak health and tendency to consumption. All was in vain, he was determined. By Linné's hard work the funds were collected, and a free passage to Smyrna was granted on a vessel belonging to the Levant Company.

With the small sum of 1,890 dalers in copper [£47 5s.] Hasselquist set out on the 7th August from Stockholm, and on the 26th November, landed at Smyrna, where he obtained quarters with his relation, Consul General Rydelius, there spending the winter. In the following March he travelled inland, and in May went to Egypt, where he stayed till March, 1751. His collections were rich, but his means were exhausted, till Linné and O. Celsius the younger, induced the Consistory to grant two more stipends, Hasselquist thus receiving assistance by this means, in all four faculties, an event which never occurred before or since. Linné redoubled his efforts, and by personal appeals collected in one week more than 7,000 dalers in copper [£175].

By this reinforcement Hasselquist was enabled to prosecute his journey. In March, 1751, he travelled to Palestine, explored a part of Arabia, and a large part of Syria, for a short time staying in Cyprus, Rhodes, and Chio, and then on to Smyrna, with a valuable store in all three kingdoms of nature. But he was immediately obliged to hurry from the unhealthy Smyrna to the village Bagda, where "our beloved Dr. Hasselquist, like a lamp whose oil is consumed, died on the 9th February, 1752, at six in the evening, to the grief of all who knew him."

To the heavy sorrow which this event roused in Linné, there came trouble about the fate of the collections and notes. It appeared that Hasselquist had incurred a debt of 14,000 copper dalers [£350], and for this, the collections and manuscripts were seized in pledge. Linné doubted the justice of the claim, but saw that something must be done to avoid a "double death, that not only had the traveller vanished, but his work threatened to vanish also, which would not be creditable to the nation among all who love science." He turned therefore to Baron Höpken, and also to A. Bäck, saying, "My heart bleeds every time I think about Hasselquist's collections. It is a great sum it is dark for me."

Linné's lamentation was not without result. In November, Bäck gave him to understand that the Queen, in consequence of his appeal and those of Tessin and Höpken, was disposed to pay the sum demanded. Linné's delight found expression in his reply to Bäck. "If the Queen redeems Hasselquist's collection, Her Majesty is a goddess and my brother [Bäck] an angel."

The following year the collections came to Drottningholm Palace. Linné, who was summoned thither, grew giddy on beholding so many novel things at one time. Later he received the scientific notes, and his estimation of them made him exclaim "God bless the incomparable Queen. All generations should praise Her Majesty, who rescued them from the fire." The manuscripts were delivered by the Queen's orders to Linné, to be set in order and published, which resulted in Hasselquist's "Iter palaestinum, eller resa till Heliga Landet." [Palestine journey, or travels to the Holy Land] to which were added the letters received from Hasselquist by Linné.

The "In Memoriam" oration was pronounced in 1758 by Abraham Bäck, earning Linné's high praise. Twenty years later Bäck did the same office for Linné himself.

Soon after Ternström's departure, the unwearied patron of Linné (C. G. Tessin) took steps for further investigations in foreign lands. After testing certain candidates, Carl Fredrik Adler was sent to the East Indies in 1748, but the result was trifling. Much better returns came from the two ships' chaplains, Pehr Osbeck and Olof Torén, who sailed in different ships from Gothenburg in 1750, and both came back on the same date, 26th June, 1752, two years later. The former's warm interest and power of observation had been often noticed by Linné. The voyage went well, four months and a half being spent in China. After his return he was invited to undertake another expedition, but as his health had been impaired, he decided to stay at home, and by Linné's influence he was appointed Court Chaplain, afterwards becoming pastor in Halland, which post he held till his death in 1805.

There is less to be said about Olof Torén. He sailed for Surat, and visited China, collecting many plants. Upon his return, he printed a short account of his voyage, and died soon after in 1753.

The first place in Linné's affections for his "Apostles" was held by Petrus Löfling, "his most beloved pupil." Born at Tolfors in Gästrikland the

20th January, 1729, he, unknown to his parents, registered in the medical faculty, and formed a lasting friendship with another student, J. O. Hagström, who thus described Löfling. "He came to Uppsala quite young, of such simple manners as might be taken for stupidity. During his first year he was intimate only with me, who, like him, had come from the fells. When he first came to my rooms, he fell upon my herbarium to search through it. This pleased me immensely, and we were together day and night. His father wished him to become priest, but he had no liking for that profession. I was senior to him by some years, so I advised him not to devote himself to logic, metaphysics or Greek, which had robbed me of much time. He made such rapid progress in botany, that he was worthy to be tutor to Linné's son. As to his character, I can frankly say, on my conscience, that his soul was graced with virtue, pious, just, loving and quick at grasping nature's many secrets. He was tall, like Kalm, of manly and pleasing aspect, and was also bright and healthy."

Among Linné's many pupils and attendants in his excursions, Löfling principally attracted his teacher's attention. He found his way to Linné's heart, and was taken into his teacher's house as companion to his son.

That Löfling gladly accepted this post was natural. He attended lectures, formed friendships with the most intellectual of the students, was early and late in the botanic garden and at meal-times plied his teacher with questions. "He lived with me in the greatest confidence," records Linné, "for he had a mind pure as gold without any dissimulation in speech or gesture. He was never effeminate nor fastidious in food or clothing; he could sleep on the hardest bench or softest bed." This evident kindness spurred Löfling on to deserve it, and in 1749 he put forward the famous disputation, "De gemmis arborum" (On the buds of trees), which showed his mettle.

This relation continued. "In the year 1750," relates Linné, "when I began my 'Philosophia botanica,' I had so severe an attack of gout as to cherish little hope of surmounting it, but as soon as the illness began to diminish, I was obliged to get my dear Löfling to hold the pen, whilst I dictated from my bed, till the book was done, and as he always asked questions on what he did not understand, at the end he was thoroughly grounded in the subject." It was decided that he should travel in Arabia and the East Indies, but the vessel sailing before the funds were secured, the project was altered to the investigation of Spain, as its flora, though known for its rarities, as a whole was practically unknown. "It is lamentable that a cultivated European land should remain in so barbarous a state as regards botany," was the remark of an Englishman, Robert More, when on a visit to Madrid; by the good offices of the Spanish Ambassador to Sweden, the Marquis de Grimaldi, it was arranged that a pupil of Linné should be sent to that country.

"Löfling was in my mind," wrote Linné, "and he was not averse from the prospect." By special effort he succeeded in gaining the degree of Magister and at once entered upon his journey, his old teacher bidding him a tender farewell. He was provided with a free passage to Portugal. After two months' voyage, he arrived in that country, and began his search for plants. He passed on to Madrid, where he fixed his headquarters for two years, his earliest task being to overcome the jealousy of the local botanists, in which he succeeded. Many letters were sent home, and many plants also, as the Linnean herbarium shows.

At the end of this period, an expedition to Spanish South America was organized, including four professors, as many skilled attendants, four surgeons and minor officials. Löfling was appointed head of the botanic department, and had as helpers, two young surgeons and two expert artists.

All went well at first; after a visit to the Canaries the expedition in May, 1754, reached Cumana, which offered a rich harvest to the botanist. A few months later he was attacked by ague, with four relapses which took away his strength. Nothing more was heard from him, till a report came from Spain that he had died on the 22nd February, 1756, at the Mission Station of Merercuri in Guiana, of a fever due to the climate.

When Linné recovered from the shock of this disaster, he determined to raise a literary memorial to the memory of his darling pupil, which resulted in the "Iter hispanicum" published in 1758 by Linné. In the preface the editor gave full expression to his deep grief at the loss of so promising a life: "Löfling sacrificed himself for Flora, and its lovers—they miss him!"

Amongst the unexplored lands to which Linné specially desired to send a pupil, was the Cape of Good Hope, with its rich and peculiar flora. Mårten Kähler was selected, and money was provided, but the opposition of the Dutch government prevented the arrangement. Finally in May, 1753, he was sent to Italy and Sicily, but met with hindrances throughout. He waited in Denmark a long time for a passage; then followed a five days' storm in the North Sea, when the cabin was flooded, and his apparatus, books and clothes were destroyed, he having to lash himself to the mast, where he remained two days and nights without food. After mishaps of many kinds, the anchor was dropped at Bordeaux, whence he wrote to Linné concerning his want of money, saying he did not dare to go further without reinforcement of his purse.

At Marseilles, reached after an escape from pirates, he was obliged to stay till the end of May,

1754, partly from poverty, partly from illness, but he sent home from there a chest of natural objects, which was captured by pirates. On his way to Naples he collected many plants and insects, but again was attacked by fever. An encouraging letter was received here from Linné, also financial help from Wargentin, secretary of the Academy of Science, and from Bäck. The new year was fraught with fresh pecuniary troubles and illnesses, but an opportune remittance from Wargentin enabled him early in 1756 to leave Naples for Rome. From here he started for home chiefly on foot, reaching Sweden at the end of May, when he was kindly welcomed by Linné. Kähler and Dr. Hallman dined with Linné, who noted that:

> One was big, the other modest;
> One was talkative, the other silent;
> One was empty, the other solid;
> One was untrue, the other true.

Kähler became Admiralty physician at Karlskroma, where he died in 1773 at the age of forty-six.

These unhappy events cooled Linné's ardour for sending collectors abroad. "The deaths of many whom I have induced to travel have made my hair grey, and what have I gained? A few dried plants, with great anxiety, unrest, and care." But this mood soon vanished, and from 1746 till Linné's death, hardly a couple of years passed without finding one or more of his pupils investigating foreign countries.

Among this later class of Linné's "Apostles" we must first name Daniel Rolander, who had come into notice by his observations on the life-history of certain insects. Linné engaged him as tutor for his son. He had some misgivings about Rolander, for though proving himself a theologian and entomologist, he was not apt at research.

In the summer of 1754, when Linné heard from Bäck that a Swede settled in Surinam was disposed to show some hospitality to a young naturalist, Linné's eagerness flamed up anew. The Swede proved to be Carl Gustaf Dalberg, at that time revisiting Sweden. Finally Rolander was recommended for the post. From Count C. De Geer, 600 dalers [£15] were received for equipment, and provided with instructions, he started at the end of 1754 or the beginning of the next year.

Little is recorded of what followed, but he was detained by illness at Amsterdam, only getting to Paramaribo in June, 1755, which he quitted in the following January. Dalberg tried to persuade him to stay longer, but he pleaded his weak health, dislike of the climate, and weariness of working in the heat.

Thus Linné's hopes of Rolander's success were blighted, the only thing he had sent to the Uppsala garden being specimens of the Cochineal cactus, referred to later; and he complained that "Rolander, that ungrateful pupil, gave him nothing of his collecting," but delivered them to Court Marshal De Geer, who afterwards presented Linné with a store of rare plants.

It soon appeared, however, that Rolander was less ungrateful than unfortunate. Before long he was seen to be disordered in mind. This showed itself chiefly in the fact that he brought home some grey seeds and said they were pearls, fine pearls, and even though their shells were fractured, he guarded them as being precious. On other topics he spoke sanely, even studying Materia medica in Stockholm, till Linné found him incompetent. Afterwards he went to Denmark to sell his pearls, which were, however, stolen, then he lived in Skåne upon charity, and died in Lund in 1793.

One of Linné's cleverest pupils was Anton Rolandsson Martin, who had a long struggle against poverty, with experience of suffering, and of disappointed hopes, but found in his teacher a faithful helper and comforter. Born in Livland in 1729, he became student in Åbo, but moved to Stockholm, and in the autumn of 1756 he went to Uppsala, where he was kindly received by Linné, who promised to make a distinguished man of him and to procure him a stipend. The Academy of Science received in 1756 an invitation from the Greenland Company to send a young naturalist to the Arctic Sea, and Linné selected Martin as the most suitable for the whaling voyage, and thus he became the first Scandinavian polar naturalist.

His departure from Gothenburg was made in April, 1758, with Spitsbergen as the goal and return was made on the 21st July; he had only landed on two islands, where no plants were in flower; so he collected birds, some hitherto undescribed, and Linné judged that he had done all for which he had had opportunity. Undismayed by his former hardships, he started over the mountains to Trondhjem; in the autumn he was at Bergen collecting both plants and animals, and then went by sea to Malmö and Uppsala. In the spring term of 1761 he graduated Candidate in Medicine, but soon after fell ill, and a leg had to be amputated. His remaining days were spent in Finland, where he lived partly on the funds from the Academy of Science, and partly on his earnings as tutor. He was always diligently observing, and many of his papers were published in the Transactions of the Academy. He died early in 1786.

Following the chronological sequence of the travellers for natural history we now come to Pehr Forskål, who, born in Helsingfors, studied at Uppsala, and then at Göttingen, where he devoted himself to oriental languages, under the celebrated Michaelis, but without neglecting botany, chemistry, physics and philosophy. On his return to Sweden, he published a political treatise, which the government considered pernicious, and his prospects in Sweden being clouded, he embraced an offer to join a Danish expedition to the East. Sailing on the 4th January, 1761, from

Copenhagen, Marseilles, Malta, Smyrna and Constantinople were visited; Alexandria being reached in October. Clothed as a peasant to escape marauding Bedouins, he roamed round Cairo, and was successful in making a good collection of new plants. He then travelled by Suez and Jeddah to Arabia Felix, finding a hundred novel species and thirty new genera. Soon afterward he was stricken with plague at Jerim, where Carl Niebuhr, sick himself, had the grief to watch him die, on the 11th July. Forskål had previously sent a small book, relating his discoveries and ready for printing, to Denmark, where it was seen through the press by Niebuhr, the sole survivor of the expedition. The genus *Forskolea* botanically commemorates the ill-fated discoverer of a very distinct genus.

A journey of quite a different character was undertaken by Clas Alströmer. Born in 1736, he was sent by his father, the well-known Jonas Alströmer, to Uppsala to study economics, under the guidance of Linné, Wallerius, and Berch. He had the opportunity in 1760-4 to embark on extensive travels through Spain, Italy, France, England, etc.; finding his letter of recommendation from Linné of immense use, and sending in return many plants, seeds and shells. He became Baron in 1778, and died in 1794.

In the year before Forskål quitted Sweden, Daniel Solander also left it. He was born at Piteå in 1733, became student at Uppsala in 1750, where he was constantly at Linné's house; he travelled in his native province in 1756 for plants, which are still to be seen in the Linnean herbarium, and the Consistory put on record their estimation of his diligence and skill.

He also was attracted to the investigation of foreign lands. The zealous London naturalists, John Ellis and Peter Collinson, had requested Linné to send some of his pupils to encourage the study of natural history in England. For this, Solander, the best of his disciples after Löfling, was selected. He came here in 1760, and after three weeks in London, reported on his good prospects; he became a thorough Englishman and never again saw his fatherland. He was appointed to a post in the British Museum, and in 1764 was installed as librarian to Sir Joseph Banks, with whom afterwards he sailed round the world with Lieutenant J. Cook in the *Endeavour*. On their return in 1771 he accompanied Banks to Iceland, a trip taken when their intention to sail again with Cook had become impossible.

Linné complained that his old pupil never sent him a single plant or insect from the voyage of the *Endeavour*, but he overlooked the fact that the whole of the collection belonged to Banks, who had incurred great expense in equipment. Further, it was hoped that Linné might be persuaded to visit England, to inspect and help to name one thousand two hundred new species with a hundred genera, and a multitude of animals, fishes, insects and mollusca.

The grateful and affectionate regard Solander entertained for his former teacher, was shown by his reception of the younger Linné in London in 1781 and 1782. Although he had no great estimation of him as a naturalist, he neglected no opportunity to further his studies, and even nursed him during a severe illness. When an article by Fabricius appeared in the "Deutsches Museum" which Solander considered defamatory of the Linnean household, he bought up all the copies he could find, and destroyed them, to prevent a misconception being spread abroad. It was therefore a heavy trial for the younger Linné, when an apoplectic stroke, in 1782, suddenly ended his countryman's life.

Forskål and Rolander, already mentioned, as tutors to the son, gained special instruction from the father; and now a third may be added, Johan Peter Falck, also an "apostle." At first meant as a companion of Forskål in his Arabian journey, it was found impossible to arrange that; he therefore was dispatched in 1763 to St. Petersburg, where he became administrator of the rich museum belonging to the Imperial Body Physician Kruse. But constant hypochondria and ill-health, due to a sedentary life, embittered his existence.

Kruse had died before 1765, so Falck was offered a position in the Collegium medicum, with superintendence of the medical garden, but he could not support the prospect of continuing in Russia. In 1768, J. J. Lerche nominated him as naturalist for an expedition first intended for Persia, but afterwards changed for Orenburg. He started on his journey, but was stopped in Moscow by a complication of disorders till his health was somewhat restored. Finally in 1772 he was found dead in his bed, from a pistol shot in his head, self inflicted.

Anders Berlin after finishing his academic studies, came to London in 1770, where he was most kindly received by Solander, and engaged by Banks as an assistant at £80 salary, which post he held till 1773, when he accepted an invitation to visit the Guinea coast, as a scientific helper to Smeathman, a young Englishman, who had gone thither a year earlier. He was delighted with the vegetation and sent some plants to Linné, but the unhealthy climate claimed him as a victim a short time later.

The next of Linné's pupils to be considered, is Carl Peter Thunberg, who had a double claim on Linné's good-will, as a zealous naturalist and as a Smålander. In 1770 he went to Paris, but stayed so long abroad, that he only came home in 1779, when Linné had been dead fifteen months.

This happened because the open letter of recommendation he had from Linné, operated so advantageously that in Holland he was induced to undertake a long voyage as far as Japan, which land was at that time closed to all nations except the Dutch; therefore to gain entrance, Thunberg had to enter the Dutch Company's service as surgeon. He first travelled to the Cape of Good Hope, spending 1772 to 1775 in research on its flora and fauna, discovering three hundred new species. After this he voyaged to Japan, touching at Java on his way. Returning at the end of 1776 with a second stay in Java, he gave seven months to Ceylon, thence to Holland, where he was offered a professor's chair at Leyden, and by England and Germany back to his native land. He kept up correspondence all the time with Linné, and supplied him with so many plants that his old teacher admitted he had never benefited so much from any other traveller. Before he died in 1828, he had occupied the chair of botany at Uppsala forty-five years, in succession to the younger Linné.

Thunberg had a worthy competitor in Anders Sparrman. When only seventeen, he in 1765 had served as ship's surgeon on a voyage to China under C. G. Ekeberg. He embarked again in 1772 for the Cape, and pushed his enquiries vigorously. At the close of that year, Captain James Cook, on his second voyage round the world, on board the *Resolution,* touched at the Cape. He was invited to accompany the Forsters, father and son, and went with them to New Zealand. Returning to the Cape after sailing six thousand nautical miles, he renewed his quest after Cape plants for eight months longer. He was back in Sweden in 1776, but was quite worn out with his labours. Nevertheless, he ventured again in 1787 to Africa, became Professor of Natural History in Stockholm in 1790 and Assessor to the Medical College in 1803, dying in 1820.

Jöran Rothman, son of Johan S. Rothman, Linné's benefactor at Växjö, after taking his degree of M.D. at Uppsala, failed as a physician in Stockholm, but was selected by the Academy out of three applicants, to investigate Barbary, starting in 1773. The result justified Linné's fears; the promises of the native envoy were not fulfilled, and the money ran short, but an excursion towards the Atlas Mountains yielded some botanic results. He returned to Stockholm in 1776, in broken health, and, two years later, died as Assessor to the Medical College.

Another of Linné's pupils was Johan Andreas Murray, who without doubt became a star of the first magnitude in botany. Born in 1740 at Stockholm, his father being chaplain to the German colony there, he became student at Uppsala in 1756, and Linné soon discerned in him one of his most diligent and clever pupils. A close friendship was formed, as between father and son, which was only ended by death. Murray took his degree as Licentiate at Uppsala, and proceeding M.D at Göttingen, became Professor of Medicine there in 1764, with superintendence of the botanic garden, and died in that place in 1791.

During a visit to Sweden in 1772, noticing an interleaved copy of the twelfth edition of the "Systema Naturæ," he begged Linné to bring out a new edition, but the reason he alleged for not doing so, was want of time; however, he put the volume into Murray's hands, with permission to print it if he felt so inclined. Murray agreed, found a publisher in Germany, and astonished his old teacher, when the book was printed in 1774, by sending him a handsome sum for authorship, received by Linné with great satisfaction. After Linné's death, he issued another edition of the "Systema vegetabilium," the fourteenth. Linné was much touched by Murray naming him godfather to his little girl Carolina.

There were many pupils who deserve mention here, but the number must be restricted to a few more. Johan Otto Hagström distinguished himself by drawing up a list of the plants eaten by various farm animals. From 1754 his time was taken up by his duties as provincial physician in East Gothland, but he produced his "Pan apum," a small work on the flowers affected by bees, to which Linné added a preface; later he wrote to the author, "I have read your 'Pan apum' eight times and I can say it is a jewel."

Another pupil was Johan Gustaf Wahlbom, who showed himself worthy of the post he obtained at an early age, as Adjunct in 1748, and M.D. in 1751; in the latter year being appointed Professor of Natural History at Lund. He increased his knowledge in Germany, and on his return home, settled in Kalmar county as provincial physician up to 1794.

About the same time as Hagström and Wahlbom, Erik Gustaf Liidbeck was living at Uppsala, and by his devotion to natural history endeared himself to Linné. He served as secretary in the Linnean journey to Skåne, and later, in 1756, was installed in the newly established chair of Natural History at Lund.

A still greater eminence in science must be allowed to Peter Johan Bergius. After studying two years at Lund, he came in 1749 to Uppsala, where, attracted by Linné's lectures, he gave himself up to biology. After concluding his academic studies in 1761, he became Professor of Natural History and Pharmacy in Stockholm, and Assessor in the Medical College in 1766. With his brother Bengt Bergius, he raised a memorial to his generosity by a large donation to the Academy of Science, which is now extant, with its library and collections, as the Bergian Garden.

Upon settling in Stockholm P J. Bergius took up medicine, and speedily became famous as a physician. After a little coolness between them, caused by adverse criticism by Linné on a paper written by Bergius on the Soja bean, the old cordial relations were resumed, and Linné greatly pleased Bergius by the gift of duplicates from his herbarium. Linné also praised Bergius's volume "De plantis Africanis," and in return received many heaths from him.

J. A. Murray's younger brother, Adolf, was born in

1751, and when only thirteen, became a student at Uppsala, where his progress was so marked, that in 1768 he was prosector in Stockholm, and in 1770, at the age of nineteen, lectured in the place of Professor R. Martin. He was promoted M.D. in 1772, on the last occasion when Linné officiated as promotor. During an extensive tour in Europe, hearing bad accounts of Linné's health, he gave full expression to his sorrow in one of his letters. Upon his return, he found the old Professor still alive, and by his influence he was instituted as the first Professor of Anatomy in Uppsala University.

Erik Acharius was Linné's last pupil. Born in Gävle in 1757, he was inscribed as student at Uppsala in 1773, and disputed on "Planta aphyteia" in 1776, the last occasion when Linné was Præses. He travelled afterwards in the New World, but his famous lichenological works belong to a later period. He died in Vadstena in 1819.

Finally may be named a man who was never a student at Uppsala, but was occasionally connected with Linné, who exercised so great an influence over him that he may justly be counted in the ranks of his pupils. This was Anders Jahan Retzius, "The giant in learning," because to him there was "nothing unknown, nothing unheard, nothing new." Thus he wrote in his "Prodronus Floræ Scandinavicæ": "At the most, I have been only twenty hours with that incomparable man, but I seem to have had as great advantages as of two years' coaching. I fancied that in the Linnean method were many defects, but in my youthful bashfulness I was ashamed to ask his help in explaining them. But those were golden and priceless hours during which I conversed with the world-famous old man!"

With this is closed the sketch of Linné's pupils and his relation to them. The words of one of our most distinguished naturalists, Elias Fries, may be quoted: "What Gustaf Adolf and his heroic band in a political aspect were for Sweden, Linné and his pupils were for natural science. Each nation has had its heroic age in the world's history; and each science which has won development has had its chivalrous period. In botany, it is the Linnean period which has left its mark, and now we may with reason, and with slight alteration, quote a poet's words:

> There is no age which lets its memories sleep:
> No land like ours, with memories so deep!

CHAPTER X

LINNÉ AS ADMINISTRATOR OF THE BOTANIC GARDEN AND MUSEUM

SIMULTANEOUSLY with the exchange that Linné made with Rosén, he took over botanic instruction, and also assumed the superintendence of the botanic garden. Keeping in mind the rich and costly gardens he had seen in Holland, England and France, he set himself to transform the little, marshy Hortus Upsaliensis to one which should vie with any of the foreign ones, and that in spite of its situation, unfavourable climate, poor soil, narrow limits, and poverty-stricken economic relations, and what is more, he succeeded.

In what condition the Uppsala garden was in 1740 has already been told (p. 189), though the dawn of a better day showed itself when the skilful gardener D. Nietzel came, and a few improvements were begun by Rosén. Linné took it as a happy omen that the enlargement of the garden was sanctioned by the Consistory. He was not long in grappling with the work, first setting out the need of a glasshouse or orangery, the soul of a garden, for raising seeds and growing plants and trees from southern lands, and for harbouring them against the attacks of wintry weather. The sum of 5,226 copper dalers [£130 13s.] was assigned for the building of a house 32 ells long [63 feet] and also for " the dwelling house of the Prefect, at present a sad spectacle, more like an owl's nest or robbers' cave than a Professor's abode." This being approved by the Consistory, Court Intendant Hårleman designed it, and in 1743 the orangery was completed with its

LINNÉ'S HOUSE IN UPPSALA (Photographed about 1870).

two wings. At the same time, the old stone house which was built by the elder Rudbeck, after the disastrous fire of 1702, with iron doors and joists, instead of wood, was pulled down, and a new one built, with larger windows and rooms, and the added luxury of plaster ceilings. The Chancellor agreed that Professor Linnæus should occupy the house rent free, and soon after he moved into it, living there till his death. The garden was remodelled according to the French fashion, with straight paths, flower quarters, borders, clipped hedges, etc., which showed a striking contrast to the old one, which was almost without plan. Hereby the garden and its buildings received the arrangement which was retained during Linné's time. Only a few slight alterations were made in 1775, in consequence of damage by a storm. By degrees more land was added to the extent of the garden, up to 1771, but the accession of plants during Linné's prefectship were so great that everything was greatly crowded. A good idea of the garden may be obtained from "Hortus Upsaliensis" of 1745, and in an interleaved copy belonging to Linné, we find instructions as to sowing seeds and the conduct of the garden, which throws additional light upon the state of things in the middle of the eighteenth century. One of the wings of the orangery was the dwelling house of the gardener, whilst Linné's work-room at the other end of the garden, commanded a view of the whole. The total number of plants rose from 200 to more than 3,000. In the year 1742, 567 plants were sown; in 1743, 500; and in 1744 more than 1,000. These additions came mostly from Linné's numerous correspondents, till in 1747, the gardener had to quit the orangery wing, which was converted into a plant-house. By 1762, Linné was able to say that no botanic garden in Europe was so rich as that of Uppsala in plants of all kinds. In 1761 came a chest of living plants from Governor Tulbagh at the Cape, and in 1769, from the same source, "a splendid collection of dried plants and bulbs." The Empress Catherina II., in 1773, sent several hundred sorts of seeds collected by Gmelin, Pallas and others in the Russian service. Linné, at an earlier date, had declared that he was body and soul in the garden, and his love for it never waned. He urged his friends to come to Uppsala at certain times, when it showed itself at its best. To the then current belief that the garden should be a nursery for medicinal plants, as it belonged to the medical faculty, Linné protested that it was a living library of plants for the public to learn their names, to follow their development, to observe their metamorphoses, and to become acquainted with their smell and taste, as a good means of noting their application to medical uses. Economic plants too were cultivated and observed, in order that native plants might take the place of imported products. He further endeavoured to acclimatize plants, such as a substitute for tea, which he hoped might be grown as readily as lilac; and also made repeated attempts to procure a living tea-plant. If he succeeded, he thought the Chinese would lose 100 tunns of gold annually, each tunn valued at £1,400. A report that tea-plants were in actual cultivation was investigated, when it was found that it was only *Salix repens*, " as different from the tea bush as a peacock from a crow." His delight was great when he learned that two living specimens had reached Gothenburg from China. They arrived safely at Uppsala, but proved to be a species of *Camellia*.

At last the day came when Linné's earnest desires were realized. Captain C. G. Ekeberg wrote that he had several tea-plants at Gothenburg. He had followed Linné's instructions, that " seeds should be sown in a pot, when starting home from China, and treated as if in a forcing house." In his reply to Ekeberg, he burst out rapturously: " But living tea trees! Is it possible? Is it the true tea tree? But I am certain it will not come unharmed to Uppsala; fate is against it. I am old, but were I sure it was the genuine plant,

I would venture to Gothenburg and bring it myself in my hands to Uppsala. I beg you, Captain, by all that is sacred, to give it the most pious care." The first sending miscarried, and Linné had to wait for the few remaining specimens. He thought it would be best for some poor fellow in Gothenburg to bring it by hand. "He could come in a fortnight, it is only 50 miles [332 English miles]. I will pay him 100 copper dalers [£2 10s.] for his trouble. If I could, I would pay ten times as much, but a large family hinders me." The wife of the Captain travelled to Uppsala in a covered carriage, and during the entire journey held the box containing the tea-plants on her knees, finally handing her precious burden to Linné, who hastened to spread the glad news to his correspondents. The young plants throve at first and even flowered, but soon showed themselves unable to withstand the climate, thus destroying all their hopes. The last time they were mentioned was in 1769; but they must have died soon afterwards, to Linné's great grief, as he wished to "shut the door by which all the silver in Europe goes out."

This was not the only loss of a plant from a distant land which Linné suffered, but on the contrary many such perished owing to the Swedish winter's darkness, and the damp of the glass-houses. He usually met such misfortunes with resignation, probably knowing that "three Uppsala gardens would not hold them all." But there was one sad occurrence which deeply grieved him, when the gardener received from Rolander a living cactus from Surinam with the cochineal insect on it. Linné at that moment was presiding at a disputation, and "the gardener who received the plant, saw it swarming with vermin, which ought to be cleaned away, which he did so thoroughly that not one insect was left." As soon as the Archiater [Linné] had arrived home, he asked if the expected cactus had come and if any maggots were on it. "Yes," said the gardener, "it was full of maggots, so I cleaned them all off." "The deuce! You have done much mischief with your officiousness," answered Linné; but what was done, was done, and Linné's hope of establishing the cochineal in the orangery was blasted. This troubled him so much, that he had megrim forthwith; one of the severest attacks he had ever experienced.

In consequence of Linné's exertions many trees of Russian and North American origin have been widely distributed, many of them being now common in parks and gardens, such, for instance, as *Vitis hederacea*, Virginian creeper; *Acer Negundo*, ash-leaved maple; *Ptelea trifoliata*, shrubby trefoil; *Dielytra spectabilis*, bleeding heart; and many more. Of special importance were many of economic value. Naturally the majority of visitors to the garden were Swedes, but eminent men coming from foreign countries found Linné an entertaining guide. Among them may be named Frederick Calvert, Lord Baltimore, who came in 1769, and was so delighted at the sight, that he not only sang Linné's praises in his "Gaudia poëtica," but also sent him a splendid gold snuff-box of 100 ducats in value [£48] and a "necessaire" of silver valued at 12,000 dalers [£300]. By 1744, the requirements for heating the stoves amounted to sixty loads of wood from the University forests, and nine furnaces were installed, in place of the five previously existing. The labourers in the garden, through Linné's influence, had their wages doubled.

For the maintenance of the garden with very restricted means, a skilful and trustworthy gardener was essential. He had such in the person of Nietzel (see pp. 153, 188), who worked till his death in 1756. Through Linné's efforts Nietzel had not only his salary raised, but received the title of "Inspector" to distinguish him and to assimilate his position to that of "Exercitie" master in the University. A further instance of the esteem in which he was held is, that when he left behind him a weak-minded widow, who had caused him great trouble, and a young daughter,

Linné took the latter into his own house, where she lived for several years. An account of this distinguished cultivator will be found in the "Gardeners' Chronicle," III. lvii. (1915) p. 359.

Upon Nietzel's death, a period of anxiety and hard work for Linné supervened. Owing to the dear times a suitable gardener could not be obtained, the stipend available being only that of a day labourer. Consequently Linné had to act as his own gardener, every day in the week, day and night. It was not until 1758 that any improvement took place.

A Demonstrator was needed for the garden, and ultimately the younger Linné was appointed, at that time eighteen years old. This action of Linné has been blamed, but though it cannot be denied that the appointment was peculiar, no better person at the time was available. He lived in the garden itself, and from his tenderest years was familiar with its administration. At first and for an indefinite period the salary was of the scantiest, whilst no small sacrifice of time and trouble was required. He had listened to his father's lectures, demonstrations, and herborizations, from the time he could walk, so that no one of his years could compete with him in botany. In the autumn he was awarded a small royal stipend, and in the following year, was to serve as an amanuensis with increased remuneration and perquisites.

The young Linné entered at once upon his duties, his father sparing no pains to encourage him as a botanist. He did not, however, seem to take his duties very seriously, this being evident from the descriptions he drew up of rare plants, and his intention to bring out a second volume of his father's "Hortus Upsaliensis" was never fulfilled. As an instance of his idea of his duties, the following list drafted in 1772 may serve:

The Demonstrator's objects must be:

1. To teach students the parts of animals and plants.

2. To teach them the terms in the vegetable and animal kingdom.
3. To draw up correct descriptions of animals and plants.
4. To discharge the herborizations when the Professor is prevented from doing so, by age, sickness or other cause.
5. Daily to see after the garden, and to report to the Prefect [Linné].
6. To compile lists of plants in the garden, and the results of sowing, with their respective numbers.
7. To have in readiness plants from the garden and field for demonstrations.
8. To keep the museum so that nothing may be spoiled.
9. To keep the accounts of the garden.
10. If the Professor should be ill, to lecture in his place on natural history and botany, in public.

To obtain a skilled gardener seemed impossible, but a substitute was found in L. Broberg, spoken of by Linné as at first having little insight, but an ardent disposition, so that he daily improved in a marvellous degree. Seeing the need of studying in other countries, he obtained leave of absence, his salary being continued for the support of his family, with a contribution for his travelling expenses. During his absence, a German named Hancke acted as deputy, but as soon as the war in Germany had ended, and he felt free from war-service, he went back to his native land. Linné then had no assistant but Löfgren, an old man, and further help had to be supplied for the wants of the garden.

During his absence Broberg visited the larger gardens in Denmark, Holland, and Germany, and on his return, Linné was so pleased with his progress that the Consistory appointed him Academic Gardener, and as such he remained during Linné's lifetime.

A controversy which caused much annoyance to the

Prefect, and much writing, was the question of the stable manure obtained from the University stables, to be used in the garden. The Royal Castle gardens in Uppsala were at that time worked by a gardener named Burman, who managed to intercept much of the manure for his own use in forcing vegetables and melons for sale, to the prejudice of the botanic garden. Linné protested against this breach of contract. After each order by the Consistory, Burman managed to evade it, and on four occasions, Linné had to appeal to the Consistory. These irregularities might have gone on further, had not Burman died in 1764. His successor tried to copy Burman's action, but was stopped in 1768.

The younger Rudbeck had begun to maintain a zoological department in the garden, which Linné decided to increase. In 1747, the Crown Prince gave a living Indian bear; in 1751, sundry birds; in 1754, guinea-pigs; and in 1757, a matchless cockatoo. In after years there came an ape, a monkey, and four kinds of parrots, also a young ourang-outang, and gold fish. New contributions still came, so that in 1769, there were eight monkeys, an Indian bear, an aguti, guinea-pigs, five parrots, musk ducks, peacocks, guinea-fowl, and other animals. The Queen Lovisa Ulrika gave a cassowary which lived long in the garden, and the King was also a generous donor.

These animals, especially the monkeys, and the observation of their mode of living, provided much amusement for Linné. Among the birds, parrots were his favourites, one especially entertaining him greatly. It used to sit on his shoulder, sharing his meals. When therefore the parrot felt hungry it would say, "Mr. Carl, it is twelve o'clock." It had taught itself another trick; when anyone knocked at the door, it would imitate Linné's voice and say, "Sti'in" [Step in] to the astonishment of the entrant who found no one in the room. The parrot sat still and silent, but when the visitor went out and knocked again, the trick was repeated. One of the garden staff, old Löfgren, had the habit when going in, to blow his nose loudly; one day he was amazed at the parrot crying out, "Blow your nose."

Linné himself bore the cost of keeping these animals until the Consistory made an allowance for that purpose.

Before leaving the subject of the garden, there is one matter worth mention, a by-product of Linné's activity, namely the alteration which he made in the thermometer. Actually the centigrade thermometer in constant use is not that of A. Celsius, but of Linné. Both of them divided the difference between freezing and boiling points into 100°, but Celsius made 0° boiling point, and 100° the freezing point, whilst Linné started at 0° as zero and freezing point, working up to 100° as boiling point. The explanation is easy; the physicist Celsius had his point of view, the botanist for plant-cultivation reversed it, as we now find the centigrade thermometer, having zero as the fatal freezing point, is of the greatest importance.

The first time in which Linné in print mentions his thermometer, is in the disputation, "Hortus Upsaliensis," which was discussed on the 16th December, 1745. Already in June of the previous year he had ordered one such from the skilful instrument-maker, Daniel Ekström in Stockholm, but it was broken on the way to Uppsala. After repeated reminders and through the help of Elvius, to his great delight in the beginning of November, he received a new one, "excellently made." A thermometer was regarded at that time in Uppsala so remarkable an instrument that the Consistory decided to order one when they saw the one belonging to Linné, which had cost 5 plåtar [15s.]. In the Uppsala observatory from the 1st April, 1747, daily observations were begun by O. P Hiorter, and from that date the thermometer made its victorious way throughout the civilized world. In foreign lands it became known as Linné's or Strömer's thermometer in distinction to that of Celsius, sometimes as the Swedish, which name, or " mine," was used by Linné.

It must not be omitted to mention that there was another one made (differing somewhat from Linné's) by a little known man, Christin, a Frenchman, two years before Linné had introduced his. It was shown at the meeting of a scientific society in Lyons on the 9th May, 1743, but was so little thought of even in the fatherland, that observations by its means were only reported in 1754, after that remaining practically unknown. It may be regarded as a fact, that Linné, during his residence in Holland, had prepared and used a thermometer graduated in his own fashion, for as it was depicted in the frontispiece to his " Hortus Cliffortianus " in 1738, it cannot be denied that Linné was earlier than Celsius, who only published his plan in 1742.

It now remains to consider the fate of the garden, the pride of Linné and of the University. If one visits the place, it is found that the original glass-house is still hardly changed as to its windows. Since it became alienated from its original purpose, it was used by the East Gothland nation, then as the University's " Sloyd " workshop, and now the Archæological Museum occupies it.

In the unchanged garden, one may trace the remains of the tanks and clipped hedges which were extant in 1840-50, but the former are now filled up and the latter cut down; but it is to be hoped that the old garden may long be kept as it is, and remain planted as a healthful spot in this ill-favoured part of the town. (At the time of writing it has been piously restored by the Swedish Linnean Society, and the Professor's house is now, 1923, a museum.)

Even in the time of the Rudbecks, the position was complained of, because of its dampness, and even in Linné's time, it not being satisfactory, he exerted himself to improve the soil and prevent flooding.

Still, it was natural that as his strength declined and age increased, the garden showed signs of dilapidation. The younger Linné relates, "that his late father became tired of the garden in his later years, and was so annoyed by the Consistory, that he ceased to deliver the seeds he received, but kept them himself, the garden thus losing its renewals, and the animals died." He set himself to renovate the place, but lacked his father's energy and enthusiasm; further, a lengthy foreign tour, and then illness, prevented him from succeeding.

His successor, C. P. Thunberg, was a diligent prefect, who induced Gustaf III. to allot to him a new botanic garden in the higher part of the town, where the foundation-stone was laid, a century after Linné's birth, to honour his memory.

Most of the plants were transferred from the old to the new garden, but some remained until recent times. In 1850 there were about forty species of undoubted Linnean origin; by 1903, the number had fallen to three, besides a black poplar near the entrance, which in its turn succumbed in 1911. Of the plants transferred, there were, in 1877-99, when Professor T M. Fries was prefect, several old laurels (bay trees), *Justicia Adhatoda; Prunus Laurocerasus*, cherry laurel; *Taxus baccata*, yew; *Cupressus sempervirens*, *Thuja occidentalis*, mulberries, white and black, almonds and myrtles; the last three dying in 1890, in spite of every care to prolong their existence.

A hop-garden was laid out by Linné at the wish of the Consistory near the university building, for brewing; elsewhere he advocated tree-planting. In 1710, for the first time in Uppsala a plan was suggested of establishing a natural history museum, when Dr. Lars Roberg stated that among other ways of benefiting the academic hospital, he had the idea of arranging the articles presented to him in a museum or natural rarities room, to which Librarian Benzelius would contribute the gigantic bones from Dr. O.

Rudbeck's whale, two thigh bones and other things given by himself. The project was approved by the Consistory. This natural curiosity chamber, eventually formed, aroused so little attention, that after Roberg's death in 1742, a dispute arose between the Consistory and Roberg's heir, Madame Eenhielm, about its proprietorship, which led the Consistory to desire that a proper chamber should be formed, after the proprietorship had been decided. After many discussions, it was determined to place these objects in Professor Mathesius's house, being sealed in 1746 by the University, and there they remained till 1752, when rent being demanded for the past years, the University declared that these objects were theirs. There was also another small collection housed in the Library (but practically inaccessible), including the Burser herbarium, and a Swedish collection formed by Gabriel Holsten.

Linné (who must be regarded as the founder of the University zoological collections) began, after he became Professor, to present objects, many of which he described in theses. In September, 1744, he had the pleasure of declaring that the Chancellor, Count Carl Gyllenborg, had presented specimens preserved in spirit; and other gifts from Prince Adolf Fredrik followed. In 1746, a shell cabinet was given by Councillor E. Petraeus. Within the next few years there came additions from the Swedish members of the East India Company, Lagerström and Alströmer. These were at first lodged in the top floor of the Professor's house in the garden, but were soon removed to the orangery, where means had to be devised to protect them from damp. Unhappily these means were adopted too late, as the inventory made after Linné's death showed that the stuffed animals were injured by moth or damp.

While Linné, especially early in his professorial career, showed himself active in promoting the zoological museum, he was also engaged in getting botanical specimens. He induced the Consistory in 1758 to buy Patrick Browne's Jamaica collection for 600 copper dalers [£15], but meanwhile he had bought it through Peter Collinson for £8 8s., and when on its arrival at Uppsala he found it to consist of more than 1,000 rare plants, he "could only marvel that the English should let so excellent a collection of rarest American [West Indian] plants go out of their country in return for 100 plåtar" [£15]. He was so engrossed with these plants, that, not availing himself of the sum provided by the University, he added the plants to his own herbarium at his own expense, and devoted himself so eagerly to their study, that he remarked, "I am forgetting friends, relations, house and fatherland."

There must also be mentioned a small collection of drugs wanted for the lectures on Materia medica, which were bought out of a small grant from the Consistory, with a hundred tin boxes for their preservation. An amanuensis was also provided; in the spring term of 1763 it was J P Falck, and in the succeeding autumn term, J. Elmgren, who filled the post.

The establishment of this museum gave rise to a small university reform. When Linné reported that Prince Adolf Fredrik had given a splendid collection of all kinds of Indian animals, fishes, insects, etc., and for their cataloguing, he asked for the loan of certain books from the library to be returned in a fortnight, as it was impossible for him to carry bottles to the library At first this request was refused, but Linné applied to the Chancellor, and in the end gained permission to borrow the volumes. But this permission did not please Anders Norrelius, the librarian, and in various ways he hindered the loan, till he was ordered by the Consistory to carry out the King's order, when the librarian at last gave way.

CHAPTER XI

LINNÉ AS MEMBER OF THE MEDICAL FACULTY, AND HIS RELATIONS WITH THE MEDICAL COLLEGE

UPON his installation as Professor, Linné became a member of the medical faculty, though it consisted only of the two professors, each having in alternate half-years to officiate as Dean. The first time he had to serve was in the spring term of 1742, when he displayed his talent and his courage for arrangement. Until then the faculty having no minutes taken regularly, Linné hastened to provide a minute-book, which he wrote up himself till 1758, when the professor who was not acting as Dean officiated. He also procured a seal, representing two serpents, one holding an egg, the principle of life, the other a skull, typifying physiology; in the field, plants as materia medica, displaying the motto, " His servamus urbes " (With these we keep cities). The plants depicted were *Frankenia*, *Rudbeckia*, *Rosa*, *Linnæa*, after the prefects of the garden, Franken, Rudbeck, Rosén and Linné.

The brunt of the work of the faculty fell upon Linné, as his colleague, Rosén, was too occupied to give much time to it, partly from his extensive practice in Uppsala, and partly because of his position as Body Physician to the King, being liable to sudden summons to the court.

Rosén's successor, Samuel Aurivillius, was Linné's colleague till 1767 His death, besides causing Linné great sorrow, compelled him to administer the affairs of the hospital single handed, until he induced Dr. J. Sidrén to share the burden. The selection of a permanent Professor occasioned much anxiety to Linné, as the three applicants seemed to him of equal merit, but it resulted in Sidrén receiving the appointment, Linné enjoying to his dying day the help of his former pupil, now become his colleague. As Adjunct, J. G. Wallerius (see p. 194) was appointed, till he became Professor of Chemistry in 1750; he was in turn succeeded by Sidrén (see above) and then by J. G. Acrel, who afterwards came into prominence on the death of Linné.

Shortly after his appointment as Professor, Wallerius sent a request to the Chancellor, that he should have a seat in the medical faculty. As the faculty demurred, it led to the decision that all students undergoing examination in philosophy, should pass in chemistry, before promotion as Doctor. The expenses of the hospital and staff were at this period estimated by Linné at 4,500 copper dalers [£225]. A dispute ensued between Linné and Aurivillius on certain points of administration, but it was the only one with a brother professor during Linné's service. Although an additional teacher was not then granted, yet before Linné died, he had the satisfaction of seeing his plans carried out, the first Professor of Surgery and Anatomy under the new rules being his pupil Adolf Murray, who filled the post with distinction till his death in 1803. The examinations in the medical faculty before the appointment of Rosén and Linné, had become inefficient, but now great changes were effected under their supervision. Discussions were, however, continued long afterwards as to the best methods to be adopted for testing the fitness of students coming up for examination, especially those closing the period of study by promotion, i.e., being granted the degree of doctor. The ceremonies attending the degrees at first being simple, became more elaborate by 1749.

One question which came into the early years of Linné's professorship, was the provision of a chemical laboratory.

As regards Linné's relations to the Medical College, an attempt was made to bring the University into closer connection. After the war with Prussia had ended in 1762, many surgeons came back to Uppsala, and others took degrees at Greifswald, entitling them to practise in Sweden; these latter aroused Linné to recommend a further examination of their powers, to prevent "unworthy" medical men from being appointed, from Greifswald, Åbo or even Lund. In 1776, a deputation waited upon the King, at Drottningholm. Linné, then an invalid, was taken in a carriage; and when with uncertain steps he entered the presence chamber, carried away by his earnestness and forgetful of the laws of ceremony, he broke out with: "That must never be, Your Majesty, it would destroy the University and Science. I could never survive such a calamity." The King, surprised at such an unexpected statement, asked Professor J. Sidrén what was meant, and was briefly informed of the circumstances, upon which the King smiled, and went forward to Linné, patted him on the shoulder and said, "That shall not happen; go home in peace and quietness." This was the last and perhaps the greatest service Linné rendered to the faculty, to which he more than anyone else imparted distinction and honour.

Another medical question which Linné followed with great interest and some disquiet, was the dispute between the Medical College and the so-called Surgical Society, conducted with some asperity in 1750-60.

Not in Sweden only was there a sharp distinction between physicians and surgeons. The operative section of medical men was regarded as of inferior training. A compromise was finally reached, by which expert surgeons should study medicine, and physicians should acquire a good amount of surgical insight. The Surgical Society's reputation sank until in 1797 it was dissolved. A pupil of Professor Acrel, named D. Théel, who applied for a stipend, was supported by Linné, but opposed by his colleague Aurivillius, and finally the Consistory awarded the stipend to Théel, who afterwards earned an excellent reputation as a skilful surgeon, and in 1772 was appointed by the King to be Supervisor of Surgery in the kingdom.

The most striking testimony to the sound views of both Linné and the Medical College, was the award of a doctor's degree to Professor Olof Acrel, afterwards known as the "Father of Swedish Surgery," although he had not graduated at a University. Linné's strong approval was shown in the minutes of the College, alike honourable to his memory, and to that of Acrel. This was approved by the Chancellor.

There must also be recounted the help afforded by Linné in the preparation of a new pharmacopæa. The existing one dated from 1686, and had become obsolete. The necessity for a new one was evident, in that the old one had many remedies taken from animals, which later physicians had discarded. The initiative was taken by Linné. Although approved by the Medical College in 1757, the work languished until taken up by Bäck single-handed in 1761, but it was not published till 1775. This contained remedies derived from native plants, which practice had been advised long before by Linné, as shown in his thesis "Plantæ officinales" in 1753.

Specially interesting was the garden established by the Medical College in the grounds of the newly instituted Seraphim Hospital. It became a botanic garden after Peter Johan Bergius, a pupil of Linné, was made Professor of Natural History and Pharmacy, and although not lasting long, it was the origin of the garden still flourishing, known as the Bergian Garden, near Stockholm.

Linné was also instrumental in framing the laws of veterinary surgery in Sweden, due to his early observations in his Lapland and later journeys. The first occupant of the special appointment was a pupil trained by Linné, Erland Tursén. An outbreak of disease in cattle, involving much loss, hastened matters, and it being resolved to dispatch a trained man to France to learn the methods there adopted in such cases, P Hernquist was chosen, upon the warm recommendation of Linné. Hernquist became known as the " Father of Swedish Veterinary Science."

CHAPTER XII

LINNÉ AS MEMBER OF THE ACADEMIC CONSISTORY

In one of his autobiographies, Linné wrote that "he never missed a single lecture, but he considered Consistorial matters as different, and further Consistory was neither his delight nor object, for he was intended for other matters." From these expressions a false idea has been derived that he left the burden of the work in that department to his brother professor. That this is erroneous may be seen from the fact, that from his installation to the middle of 1776, when bad health forbade him to leave his house, he was present at one or the other Consistories no fewer than 1,902 times, long or short as the case might be. Of course he could not attend when he was away from Uppsala on scientific travels, or when hindered by sickness, or presiding at disputations. One single case of "school-sickness" in 1751 occasioned so much remark, that Bäck, uneasy about it, enquiring of Linné, found that he had had a megrim for twenty-four hours, but was now quite well, and was busy on his "Species plantarum," having finished "Polyandria." Assiduous work compelled him in 1768 to beg leave of absence, readily granted by the Chancellor. On other unavoidable absences, the Academic Secretary was despatched to ascertain his vote by word of mouth. Even in 1777, when he was feeble, both in body and mind, he was able to send his vote by the same means to the Consistory. So energetic and quick-witted a man was not likely to neglect his part in deliberations, and he did not refrain from giving his opinions frankly.

Amongst the important questions which during Linné's time had to be considered by the University, undoubtedly were those concerned in organization, methods of instruction, examination, the professors' duties, and other related matters, which were of the foremost importance. Linné took the greatest interest in them, as his entrance into the professorial chair took place when far-reaching changes were in contemplation. He wrote, "Now our laws are given to us: 1. No professor to go a mile [nearly seven English miles] outside the town without the Chancellor's permission. 2. If any are hindered from attendance for a day he shall pay six plåtar [eighteen shillings]. 3. Anybody beginning a day later than the beginning of term, shall lose a quarter's salary. 4. To continue lecturing from 28th January till 23rd June, from 1st September to 20th December. 5. To count the students, and note those absent. 6. To examine the pupils. 7 Each month to give an abstract of his teaching." It was also suggested that the national clubs should open, and the medical faculty be called up, and closed, as parts of the philosophical faculty.

Linné looked upon these novelties as tending to the alteration of universities into gymnasia. For his part he had lectured as much each term as the current regulations required, and further had not spared time nor trouble for the benefit of the University, but had done all voluntarily with heart-felt pleasure and competition with his comrades, but if he were compelled, it would cost more now to lecture once, than in a week formerly. Ambition can drive one, but with these rules all stimulus would be taken away; neither Haller nor Boerhaave were driven to their eminence, but attained that by inducements.

Linné also disapproved of the lengthened terms, by which the professors were robbed of the time they required, partly for the study of new works, partly for authorship; the latter was especially important as he regarded himself as of greater service to the public by his writings than if he lectured one day in the year, or every day. His dissatisfaction was so great, that he declared, "If I had no family, I would long since have accepted the English invitation (though little do I love the nation), but Oxford is still open to me."

The dreaded inconveniences, however, proved to be fewer than was feared. Certainly he and his colleagues had to give in monthly reports of their lectures and term lists of their hearers, but these caused them little trouble. The new regulations were gradually relaxed or abandoned, as experience showed their unsuitability or the impossibility of their strict enforcement. Various events also prevented the new laws being observed, such as journeys to Stockholm, partly by order of the King, partly for University needs, etc. Then too, in the case of so indefatigable a lecturer as Linné, the regulations could hardly apply; and further, it was impracticable for the Chancellor to peruse the lists always, even if marked to show uncommon diligence on the part of certain students. Thus the University reform of 1740 fell into desuetude.

Amongst the discussions in which Linné took part were those on new posts in the University, such as those for chemistry, physics, and metallurgy, but these need not detain us, as they belong to academic polemics.

The question of disputation was also discussed; before theses were printed they had to be submitted to the Consistory, and with an abstract sent to the Chancellor, who decided whether they should be printed or not. According to the constitution, all disputation had to be in Latin, but in 1758, one in Swedish was permitted, but in the main Latin was obligatory.

A political contest arose upon the death in 1747 of the old Chancellor, Carl Gyllenborg. At a meeting of the Consistory on the 14th February, Professor D. Solander, of the "Caps" party, announced the Chancellor's death, and at the same time intimated that a new Chancellor should be chosen, in the belief that his party would gain an easy triumph. The "Hats" present (Linné among them) demanded delay, first out of regard to their recent loss, secondly, for a few days reflection, but in vain. The "Caps" gained their point by eleven votes to five, naming Samuel Åkerhielm as the Chancellor. Linné's view was that the Crown Prince should become Chancellor, as best for the University, but failed to carry the motion. The victors were not slow to act; a letter was dispatched to Åkerhielm, who was willing to accept the post, and another letter was addressed to the King asking for confirmation of the vote. On the 24th a further meeting was held, and the letter to Åkerhielm sent off, but on the next day, before a reply came, a new vote was taken, and the Crown Prince was unanimously elected, a letter being sent to him which was acknowledged. What was the reason for this rapid change?

The Council of State, in which the "Hats" predominated with Count Tessin as their chief, had been informed by Linné how matters stood, and a communication to the victorious section of the Consistory being made, caused them hurriedly to throw their new chief overboard. The King had meanwhile appointed Adolf Fredrik as Chancellor, and Åkerhielm wrote thanking the Consistory for its flattering offer, which, however, he found himself unable to accept in opposition to his future King. The installation of the Crown Prince took place with the customary ceremonies.

Four years later the Crown Prince ascended the throne, and after considerable negotiations, C. Ehrenpreus became Chancellor, as Count Tessin was too much occupied with affairs of State, to undertake the academic duties.

The minor questions in which Linné was involved as a member of the Consistory cannot be told here, but it may be put on record how that body enforced discipline amongst the students, checking their exuberance at weddings, or dinners of the various nations, sternly forbidding nightly disturbances, breaking windows, and collisions with the town police. The Inspectors of the Nations were warned to keep order amongst their members. The grosser instances of rough behaviour came specially before the Consistory, such as a disturbance (or almost a riot) at a wine-seller's, named G. Kähler, which was characterized by stone-throwing, sword strokes on the ground, etc. Kähler was fined 500 dalers [£12 10s.], two students condemned to a week's bread and water, and one was ordered to lose his sword, because he at night had struck sparks with his sword from the stones in the street. At another time, a student had abused the fire watchmen, called them "sausages," and acted as if he were hungry and was eating one. As the accused had previously been condemned to death (though the sentence was commuted to eight days bread and water), he was now sentenced by the Consistory to perpetual relegation in disgrace, i.e., sent down.

The first half of 1772 was noteworthy for disturbances among the students; one, who was convicted for abuse of a fire watchman, unseemly language, blows, and stone-throwing, was condemned to "lose life and goods"; two of his companions who resisted his arrest, were sentenced to twelve days' imprisonment and to be fed on bread and water, many similar cases being recorded.

It was during this tumultuous time that Linné in the later half of the same year, 1772, became Rector, for the third time. It was noteworthy for the fact that no student was charged, no card-playing took place, and there was no masquerading nor disorder. Nobody had ever experienced such a quiet time, the reason being the respect and love which the students cherished for the old Linné. After he had relinquished his Rectorship with the accustomed ceremonies, all the Nations deputed their chiefs to thank him, and to ask him to print his speech, the Chancellor closing the proceedings by his recognition of the praiseworthy behaviour of the young students.

But all pleasant hopes of reformation were doomed to fail; in the next half-year the disturbances were in full swing again.

Another occurrence which gave great pain to Linné, was in the case of Prosector J. G. Rothman, son of his old benefactor at Växjö. He had been allowed to travel abroad, but not having returned two months after his leave had expired, did not respond to the admonitions addressed to him. He was at last condemned to be dismissed from his post, for neglecting his duties.

Linné was appointed Rector, as previously stated, three times, for half a year each time, namely in 1750, 1759, and 1772, and each following half-year as Pro-rector, and President of the Lesser Consistory. During Linné's third Rectorship, Gustaf III.'s revolution occurred, ending the Era of Liberty.

Other duties discharged during his professoriate, were superintending the granting of stipends or scholarships, inspection of methods for extinguishing fires in his own neighbourhood, cataloguing the so-called Cabinet of Arts, the direction of the academic poor box, and approving of timber to be felled on the University property.

CHAPTER XIII

LINNÉ'S RELATION TO THE SCIENTIFIC COMMUNITY—AUTHORSHIP AND SCIENTIFIC CORRESPONDENCE—INSPECTORATE OF SMÅLAND'S NATION

ON the 23rd September, 1741, Linné informed the Academy of Science, that as he intended to go to Uppsala and remain there, he besought the Academy to permit him to continue his contributions to its Transactions; the President, J. Benzelstierna, assured Linné of its good-will and its concurrence in the desire announced, ending by wishing the Professor a happy journey.

It might have been supposed that with this Linné would have ceased to be the soul and mainspring of the Academy, but this was not the case. It is true, that it was rarely he had the chance of attending the meetings, but his warm interest in it was shown by the lively correspondence kept up between him and the secretaries, P Elvius and P W. Wargentin, until old age and illness interfered. From this time he was only present on seventeen occasions, the last time being in 1774, but his reports on many kinds of papers were numerous and frequent, and his advice was freely given, either in criticism or encouragement. During his residence in Uppsala, he sent in no fewer than forty-two important papers, the total published in the Transactions amounting to fifty-two. Many of his pupils contributed articles at his instance. The Academy had been endowed with a capital fund by Court Intendant F Sparre for two annual awards for papers tending to the public benefit, and Linné had to adjudge the merits of the competing papers. In 1762 the question proposed was, "How caterpillars which do harm to fruit trees by devouring blossoms and leaves, can best be destroyed." Eleven answers were received, several being from Academicians, such as De Geer, Bäck and Bergius; the first prize was assigned to Torbern Bergman, whilst four other replies were each printed and rewarded with a silver medal.

When the notes containing the names were opened there was one signed "C. N. Nelin, N. Minist.," an unknown person, and it has been conjectured that under this name Linné concealed himself; but this cannot now be ascertained. Linné in 1765 sent in a memorial urging support for C. A. Clerck, that he might be enabled to publish his specially valuable work on rare insects, as the author was by sickness and economic stringency in want of such help; the result being that many of the Academicians readily became patrons of the book. In August Linné wrote to Wargentin, that the industrious Clerck had closed his eyes in death. "May God induce the Academy to continue to value his work, so that it may not come to nothing, for Science has never seen more elegance." After Linné's active support of the Academy, it was not surprising that it should desire to have a portrait of its illustrious member, and Per Krafft the elder was commissioned to paint it during a stay in Stockholm; Linné declared it could not be bettered, his family and contemporaries also agreeing that it was admirably like, and it is undoubtedly the favourite presentment of the great Swedish naturalist of all those which are extant. (It is shown in the Frontispiece of this volume.)

Besides the Academy of Science, there was another learned society of which Linné was a member, almost ex officio, the Royal Scientific Society at Uppsala. He had been elected in 1738, but his first attendance at a meeting was in 1741, when he had settled in Uppsala. The Secretary was the famous Anders Celsius, but during his long travels abroad, matters had fallen into arrears. Celsius dying in 1744, Linné acted as his deputy, until later in the same year he became definitely the Secretary. Among the perquisites of the office, was free postage, granted by the King, and this explains Linné's request in a letter to Arduino, in 1764, that all letters may be sent to him as Secretary, so as to spare him postage at about a ducat each letter. [Nearly ten shillings.]

The state of the Society was at this time unsatisfactory. Most of the members abstaining from the meetings, the papers for the "Acta" were few and dull, the accounts and property diminished, and the two banknotes which represented all the effects of the Society disappeared. It was Linné's problem to bring affairs into better order, and vigorous steps were taken. It was determined that all who had not sent contributions in two years to the Acta at the next New Year should be removed, and a fine of ten copper dalers [five shillings] imposed on each absentee from a meeting.

The result was that in 1744, a new volume was brought out, with a paper by Linné on Orchids, and another on Belgian fishes by Gronovius; the following years showing improvement, though not permanent. From 1750 to 1755 nothing was published, and Linné induced friends abroad to join, but finally gave up the secretaryship in 1765, Carl Aurivillius succeeding him. Eight years later a new volume came out containing two of Linné's contributions; and John Ellis's paper on *Dionæa muscipula*.

In 1762 he was appointed one of the eight foreign members of the French Academy, the first time a Swede had been selected for that honour. The foreign societies which had thus distinguished him, were: Germany (Academia Naturæ Curiosorum, 1738), Montpellier (1743), Florence (1755), London

(Royal Society, 1753, and Society of Arts 1762), Trondhjem (1766, the first foreigner chosen), Celle (1767), Philadelphia (1770), Flushing (1771), Rotterdam (1771), Edinburgh (1772), Bern (1772), Siena (1773). " I am weary of corresponding with so many " was his not unnatural remark.

The Royal Bible Commission in 1773 appointed Linné as one of the Commissioners, and he went through the Swedish version to see where it needed correction as to Botany or Zoology. The few emendations he induced his fellow-commissioners to adopt did not compensate for the time he devoted to the service.

Dr. J. M. Hulth's admirable and accurate bibliography cited later, frees one from a detailed enumeration of Linné's works, but a rapid survey will be expected here of the more important books. In 1735 his " Systema Naturæ " came out, as already mentioned on p. 142 followed by a second, sixth, tenth, and twelfth editions, the intermediate numbers being due to other editors. The thirteenth and fourteenth were styled " Systema vegetabilium " and edited by J A. Murray. The tenth edition (1758-9) forms the foundation of binomials for zoologists. Then followed " Genera plantarum," with six editions, and in 1753 his " Golden Book," " Species plantarum," which he began in 1746 and laboured night and day upon till 1748. Then he paused, and as he confessed to Bäck, he " wanted to show his competence to the world, had he only had time to complete it."

This feeling did not last long, and a year later he informed Bäck that he was progressing, that he reached *Poa* in a week; five months later he had reached Icosandria. Early in 1752 he was engaged on Syngenesia, and in August of the same year, he thankfully recorded that he had finished writing the whole book. He considered it his best work, embracing eight thousand species and many varieties, and it is a book which botanists sorely needed and fully appreciated. It retains a foremost place in every botanic library as the starting point of the modern usage of specific names in place of a long Latin descriptive phrase, a single word being sufficient to denote the particular plant meant. Linné took no special pride in these "trivial" names as they were first called; he admitted that he was not the first to employ them, and that it was only like putting a clapper in a bell. The second edition came out in 1762-3, and two Mantissæ in 1767 and 1771.

Another work of far-reaching influence was his "Philosophia botanica" which he dictated to his pupil Löfling while recovering from a severe illness (p. 233). It was this volume which J. J. Rousseau declared had "more wisdom in it than the biggest folios, in it there is not a single useless word," and L. C. Richard said that though each winter he read it through, in his seventieth year he found it new and fascinating. It was printed abroad ten times and translated into German, English, French and Spanish.

Specially noteworthy is his "Flora suecica" 1745, of which an enlarged edition came out in 1755; it was not a mere dry catalogue of plants, but embraced their properties and application. A similar book was the "Fauna suecica" with 850 species in 1746, and in the second edition with nearly 1,600. The following can only be named—"Flora zeylanica," 1747; "Hortus Upsaliensis," 1748; "Materia medica," 1749; "Bibliotheca botanica," Ed. II., 1751; "Museum Tessinianum," 1753; "Museum Reginæ Ludovicæ Ulricæ," 1764; Prodromus Musei Regis Adolphi Friderici," 1765; "Clavis medicinæ," 1766. (See Appendix: Bibliography.)

Some of the above may be regarded as the product of his work as a teacher; but the following are still more closely connected with that other sphere of his work, namely the academic disputations which bear his name. According to the regulation which then prevailed, and even to 1850, everyone who would undergo the examination for Candidate in the faculty of philosophy, such as medical men, must dispute *pro exercitio*, and afterwards before promotion, *pro gradu*. The former disputation in most cases was entirely the work of the Professor, who occupied the chair as Præses. To ascertain how many of these disputations, 186 in number, came from Linné's pen, is now impossible, but as regards a few, from the author's own words, as well as the subject, such as Löfling's "Gemma arborum," Söderberg's "Pandora et Flora rybyensis," Tillæus's "De varia febrium," were the outcome of the respondents' own studies and observations. But even these have Linné's stamp, his *imprimatur*, as he read them through, before completing, correcting and printing them.

How this was done in most cases is known from the statement supplied by his pupil J. G. Acrel. "All disputations he wrote by dictation, partly in Swedish, partly in Latin; to put these in order was the function of the respondent, and although he did not worry himself about the Latinity, he let his pupils know whether their task was well done or the reverse. To draft such a disputation required hardly three hours, as it was nothing else than a lecture on the subject which the respondent indicated." Of course all could not be so easily performed, for it often required considerable trouble and scholarship from the respondent, though the most important part was to turn out a passable Latin version. With this it was usual for him to assume the "authorship, with a flattering mention of the Præses's" learning and acuteness.

The worthiest of these disputations, together with academic orations and programmes, Linné collected and published under the general title of "Amœnitates Academicæ," which came out during his lifetime in seven volumes, 1749-69; three later volumes were edited by Schreber with a new edition of the former seven.

All the foregoing were of purely scientific import and were consequently written in Latin, then the universal speech of the learned. But Linné also published a considerable number of works in his mother tongue, among them being his accounts of his travels in the southern parts of Sweden, as well as smaller popular writings, such as deprecating the general use of spirits, appearing in the media of people's almanacks. They were written in a simple style, almost childlike in their naïve expression, but with touches of genius. Nevertheless, many of his writings remained unprinted during his lifetime, a considerable number reaching publication in recent years, many of them in the possession of the Linnean Society and elsewhere. His own important works were interleaved and copiously annotated. He gave much time to a "Lexicon" ultimately published by J. M. Brurset in Lyons as a "Dictionnaire portatif d'Histoire Naturelle." His last work was intended as "Mantissa" III, when illness finally put an end to his labours.

Closely connected with his authorship was his scientific correspondence; practically the whole of the letters he received are preserved in the Linnean Society's archives, and are in course of publication, but many of Linné's own letters despatched to distant lands, have certainly been lost.

Among his most active correspondents may be mentioned J. Burman, J F Gronovius, and A. van Roijen, with François de la Croix de Sauvages, Professor of Medicine at Montpellier. Our own countrymen P. Collinson, Thomas Pennant and John Ellis in London, and on the Continent, J. G. Gmelin, for many years in the Russian service, A. Gouan of Montpellier, J. E. Gunnerus, Bishop of Trondhjem, Baron N. J von Jacquin of Vienna, J. V Rathgeb, Austrian Minister in Venice, and D. Vandelli, professor in Lisbon.

Special mention must be made of Albert v. Haller,

who for many years was intimate with Linné till the lapse of years caused coolness and even estrangement between them. The first sign of this breach of friendship was caused in 1746, by one expression in "Flora suecica" to which Haller took exception, in spite of all that Linné could do to disarm bitterness. Haller's self-love was deeply wounded; but even as late as 1760, Linné persevered with his endeavours to placate his former friend, though vainly. Linné kept to his principle not to embark in disputes, but to let others decide between them. "After we are dead, children who now are playing will be our judges;" and he refrained from reading Haller's attacks, although in his old age, he was stirred to the depths by Haller publishing confidential letters and pointing out errors in Latinity.

Now what was the reason of Haller's implacability? He accused Linné of aping Adam by naming all animals afresh, and of being an autocrat in botany and zoology, while Linné was lamenting that Haller regarded himself as an infallible Pope, his restless misguided disposition driving him to decry Linné's merits, and refusing to hear him praised. "Judging impartially," says a most competent critic, Dr. O. E. Hjelt, "in the lamentable severance of two such distinguished men, Haller seems to deserve the greater blame, and in spite of his splendid powers and extent of knowledge, he was at times subject to the demon of ambition. In the wonder and respect which were accorded him in his lifetime, in the excessive praise and flattery which he constantly received, lay a temptation, stronger perhaps than for any other person." It was perhaps his misfortune to live at the same time as Linné and to work in the same field. "They resembled," said Bäck, "in botany, Cæsar and Pompey. One, our Linné, suffered no equal, and the other, Haller, suffered no greater one, or the reverse."

Linné had a constant habit of naming new genera

after many of his correspondents, pupils or older botanists. He considered it a great mark of distinction, and those who received it from him, took it as a most flattering favour. It has been suggested that in thus bestowing names he, in some measure, detected a spiritual likeness between them and their names. In a certain degree it was so, as for instance, he gave the name *Bauhinia* to a genus with a peculiar two-bladed leaf, in memory of the eminent brothers Bauhin; then *Commelina* with flowers having two large and one small perianth-segment, from three brothers Commelyn, two eminent and the third insignificant as naturalists; whilst Plukenet, known as displaying bizarre ideas, has his name bestowed on a plant with very irregular flowers. In many cases there is not the slightest reason to allege this occurrence, and before everything, one must strongly dissent from the repeated statement, that Linné gave the names of his friends to beautiful and stately plants, whilst he branded his opponents by giving their names to ugly or insignificant ones. Linné considered no plant ugly or insignificant; even the humblest was regarded the same as the stateliest. *Boerhaavia*, *Forskolea*, *Loeflingia*, *Kœnigia* and others were so called after persons for whom he had the heartiest good-will, though they are unpretentious, while *Adansonia*, bearing the name of an opponent, is one of the grandest trees in existence. Another misapprehension is with the genus *Buffonia*, at first and by accident published as "*Bufonia*," and therefore acclaimed as a reference to Bufo, a toad—an entirely unwarranted assumption.

He hated unnecessary words, and he himself wrote briefly and impressively, all his works showing system; his style was original, usually succinct, without a needless word in a description, often compressing into two lines more than his predecessors did in page-long descriptions. His striving after brevity increased as he grew in years, so that the purely scientific works of his later period are wanting in the poetic touches which adorn his youthful "Flora lapponica." "*Nulla dies sine linea*"—"No day without a line," was one of his mottoes, and he acted up to it.

In 1755 he was surprised to get a letter from M. Manetti, Professor of Botany in Florence, who had previously written in opposition to him, stating that the more he studied plants themselves, the more he became convinced of his previous mistakes, and as a pledge of this he asked Linné to accept the diploma as member of the Florentine Society. The Pope, Clement XIII. (Rezzonico) had forbidden the introduction of Linné's writings into the Papal States (because he had divided the arrangement of animals in a different way from Moses), but in 1774, through the influence of Cardinal de Saladas, Clement XIV. (Ganganelli) instituted a new botanical professor, Minasius, giving him orders to put forth Linné's views in his lectures.

During the last twenty-five years of his life, Linné hardly ever alluded to his opponents except in a playful way, showing that he was not deeply hurt. Even Siegesbeck, who more than anyone else had wounded him, was afterwards spoken of by him with pity. There is a story extant that Baron Sten Bjelke, who was procuring seeds for Linné in Russia, wrote on a packet of seeds of *Siegesbeckia orientalis*, instead of that name, *Cuculus ingratissimus*" [the most ungrateful cuckoo]. The packet happened to fall into the hands of Siegesbeck, to his great annoyance, and it cost much trouble to smooth over the irritation thus caused.

Sickness and vexations of various kinds tempted him to follow the example of a colleague, to live in peace and quietness, or, to devote himself to medical practice, which was better from an economic point of view. Many times he compared his own constant activity to that of Professor Mathesius: "He has rested his body, I have murdered mine; he has gained all that I have done, but spared his body and mind." Such a period of depression occurred in 1748, when he threw his pen away and declared that hereafter he would only publish a few dissertations. One of these disturbing causes was a letter from Hårleman "which nearly killed him, and his sleep was broken during two months after," because he had been accused secretly by one whom he had greatly helped. Another "death-blow" was that the State Chancellery issued an order that no Swede should print or publish anything abroad, in pain of a fine of 1,000 silver dalers [£75], which was clearly aimed at Linné. A third annoyance was that Linné had a disputation "De curiositate naturali in laudem Creatoris" [Of a natural curiosity in praise of the Creator] which his best friend Halenius publicly opposed. After this, Linné never trusted in any priest, to whom he had previously shown a disputation. A further light is thrown upon this affair by the theological minutes of the 30th June, when it was alleged that the disputation contained statements which did not accord with sound doctrine; but the decision being made that it was not injurious to theological truth, the faculty decided that the Dean should pronounce that verdict before the University, and so ended the inquisition. Linné practically treated this occurrence as a trifle, not alluding to it in his autobiography, nor even in his confidential letters to Bäck. Although he expressed himself dissatisfied with the conduct of the affair, yet he soon resumed his customary methods.

It has been previously related (p. 112) that Småland's Nation in 1734 wished their countryman "much success in his travels and enterprises." Nearly seven years later, he was installed as professor, and naturally the Nation did not hesitate to congratulate him on the honour thus acquired, resolving to make him a present in the shape of a silver drinking cup, costing 313 dalers 16 öre in copper [£6 17s.] which was gratefully received.

The connection between Linné and Småland's Nation thus renewed, was still further confirmed, when on the 24th November, 1744, after the death of A. Celsius, he was elected Inspector, the other four competitors being Winbom, Solander, Ekerman, and Ihre. He accepted the honourable office gladly, and for the third of a century he discharged his duties in so fatherly a fashion, that the relations between them were never clouded. All members were expected to attend the gatherings several times each year, Linné himself as Inspector being present at no fewer than 116, and as the funds of the Nation did not permit the hiring of an assembly room, these meetings usually took place in the house of the Inspector. Usually an oration was given by one of the seniors, characterized as "beautiful" or "neat," listened to with general delight, and sometimes recorded to the Nation's honour. These were on various subjects, national events being naturally included. The assumption of the Crown Prince Adolf Fredrik as Chancellor in 1747 was specially celebrated, the proceedings lasting from nine in the evening till three the next morning. Linné's wife and many distinguished ladies were also present. In 1769 another rejoicing took place on account of the successful inoculation for small-pox on the Royal children. Occasionally these gatherings were in honour of some esteemed member, as in 1765, of Andreas Neander, celebrating his long service as university book-keeper, an address being presented to him commemorating his piety, diligence, virtue, and administration.

The Inspector had also to warn the members against cards, dice, and other practices, of bad habits such as the haunting of beer shops, etc., to warn them that lectures must not be missed, and to exhort them to observe the regulations against fire.

During his long Inspectorship, only once did a case occur of a Smålander being summoned to the Governor's court. The old penalism (pp. 29, 109) was abandoned upon Prince Adolf Fredrik becoming Chancellor, but a modified form of "service" on the part of new members still obtained. The seniors were admonished to train the novices in good ways, such as to remain at home at night, etc. The total number of members inscribed during Linné's Inspectorship was 308, many of them becoming distinguished in after life, such as C. P Thunberg, D. Rolander, P. J. Bergius, C. M. Blom, S. A. Hedin, the eminent Latinist Hakan Sjögren, Jonas Aspelin, J. Hagsdorn the Orientalist, J. D. Roberg, and many more.

As testimony to the affectionate regard in which the Inspector was held may be mentioned the "honoraria" which was given him. The first sum, which had been intended for Anders Celsius before he died, was handed to the new Inspector in 1744. Such gifts took place every three years, and in 1778, after the death of Linné, it was awarded to his widow. A special visit was paid to Hammarby to inaugurate the stone edifice which Archiater Linné had built in a fortnight on a small eminence for his cabinet of natural history. (See p. 328.)

During the last two years of Linné's life, his health had so failed as to make it impossible to hold a meeting, even in his absence, yet the good feeling which he had implanted continued, and the respect and gratitude to their fatherly friend never failed, nor has it since, the Nation recording with pride his services as member and Inspector.

CHAPTER XIV

LINNÉ'S BENEFACTORS AND FRIENDS

THERE can be few persons of whom it can be said as of Linné, that he had no personal enemy, and still fewer, who, through all the scenes of life, have been so fortunate as he to meet with warm-hearted benefactors, who, impressed with wonder at his genius and splendid scientific works, felt themselves bound to him with affection and respect by his sympathetic personality. Among those whom he had nearest to his heart may be named Carl Gustaf Tessin and Abraham Bäck.

How Linné had to thank Tessin for his first successes in Stockholm has already been told (p. 176). His eagerness to work for Linné's advancement was no hasty flame, soon slackening, but continued to burn as clearly all his life. It was he who was responsible for Linné's appointment as Professor, and it was his influence at Court which led to Linné's receiving the title of Archiater and the Order of the Polar Star. The admiration and affection he cherished for Linné were so great and so generally known, that some of their friends in 1746 resolved to honour them both by striking a medal with Linné's bust on the obverse, and a Latin inscription on the reverse, which stated that C. Ekeblad, A. Höpken, N. Palmstjerna, and C. Hårleman, dedicated the said portrait to Carl Gustaf Tessin and immortality.

Not satisfied with the ordinary expressions of gratitude to "My most gracious Master," "My great Apollo," "Mine and Svea's welfare," Linné felt himself obliged to give his feelings public utterance. This he did by dedicating to Tessin many of his most important works, beginning in 1740 with the second edition of his "Systema Naturæ." This was repeated in all the editions he himself issued, declaring that if in that work anything should be found tending to the advancement of science, one had exclusively to thank Tessin as the creator of his success. The latter, on his side, when the tenth edition appeared in 1758-9, showed his appreciation by having a specially beautiful medal struck, with Linné's bust as before, and on the reverse the three Swedish crowns, intended to symbolize the three kingdoms of Nature; the first having heads of animals, the second having flowers and fruits, and the third being charged with crystals and stones, with rays of light streaming upon them from above, and for the inscription "Illustrat."

What in the course of years brought these great men of different surroundings into close relationship, was the interest Linné awakened in the statesman, even amidst the turmoils of political strife. Afterwards, when the great days of his power had passed, and (in consequence of economic discomfort) Count Tessin lived quietly, separated from the great world's strife and endeavour at his beloved Åkero, their friendship continued as a never drying spring of trust and refreshment. From his early days he had eagerly collected books, coins, etc., but all these were thrust into the background for his collections of natural objects, chiefly minerals and petrifactions, which he had with great sacrifice of time and money amassed at home and abroad. His museum became so rich, that Linné was induced to superintend, to put in order, and to compile a catalogue with numerous plates and remarks, which was printed in 1753 in folio, entitled "Museum Tessinianum." Although drawn up by Linné, it was dedicated to him by Tessin "as he alone should have the honour, and I owe to him for this all my gratitude." True, holidays were the occasions on which Linné visited him at Åkero, when the two veterans, each esteeming the other, wandered round the castle's beautiful neighbourhood, or busied themselves in the mineral cabinet, adorned with Linné's portrait.

During the thirty or more years that Linné and Tessin were thus united, the latter's condition had undergone a change. After being the absolute ruling chief of the "Hats," and one of the most powerful men in the realm, he had been induced, partly from choice, partly from necessity, to withdraw entirely from public life. Many who previously bowed before him afterwards regarded him with indifference or contempt, but Linné was not one of these. He deeply regretted his benefactor's adversity, and with inward respect he still reverenced his personal greatness. In a New Year's letter in 1762, shortly after Tessin's fall, Linné wrote: "After twenty years sailing on a raging sea, envy awakened tempests, but he happily came with ship and cargo into a quiet haven," and "The Children of Israel, who every year celebrated the day when God's Almighty arm by the hand of Moses, delivered them from Egyptian bondage, have taught me every year to celebrate the day in 1738, when the grace of God raised me, through your Excellency, from my congenital poverty to an advantageous position, in which I can provide for and shelter myself and mine. You are deserving, therefore, my praise, honour and gratitude, which my children will continue after me, so long as they are upon this earth." In 1770, he similarly gave expression to his feelings of enduring thankfulness to the author of his advancement. The last letter reached Tessin on his death-bed, and a week later he quietly passed away, aged seventy-five.

The Count had always owned that he was no botanist, but his Countess, Lovisa Ulrika Tessin, was an admirer of that science, which fact Linné learned at the end of 1752, through Bäck. It was at her instigation that Linné's pupil J J. Haartman published his teacher's "Indelning i Svenska örteriket" [Introduction to Swedish Botany], in 1753, dedicated to his patroness, and as she loved the cultivation of flowers, many rarities found their way from Uppsala to the garden and plant-houses at Åkero.

Next to Tessin, the Crown Prince, Adolf Fredrik (afterwards King), and his consort, Lovisa Ulrika, must be mentioned as Linné's powerful supporters. The first meeting took place in 1744, when the Prince paid his first visit to Uppsala, the Rector and four Professors (among them Linné), being deputed to give them a most humble greeting on their reaching the boundary of the county Two years later, the Prince came again, when Linné was the recipient of a special mark of distinction, His Royal Highness bestowing upon him two gold medals—many of the academic notabilities only receiving one. Soon after he dedicated his "Westgöta resa" to the Princess.

This intercourse with Royalties was not transient, but became permanent. In Sweden there was a practice of collecting, and as many new curiosities were constantly arriving in the kingdom, influential persons were continually increasing their stores. There was therefore nothing surprising that the Princess should share the prevailing taste, the beginning being made by the purchase of a splendid collection of shells and insects, bought in Holland, thus laying the foundation of the natural history cabinet at Drottningholm. Linné was summoned to describe them, in April, 1751.

In that year Adolf Fredrik ascended the throne, and the Queen's conversation with Linné awaking the King's desire to form a similar collection of his own, three years later, Linné declared that "His Majesty's cabinet has become the largest in the world; it would be hard to add anything to it." Linné was commissioned to describe these treasures, and he entered upon the task with so much briskness, that he wrote to Bäck, "I have been writing night and day in His Majesty's cabinet, so that my eyes smart, and I can hardly shut them." At length the work was finished and Linné's duties at Ulriksdal, where the cabinet was housed, were ended by the publication of the book, which was to "perpetuate his name as the battles of Charles XII.'s did in the reign of that monarch."

Naturally after the "Museum Regis" came out, "Museum Reginæ" must follow; and Linné began the work with the same feverish haste he was accustomed to show. The "Prodromus" was printed in 1764, and contained the accounts of shells and insects, but not those of the corals, crystals, or the rich gathering of metallic ores. The reasons for the delay were many, partly because of the issue of other and voluminous works, and partly because of the continual increase by purchase or gift.

Linné received many signs of Royal favour; namely, the appointment as Archiater, and the decoration of the Polar Star—these were mainly due to Tessin, but his ennoblement was a special mark of grace from the King. So far back as 1753, Linné received a gold ring with inset ruby from the Queen, who further delighted the Professor by asking after his son, and if he showed any love for natural history; and she promised that when he was older, he should travel over Europe at her expense, which greatly delighted the father. In 1765 he wrote to Bäck, that His Majesty had made him a splendid gift of sixteen big chests containing plants from South America, preserved in spirit, admirably preserved.

From his youth, Linné was a Royalist, so it is not surprising that he dedicated his "Species plantarum," "the fruit of my best and most of my life," to the King and Queen.

The same affection and admiration for Linné which were manifested by Adolf Fredrik and his Queen were shown also by their son Gustaf III., both as Crown Prince and King, but on account of Linné's advancing age, the display of it was not so often made. One event which gladdened Linné was that in 1769, the Crown Prince, Gustaf, visited Hammarby to inspect Linné's museum. In 1774, Linné had experienced the "first death-messenger," an apoplectic stroke, and by the close of the year found himself weak and without wish to work, but at Christmas, His Majesty sent a collection of plants from Surinam in hogsheads of spirit. Linné at once left his bed and received new life in examining and describing the plants, about two hundred in all. The next year the King travelled from Ekolsund to Uppsala to visit Linné and no one else, and stayed with him the whole afternoon. They never met again, but later on it will be shown that Gustaf III. did not forget his old subject.

An admirer of Linné, who had at least intended to raise a monument to his honour, was the Margravine Carolina Louise of Baden Durlach, born Princess of Hesse-Darmstadt, who proposed to issue 10,000 plates illustrating all the plants in "Species plantarum" at a cost of 90,000 ducats [£44,745]. A beginning was made without Linné's knowledge, but he first heard of it from the Swede, J. J. Bornståhl, an eminent orientalist (1731-1779), who stated that a skilful French engraver was at work upon it. "All the Veronicas are ready, quite beautifully done, for the Princess has sufficient insight in this. She is not only skilled in botany, but also in art; she examines every plate, corrects every fault, and alters every variation; then she colours all in life-like tints, so that this work becomes the most accurate and splendid in botany." A plate was sent to Linné as a proof, with the promise that the Princess would carry out his corrections and remarks. She even said that if he and his son would come to Carlsruhe they should be well lodged, and have all the comforts of Hammarby. The same Princess had amassed an incomparable museum, but had nothing in it from Sweden. Linné wished to reward the young Princess's enthusiasm, and sent to her a drawing of a handsome, undescribed, exotic plant, to which he had affixed the name *Carolina princeps*, which immensely pleased her. In spite of this, the work soon stopped, presumably on account of its great cost; and the Princess died in Paris in 1783, at the age of thirty-two.

Particularly helpful in furthering Linné's plans with regard to the institutions, whose administrator they were, were the Chancellors of the University. The help of Count Carl Gyllenborg's powerful influence obtaining his chair at Uppsala has already been recorded (p. 196) as well as the help he gave in the matter of restoring the botanic garden and museum. Linné realized that in him he always had a steadfast supporter, and it was therefore with great grief that he learned of his patron's fatal illness in 1746. Time after time he wrote to Bäck concerning him; but his condition becoming worse, Linné wrote, " Almighty God help the good old man, who has done so much good to mankind. If the University should lose him, it will never have another Count Carl in our time, and hardly in our children's," and a month later he lamented, " With the great Chancellor I have lost immeasurably." He had dedicated to him during his life, " Flora suecica," and the second edition of it was dedicated to his memory nearly ten years after his death.

Still more was he drawn to Count Anders Johan von Höpken, with whom he had close connection during his residence in Stockholm. Linné entertained a sort of veneration for him. " If we only had him as Chancellor," was his wish in 1753, and that desire was fulfilled in 1760. How he discharged his high function may be learned from this, that during his four years of office, Linné dedicated to him no fewer than three works, " Fauna suecica," 1761; " Genera morborum," 1763; and " Genera plantarum," Ed. VI., 1764; and he had previously dedicated " Philosophia botanica " to him on its issue.

The short time which Höpken had as Chancellor prevented any deep impression being made by him in University matters, but it was with Linné that he chiefly consulted on such subjects.

At the time when Linné felt worn out with the duties of his chair, Höpken confessed to him that he too was weary of the Chancellorship. In 1763 he wrote, "Would that all the Archiater's colleagues were like him, more devoted to science than to schemes and intrigues," and later, "among the advantages he had enjoyed during office, he counted his acquaintance with Linné."

When Höpken withdrew from political and academic strife to his estate of Ulfåsa, the correspondence between him and Linné became closer, as shown by many warm expressions in Höpken's letters. In 1774 he wrote, "I long for Uppsala, not indeed for its own sake, but that I might have the pleasure of talking with the Archiater."

Linné's most intimate friend, however, was Abraham Bäck (1713-1795). They first became acquainted in 1740, when both were candidates for the Chair of Medicine at Uppsala, and thence began a friendship so warm, intimate and unclouded, that a similar instance can hardly be found. Their truly brotherly love showed itself in many ways; in 1774 Linné noted, "Each time he stayed in Stockholm, he lodged with his best and truest friend, Archiater Bäck, as with his bodily brother." The evenings and much of the nights were spent in talk, earnest or playful, and so refreshed was Linné on such occasions, that he could write, "Ever since I was at Stockholm I have been livelier, better for work and quicker; before, I was depressed and melancholy, and could do nothing." Bäck had contrived the visit to be so enjoyable, that he willingly tried to tempt Linné to come again soon.

It was not only the personal meetings which made the friendly ties between them, but the intimate correspondence during thirty-five years, in which each confided his pleasures and sorrows, hopes and fears, intentions and happenings to the other. About 520 letters are extant from Linné, though many have probably been lost. On the other hand all letters from Bäck except fifteen of early dates, have not been traced. Probably this was due to the fact that all the correspondence was entrusted to Bäck for his funeral speech on his beloved friend, and presumably he kept his own letters when he returned those of Linné.

These letters show in the liveliest manner the thoughts which were occupying Linné's mind at the time of writing. It has already been stated (p. 193) that Bäck was a competitor for the post which Linné obtained in 1741, and after that date, he undertook a journey abroad, returning in 1745, when he became Assessor in the College of Medicine. In 1748, he bought a court practice for 14,000 dalers [£1,050], "far too dear," thought Linné, "it grieves me and the whole modest world." The next year he became body physician, and in 1752, President of the College, to the intense delight of Linné, who would therefore "break the necks of some bottles of wine and drink from them, even if that gave him a megrim."

Bäck's position as acting body physician at court made him an intermediary in such matters as purchases for the Royal Collections, or Linné's visits to the palaces of Drottningholm and Ulriksdal. Linné invited Bäck to show to the Royal family the movement of the stamens of *Berberis* when touched with a pin, "in the same way as [Bäck] felt the pulse of a damsel after the method of Paracelsus; see and marvel at it!"

The most diverse matters were handled by Linné in his letters, including his friend's affairs. He took the greatest interest in the news that Bäck had become engaged to a certain lady, and hastened to convey his hearty congratulations, but more trustworthy communications showed that this engagement was mere rumour. However, at the end of 1754, Bäck really became betrothed to Anna Charlotta Adlerberg, the wedding being celebrated in March the year after. Linné, not being able to be present, meant to have had a quiet celebration at home, but the apothecary, D. W Böttiger, persuaded him to be present at a formal collation, when healths were drunk and the festivities lasted till one o'clock next morning, when Linné and his wife went home. Soon after, the newly wedded pair visited Linné at his Uppsala house.

This change in Bäck's mode of life did not in the least disturb the old friendship. On his Stockholm visits Linné was so hospitably received by Bäck and his wife that he soon came to call the latter " Sister "—having long before called Bäck " Brother "—the intimate form of address in Sweden. It was therefore with the keenest sorrow that he heard of her death in 1767, and a long and most sympathetic letter to the bereaved husband testified to his deep feeling at Bäck's loss.

Affection for the parents continued also for their children. On the first being born Linné was invited to become godfather, which position he gladly accepted, and afterwards frequently referred to the little maid, who, however, died early. When a son was born, he was baptized Carl Abraham Bäck, and in due time he had as his tutor one of Linné's pupils, D. H. Söderberg. Unhappily, the son, inheriting his mother's tendency to consumption, died in 1776, at the age of sixteen. At this time Linné was laid aside by repeated strokes of apoplexy, and his attempt to console his friend only resulted in a few sentences, " Farewell. I am Brother's, Brother is mine, constant to death, Broth——" this pathetic fragment closing the long friendship. By some mischance this letter, addressed to " M : sr Abrah. Brách," was never despatched, but being found amongst Linné's papers after his death, was sent to Bäck by the younger Linné thirteen months after it was penned.

Bäck's remaining children enjoyed with their father the constant affectionate attention of Linné. Many times he urged their visits to Hammarby in order to enjoy the fresh air: " Let the poor little ones who have no mother, enjoy themselves with my girls this delightful summer." Bäck had the pleasure before his death of seeing two of them grown up and married.

In 1772 Bäck invited the entire household of Linné to come to Stockholm for the coronation of Gustaf III.; his wife and three younger girls went, but Linné was too busy to accompany them. On their return he wrote that they would never forget the time they had had, nor would he during " his short days." Needless to say that Bäck was a welcome guest at Uppsala when he did come, and in 1775 he paid what proved to be a farewell visit, for the old friends never saw each other again.

Among Linné's friends from his student days were two with whom he remained in close companionship till death closed the bond. Both attained high office in the church, and played a considerable part during the Era of Liberty These two were Johan Browallius and Carl Fredrik Mennander.

It has previously been recorded that the former had an influential share in Linné's career (p. 107). In 1737, he was appointed Professor of Physics in Åbo University. His love for natural history was maintained even when in 1746 he became Professor of Divinity, and in 1749 Bishop of Åbo, which latter office he filled most successfully. An opponent of Linné was J. G. Siegesbeck (p. 170), who wrote condemning the sexual system root and branch, while Browallius took up the cudgels in his friend's behalf. In his " Examen epicriseos in systema plantarum sexuale Cl. Linnæi " (see p. 183), he doughtily defended Linné's arrangement, no reply to this being forthcoming.

Another debate on " Vattuminskingen " or decrease in water level, started by Anders Celsius in 1743, and maintained by O. von Dalin, the historian, was attacked by Browallius on the ground of his own observations, which conflicted with Celsius's statements. Linné, on the other hand, took Celsius's side, till convinced by Browallius to the contrary. His name is perpetuated by the genus *Browallia*, concerning which there has been some misunderstanding.

How this arose seems unknown, although it was current in 1835, and exposed by H. E. Richter in his "Codex Linnæanus" that year, but also recorded by Augustin in the "Botaniska Utflygter," 1 (1843), p. 150; the most blatant exposition known to the writer is that related by "X," in the "Gardeners' Chronicle," III. x. (1891), p. 188, thus:

"The great botanist Linnæus had amongst his numerous acquaintances a certain friend named John Browall, who was very humble in his relations with Linnæus, and, having adopted his new sexual system of botany, wrote an article against Siegesbeck defending that system. Linnæus, in acknowledgment of his friend's services, dedicated to him a genus of a single species, naming it *Browallia demissa.* Shortly afterwards, Browall, having been made Bishop of Åbo, assumed the pomp and dignity of a great magnate, and Linnæus, having discovered a second species of this genus, named it *B. exaltata.*

"This excited the wrath of Browall, who proceeded to write pamphlets against Linnæus, denouncing him in the most severe language. Later on, Linnæus discovered a third species differing slightly from the original outline of the genus, which he named *alienata.* The two men were never afterwards reconciled to each other, and thus we have preserved in the nomenclature of this genus an historical incident to which future generations of botanists will look back with considerable interest."

The genus was instituted by Linné in his "Genera plantarum" in 1737, but first provided with the specific name *americana* in 1753, when the

"Species plantarum" came out. In 1759 appeared the tenth edition of the "Systema Naturæ" with a revision of the plants, the genus *Browallia* then having three species, *B. alienata*, *B. demissa* and *B. elata;* all these being first published four years after the death of Bishop Browallius in 1755, who consequently never knew this enlargement of his genus, and therefore the whole of the legend is shown to be baseless.

The other friend, Mennander, having left Uppsala for Finland in 1737, became Master in Philosophy, taking the place of Browallius in 1746 as Professor of Physics. He then passed to the faculty of divinity, and from 1757 to 1775 was Bishop of Åbo. How warm was Mennander's liking for natural history was displayed by his support of P. Kalm and A. Martin as naturalists in the Finnish University.

Frequent letters between Linné and Mennander, which correspondence began when both were students, were continued during life, till in 1775, the latter becoming archbishop, removed to Uppsala, but a few months later Linné was so broken in health and spirits, that he could no longer enjoy conversation with his trusty friend.

Another friendship formed during Linné's Stockholm residence, was with the then Captain Augustin Ehrenswärd, one of the earliest members elected into the Academy of Science, who in 1740 became its secretary. As both belonged to the "Hats" party, the relationship between them was specially cordial, a testimony to it being the fine copper-plate portrait of Linné *amica manu* in the same year prepared by Ehrenswärd. Their paths separated when Linné removed to Uppsala, each going his own way, but afterwards when Ehrenswärd came to Finland to fortify Sveaborg and other places, and to form the Swedish fleet of galleys, the friendship was renewed. From the Uppsala garden he introduced the cultivation in Finland of the grass *Glyceria aquatica*, but his sanguine hopes about its success were not shared by Linné. Applying for a tutor for his son, "Will you," he wrote, "provide me with one, who has not studied divinity, oriental languages, nor metaphysics, but has given his mind to other important sciences? I would gladly choose a botanist." The death in 1772 of the then Count and Field-Marshal Ehrenswärd severed the connection between these veterans.

Another friend who has been mentioned before was Court Intendant Baron Carl Hårleman. The friendship began when plans for the restoration of the Uppsala garden were prepared by Hårleman, and continued when the castle, the cathedral, and similar works were taken in hand. It was by his initiative that Linné was commissioned, for economic reasons, to undertake his Skåne journey. Linné wrote to Bäck in praise of his friend's endeavours for the economic improvement of the country, and it was to him that in 1749 Linné's "Materia medica" was dedicated. Hårleman's death in 1753 deeply grieved the survivor. "God help us, who now will be our Hårleman?"

A comrade in the Academy of Science was Baron Carl Sten Bjelke, Assessor, afterwards Aulic Councillor in Åbo Court of Appeal. He was devoted to botany, having an uncommonly accurate knowledge of grasses. In 1744, with Kalm, he travelled in Russia, and in consequence, Linné received from him a rich collection of dried plants from that country, more than two hundred in all, those from Siberia being nearly all new and undescribed. How grateful Linné was for this gift appears from his remarks dated 1745. "Most East Indian plants are now in my hands; all of Ceylon, more than two thousand in number, I am examining without ceasing. There is a genus *Bielkea*, a rare grass, which shall thank the Baron for his love for botany, and shall make him known on the sun's rising upon his return to Sweden." Bjelke settled at Löfsta on his estate in the parish of Funbo. There he busied himself in the introduction of new supplies for fodder and food, Linné being often a welcome guest there, but the pleasant intercourse was ended in 1753, the Baron dying in that year, aged forty-four.

Similarity of pursuits united Linné with Charles De Geer, who became known by his accurate and sensational observations on the habits and development of insects. In 1739, when only nineteen years old, he became one of the first members of the Academy of Science, and at the age of twenty-eight was a corresponding member of the French Academy. Later, in 1750, he was appointed Court Marshal, gaining the title of Baron in 1773. His valuable museum and library at Leufsta readily tempted Linné thither, and in return, De Geer, sometimes accompanied by his wife, visited Hammarby or Uppsala.

Nils Rosén von Rosenstein must also be reckoned amongst Linné's most intimate friends. It has already been recorded (p. 123) that strained or cold relations existed between them in earlier years, but they eventually became good comrades, with mutual regard for each other's attainments, working in harmony for elevating medical study in the University. How this close intimacy arose was related thus by Linné: "In May, 1764, I was attacked by a dangerous pleurisy, from which I was rescued by Rosén's faithful services, whereafter I entertained an incredible friendship for him." That this was no hastily kindled feeling of gratitude, as quickly cooled, is shown by Linné's references in his letters to various friends. The year following Linné's illness, Rosén being severely attacked by the so-called Uppsala fever, it was then the turn of Linné, who attended him night and day for two months with the happiest result. Rosén afterwards removed to Stockholm, but when in consequence of repeated illnesses, he presaged his approaching death, he came back to Uppsala, which he reached in a very weak state, in June, 1773. Thenceforward Linné was found in constant attendance till the day before his death, entertaining him with the account of new discoveries in Natural History. "I can never forget," says one of their conjoint pupils, J. G. Acrel, "the conversations of these two great men on certain medicaments, their use and application, Rosenstein on his sick-bed with an old man's experience, but with the fortitude and continuance of one of middle age." On the 16th July of the same year, Linné journeyed from Uppsala to wait upon the King at Ekolsund, and on his return the next day, found that Rosén had ended his days. His sorrow for the departed was moving. He hastened to the house of death, standing by the bed where his dead friend lay, and burst out with, "Here has a whole university closed its eyes." It had been the wish of the University that Linné should deliver the memorial oration for the "man of strictest probity in medicine in Sweden," who had "both learning and experience," and it was only with the greatest trouble, in view of Linné's age and greatly weakened health, that he was induced by his friends to forego that duty.

Among the many who might be reckoned as intimates during Linné's professorship, a few may be briefly mentioned. During the early years after his removal to Uppsala, there were two, already recorded, namely the eminent astronomer, Professor A. Celsius, and his powerful advocate in the Wallerian disputation, Professor Magnus Beronius. Both friendships were of short duration, as the former in 1744 quitted worldly scenes, and the latter in 1745 went to Kalmar to become Bishop. In 1764 he came back to Uppsala as Archbishop, but died in 1775, two years before Linné, so that he was unable to deliver the funeral oration as he had promised. Two other professors in later years were on the most familiar terms with Linné, namely the celebrated mathematician, Samuel Klingenstierna, and the not less famous orientalist, Johan Ihre. The former was almost a constant member of Linné's household, even going so far as sharing linen and garments at Hammarby. The latter was a neighbour, his estate being Edeby close by. Linné showed them both much sympathy, sharing their sorrows, when the former lost his wife, and, through drowning, his dearest son; while the latter, Ihre, also became a widower.

Yet another one must be mentioned, who during a long succession of years was closely connected with Linné in his authorship; this was Lars Salvius, the publisher of the greater number of Linné's printed books. The first connection between them seems to have been in May, 1745, when an author's honorarium of 18 copper dalers [nine shillings] for each printed sheet of the "Flora" and "Fauna suecica" was granted by Salvius, and proved so satisfactory to both, that Salvius undertook the issue also of the tenth and twelfth editions of "Systema Naturæ," "Genera plantarum," Editions V. and VI., both editions of "Species plantarum," "Hortus Upsaliensis," "Materia medica," "Amœnitates" and others, and the "Mantissa" of 1771, as long as Linné's strength held out, but Salvius died in 1773. There was no unpleasantness between them, though in later days Linné complained that there was "bad paper, worn-out type, and carelessness as to corrections," but on the other hand, "he always paid as promised, and always promptly." He also acted as commissioner for Linné, in distributing letters and packets of the greatest importance. Agreements about the issue of new works were verbally made, and written contracts appear never to have been drawn up, Linné observing, "My days are running out, and what I have to do must be done quickly. We are both old, and I believe equally old."

CHAPTER XV

LINNÉ AS A PRIVATE PERSON AND HIS FAMILY RELATIONS

THE portraits of Linné show that as he advanced in life his appearance changed considerably. He speaks of himself as: "Linné was moderately big, rather short than tall, more lean than fat, with fairly muscular limbs and prominent veins from childhood. . . Large head, the back of it with a transverse depression along the lambdoid suture. Forehead moderately high, wrinkled in old age. Hair neither straight nor curly, in childhood flaxen, afterwards brown, ruddy about the temples, grey in old age. Eyebrows brown. Pale in face. Eyes brown, very sharp, lively, gladsome; sight excellent, descrying the smallest object. Nose straight. A little wart on the right nostril, and a somewhat larger one on the right cheek. Teeth bad, decayed from severe toothache from youth to fifty years of age, entirely toothless before sixty. No ear for music. Weight in 1734 9½ lispund, or Stockholm's weight [178 lbs. avoirdupois]. Walk very easy, quick and lively."

"He was not luxurious, but lived moderately and was no toper. Housekeeping he left entirely to his wife, occupying himself solely with the productions of nature. He was neither rich nor poor, but lived in dread of debt"; his works were written not for gain but for honour. He slept in winter from nine till seven [in old age from eight to eight], but in summer from ten to three.

To his own account of his appearance and habits the following may be added, written by one of his pupils, J. G. Acrel, who, during the later years of life, often met him. "Linné was short and squarely built, not fat but muscular, the back of his head unusually big. In a not unpleasing countenance he had quick and fiery brown eyes, somewhat short-sighted and blinking, more from habit and work on delicate objects than naturally; in the course of years he developed deep wrinkles round the eyes, through muscular contraction. Went somewhat bent as to body, but otherwise when young had an easy gait, which gradually disappeared, so that at fifty he began to shuffle his feet forward in place of lifting them."

As to his clothes, the same authority states that he was neat, but never magnificent; sometimes he was fine, but always wore shoes; in his house he was mostly dressed in a short dressing-gown and velvet skull-cap. He was moderate in food, extremely abstemious in drink, but he drank coffee and used tobacco with avidity, almost to excess. In winter time he slept from nine to seven, but in summer from ten to three, when he began work, in the garden or in the fields; not infrequently he even in winter rose, lit his pipe, and sat down to work, till he was tired. His books did not produce so much money as reputation, though they certainly contributed to his moderate income. From the time when he removed to Uppsala as professor, he never practised medicine, except the giving of some simple advice to a poor person or dependent.

There is little to add except one or two things from his letters. As to diet, he wrote to Bäck in 1772. "I never taste brandy, except when I am away, and then only a sup with plenty of water, often not a drop for a whole month. I have truly not drunk a tankard altogether in six months, for fear of acidity. I take three cups of tea each morning and two cups of coffee each afternoon." He was equally abstemious as to wine. "My motto is: I never drink wine at another person's house, for I will never treat others to wine." An exception was at Bäck's, where he enjoyed excellent wine. In conclusion, an extract from a letter of the younger Linné in July, 1778, to Bäck may be added. "My late father never worked unless he was in good humour. He rose early in the morning, when he awoke, lit his fire, and sat down to work, but as soon as he felt the least tired, he left his work and rested while he smoked a pipe of tobacco, or else threw himself on his bed, for he had the knack of sleeping at once, and snoring; in a quarter of an hour he was up again, alert, and returned to his pen. This happened several times in the day, till about four in the afternoon, when he would have some society to clear away what had till then occupied his thoughts. He readily joined in a laugh with comrades or pupils as he 'could digest a little of anything,' adding, 'It was well said by the old Romans, that one ought to have an understanding throat, such as mine, for I can digest anything, and I can also go without.'"

Another of his peculiarities must not be passed by, namely, a certain amount of absent-mindedness, as when he visited Bäck, he would often leave behind him a sword or a nightcap, manuscript or letter, while taking away something belonging to his host, and would excuse himself in his next letter.

His character may be summed up thus: In his youth he was merry and glad, in his age never surly; easily moved to joy, sorrow, or anger, and was soon placated. Not hasty in judging, he held fast to his opinion. He had an excellent memory till he was sixty, when proper names began to be forgotten; to learn a modern language was never his custom; he did not begin tasks which he did not finish; indefatigable in observing everything, he never went into or out of the garden without noting something.

He was far from quarrelsome, would not willingly dissemble or deceive, hated everything that tended to pride, stood firm to his promise, and could not easily be disturbed from it. He was not inquisitive; he always had respect and admiration for his Creator, and sought to deduce his science from its Author, and had inscribed over the door of his bed-chamber, "*Innocue vivito*; *Numen adest*" [Live innocently; God is here"].

Such is the character which Linné has left of himself. As no one can be regarded as a competent judge in his own case, it will be well to adduce the testimony of others.

As to his merry and glad disposition, all accounts agree that he delighted to see happy faces round him. "In society he was cheerful and gay and would gladly listen to anecdotes with enjoyment," says his pupil Hedin, and Acrel adds, "In youth he was frolicsome, in middle age ever mirthful, quick and ready in words, and in jolly society shared laughter with the others, which lasted to his latest year." He did not talk much, but enjoyed hearing others talk, and sometimes struck in with special short interesting remarks. Weary of work, he felt the need of refreshing laughter, and therefore amongst his pupils, Tidström was a favourite, whom he had as a guest at Hammarby at Christmas, "simply to laugh and be free from care." He particularly wished young people to enjoy themselves, and it is related that when the newly appointed Archbishop Mennander forbade the accustomed dance assemblies, the young girls of Uppsala got the old Linné to obtain permission for their retention.

He quickly flamed up, but soon abated his anger, and it was well known among his friends and pupils, that he was entirely free from rancour. His habit was never to delay anything which ought to be done, and to note down his thoughts at once, on paper in pithy expressions, which needed to be somewhat pruned. This appears evident, that judgments on the same person or event sometimes were recorded in a short time from each other, as not unimportantly diverse. As soon as the roused feelings had subsided, a calm ensued, and he expressed in a letter written in 1754, "I have learned in forty-seven years of life that if others are let alone in peace, the world goes on without disturbance, but as soon as commotion is made, friction arises, and a little hornet will often produce a dangerous wound."

His compassion towards the unhappy was very striking, his eyes often filling with tears on hearing of sad cases, especially when it was little children who were afflicted.

Then his statement that he was not hasty in judgment must be taken with a certain modification so far as it concerns scientific questions. A striking instance of this was his suspicion that the prohibition of printing Swedish authors' books abroad was aimed directly and exclusively against him. Among those who took part in this prohibition were Tessin and Ekeblad, and it needed but little reflection to realize that these friends and admirers would not have willingly caused him harm or anxiety.

"He did not possess the art of dissembling," records Acrel, "his face showed at once whether a person pleased him or not, the surest proof of his dislike being his silence, but he hated most, quarrels and coarse answers." Many instances of his straight forwardness might be given, but two such may suffice, preserved by tradition in the family. When Queen Lovisa Ulrika, whom he so highly admired, expressed her wish to receive one of his daughters at Court, he answered with a positive refusal. The Queen being startled, asked if he could not entrust his daughter to her care; he readily assented, but said that he thought the matter had not been rightly understood. Another time when he was staying at Drottningholm, to arrange the Queen's Museum, he was called in to play blindman's bluff with the courtiers. In such cases it was strictly against etiquette to catch the Queen, but Linné, who thought he could better employ his time than in playing, took care, when he became blindman, to see a little, and then caught the Queen, as soon as he could. She cried out at once, "It is I"; but without in the least allowing that to deter him, he clapped her on the head, and said, "Clap, woman, sit on the bench," as the custom was when anyone was caught, and added, "Those who play, must put up with play"; and afterwards he ceased to play blindman's buff.

The first public distinction he received was the title of Archiater, bestowed in 1747, without his knowledge or request. That he should be gratified at this mark of appreciation is easily understood, though other Professors of Medicine, Rosén amongst them, had previously been thus honoured, and soon after, a barber-surgeon was distinguished in the same manner; but Linné looked upon it as a recognition of his deserts. On the other hand, he felt far from pleased at having to pay for the title the fee which was annually demanded of him. Upon a request he put forward, he was freed from this, and later, his previously paid fees were returned to him.

As regards his being dubbed "Knight of the Polar Star" in 1753, it was certainly in his time a very unusual mark of royal grace to any Doctor, Archiater or Professor; and was correspondingly appreciated by the recipient. His ennoblement was such as was frequent at the time when prominent professors found a place in the House of Nobles. His turn came in November, 1761, when he assumed the change of name as "Von Linné," though the grant was antedated to August, 1757 Next the question arose as to the arms he should bear. Quick in application as he always was, he suggested them thus: "My little *Linnæa* in the helmet, but three fields in the shield, black, green and red, the three kingdoms of nature, and upon it an egg cut in two, or a half-egg to denote nature, which is continued and perpetuated in the egg." This design was forwarded to the Governor, Baron Daniel Tilas, who, as State Herald, had to frame or define such matters, but it was rejected. In its place another was designed, in its three crowns recalling the Tessin medal. Linné was not pleased with it, and to gain his end he applied to Wargentin and his wife, begging their help, to get rid of Tilas's "absurdities," but this appeal not being sent before a new letter from Tilas had come with a final decision, Linné felt himself obliged to acquiesce, afterwards finding that the blazon "was truly honourable and beautiful, more than I deserved," and thus it was settled.

It may be asked if Linné was proud at changing his name and status with the addition of "von." In that case it must be answered, that if he were so, he at least managed effectually to conceal it. It is certain that he showed no haste to enter the House of Nobles, for he did not pay the fee due till 23rd April, 1776. These payments were for the patent (Charta sigillata), 200 dalers, for the Chancery fee, 150 dalers in silver [£17 10s. together], and 25 and 40 dalers [£1 12s. 6d.] for soldier's fee. Linné quite as often signed his name without as with the "von," as may be noted on page 307

With all his actual greatness in many respects he was as pleased as a child at his hard-won distinction, but between this and pride is a long step. He wrote to Bishop Mennander in his autobiography, "I cannot send personal details for self-praise is offensive, and self-love may creep in here and there. Be so good as to alter or preface, as you may see fit in your wisdom." What he really thought of these honours we see by his calling them "empty nuts," and in his "Nemesis divina," where he expresses his deepest thoughts, they certainly witness to no pride when one reads, "There is no greater character than to be an honest man," and in another place, "I gave myself no rest day or night. What had I for it? Call it wind, which is annihilated by another. Titles are wind: Noble, Knight, Archiater"; and again, "What is greatness, when the wheel of success turns? What is wisdom? To realize one's own ignorance. What is power? The foremost place among fools. What are riches? Guardianship for other fools. What are clothes? Parts in a comedy to frighten children."

There is more truth in the claims that he heard and read praises of his writings and views with satisfaction, and disliked to find that questions which he had

1 Carolus Linnæus Smolandus.

2. Carl Linnæus

3. Carl Linné Carl Linné

4. Carl Linne

1. SIGNATURE ON MATRICULATION AT UPPSALA, 23 SEPT., 1728.

2. SIGNATURE IN 1755.

3. SIGNATURES IN 1765.

4. THE LAST SIGNATURE KNOWN, 1777.

thoroughly settled, were sometimes denied or assailed with bitter words. It must be admitted that few men of eminence exist of whom that cannot be said. Some have asserted that he wished to elevate himself to a position as absolute in natural history, and have cited such instances in his "Species plantarum," where he omitted mention of plants discovered by others, or transferred them to other genera, when he had had no opportunity of examining them himself. There is an element of truth in this, but we must consider how many doubtful species were described wrongly by the old authors, to explain his unwillingness to recognize all their statements. Acrel assures us that "no one could be better pleased than he at being corrected, and the disciple who by research could convince him of a mistake, became his best friend."

The strongest ground upon which people taxed him with pride, is derived from his remarks on his own merits and work in Afzelius's "Egenhändiga anteckningar af Carl Linnæus om sig sjelf" [Remarks under Linné's own hand], but these were jotted down for a special purpose. At that time, much importance was given to the memorial oration after death, and it was not unusual for the man himself to write and leave biographies which would spare the speaker special trouble. This was the case here; Linné had provided his friends Bäck and M. Beronius with what he wished them to say How could he be expected not to give a short summary of what had been expressed in many writings on the most varied occasions; if he had omitted, denied or lessened these praises, would not that have been transparent hypocrisy?

Another accusation was of avarice; but nothing can be less admitted than that. Considering his generous goodwill to pupils, studious fellow-countrymen, and other needy people, the support that he gave to scientific travels, and the purchase of costly books, it is clear that he thought too little of his own advantage or his family requirements. How such an opinion could arise was due to the peculiarity of his fondness for ducats, especially Dutch. When he received any, he hid them carefully in the drawers of his writing-table. "This was only a freak of his," relates his pupil Hedin; whilst Bäck declares, "The noblest of metals gladdened his eyes, and why should he not collect them, as naturally as other things?"

In another aspect he was more greedy than his pupils wished, and that was regarding the plants in the garden; but that was due to the large number of his pupils. If he had permitted every one to make a herbarium of the plants there flowering, the garden would have been stripped, to the injury of himself and the institution. He was greedy in adding to his collection, and thankfully received any contribution to it. He considered that he had a right to receive the plants collected by his pupils, at home or abroad, and valued them highly; if he missed such confidence he could not conceal his displeasure.

Without doubt the most noteworthy trait of Linné's character was his ardent piety. He rarely missed attending church on Sundays and holy days, and on those mornings his daughters or granddaughter would come into his room, and sing "Papa's Song," probably of his own composition. This gives a clear impression of the rules of life which he set before himself, and carefully followed. Amongst his memoranda, never meant to be read by any but his children, he, in his simple, childlike piety, gave expression to his feelings of humility, respect and thankfulness to Him, who had so wonderfully and happily directed his way unto his old age. Thus he wrote:

"God has conducted him with his own Almighty hand;
" has let him grow up from a trunk without root, planted him in a distant, splendid spot, let him grow to a considerable tree;
" has given him so ardent a mind for science, that it became the most desirable aim in life;
" ordained that all suitable means should be available in his time, to aid his progress;
" so directed him that his failure to win what he wanted became his greatest advantage;
" caused him to be taken up by patrons of

Science, even by the highest in the King's palace;

God gave him the best and most honourable duty; precisely what he most desired in the world;

" gave him the wife he most desired, who kept house while he worked;

" gave him children, who were good and virtuous;

" gave him a son as successor;

" provided him with the greatest herbarium in the world, his greatest delight;

" bestowed goods and other possessions, so that there was nothing superfluous, nothing wanting;

" honoured him with a title of honour (Archiater), Star (Knight), Shield (Nobleman), name in the learned world;

" preserved him from fire;

" preserved his life beyond sixty years;

" let him gaze in His secret Council Chamber;

" let him see more of His created world than any mortal before him;

" bestowed upon him the greatest insight into the knowledge of Nature, more than anyone had hitherto enjoyed.

"The Lord has been with him whithersoever he went, and cut off all his enemies and has made him a great name, such as the greatest on earth. —2 Sam. vii. 9."

Thus clearly does his view appear in his "Nemesis divina," that God punishes transgression. A bad action must bear bad fruit, and punishment follows, often in the most striking manner. He did not leave an ordered account of this doctrine, but collected certain facts, which he considered supported his views, and as he expressly said, to warn his son against the sins which Nemesis would specially avenge. He often adverted to this in his lectures and writings, such as the preface to the later editions of his "Systema Naturæ." Though these accounts were not published in his lifetime, they have been printed twice in recent years.

This sincere piety of Linné did not impel him to a slavish assertion of everything which the orthodox divines of his time regarded as indisputable articles of faith. It is true that he, like most naturalists of the day, looked upon the Bible as the main source of information, even in natural science questions, but he sometimes allowed himself to interpret passages in his own way, although they were not in accordance with the church's doctrine. Thus, he doubted the existence of a universal flood, or that only six thousand years have passed since the creation of the world, also that the elements existed before the Mosaic account narrated their creation. Though he did not print these views, they could not remain unknown, and many priests and laymen thought him a secret atheist, or at least heretical in some degree.

Before closing this account of Linné's personality, his political interests and views may be mentioned. Of the two political parties, he sympathized most with the "Hats," but regretted the persecution which prevailed. Only once did he awake from his political apathy, and that was when the revolution under Gustaf III. caused high hopes for the prosperity of the country. He called his pupils together, and after delivering an oration, all sang "Jubilæum" on the occasion, Linné joining in most lustily.

In 1743, Linné removed to the dwelling-house erected in the botanic garden, and he felt himself then more than happy. In one of his autobiographies he wrote: "Now had Linné honour, the office he was born for, sufficient money, partly from his marriage, a dear wife, pretty children, and an honoured name, he lived in a palace built by him close to the University, and he completed the garden. What more can a man desire who has all he wants, though it is impossible for it to continue. So many stones, which are in his collection, so many plants in his herbarium and garden, so many insects which he collected and set on pins, so many fishes which he had glued on paper like plants, all besides his own library, were his pastime."

A description of the house may be given; on the walls were hung the portraits of the two Rudbecks, so lifelike, that they could not be excelled, two worthy predecessors, beside drawings of the greatest botanists, Tournefort, Ray, Morison, Rivinus, Vaillant, Boerhaave, Burman, Plukenet, Breynius, Columella, Jungermann, Koenig, Simon Paulli, Camerarius, Tilli, C. Bauhin, Sloane. It is known that at first the family had the ground floor, and part of the first floor, the other rooms being used as a museum, and for private lectures. In the corner room towards the garden, he had his library where he worked, and where he could also keep a watchful eye on the workmen and the visitors in the garden. As the family increased, and more room was wanted, the museum was moved to the orangery, even the mineral collection finding its place there. This was done by leave of the Consistory, who stipulated that it should not be mixed with the University collection.

As to the arrangements in the house, Linné did not trouble himself, his time not permitting any interference; therefore he regarded it as God's gracious gift that he had a wife who kept house while he worked. In the course of years she altered considerably, both in appearance and disposition. The smart lively young wife became a big, corpulent, rough matron, with coarse features. During her early years in Uppsala, she seems to have devoted herself to enjoyment, had dances in her house and, above all, was addicted to card-playing. When Linné was absent in Stockholm, nearly every evening there was a card-party, including the wife of Professor Klingenstierna and Magister Berge Frondin, when the play became so jolly and free that Linné, who did not entirely lack suspicion, evinced a certain degree of jealousy That this feeling was not justified and soon passed away, is shown by a friendly letter to the suspected young man, who continued to be a frequent guest in his house.

As years passed on, Madame Sara Lisa became an altogether prosaic, but able house-mother. The education which she received in her parents' house was without doubt scanty; and though at this time all ladies' letters, even in the highest circles, showed little ability in the use of their mother tongue, Linné's wife displayed her entire ignorance of spelling and syntax of Swedish usage. Hence arose dislike for more intellectual employments and a liking for dissipation. It was truly a sacrifice when she, on Linné's earnest invitation to Bäck to pay them a visit at Uppsala, promised her husband she would refrain from all pursuits during the visit likely to displease him. Card-playing was excepted, but afterwards, when Bäck returned an evasive answer, she declared herself ready to give up even cards, " if only my Brother will come."

Even if she could not, nor even tried in the slightest degree to interest herself in her husband's views and occupations, she was an industrious, strong, able and domineering house-mother, and a useful manager when the increasing family demanded more outlay and expense. In the house there was shown a liberal hospitality, especially during the two great events in academic life, at each promotion (granting degrees), the house being crowded with guests, who did not make short visits, but stayed for many days. The food was certainly simple, but it was good and abundant, both for the family and the servants. Furniture, as the years progressed, became handsomer, and to judge from what has come down to the present day, the Linnean household bore the stamp of elegance and comfort, though in some degree it was simple and unpretentious. Clothes for everyday use were plain enough, but the girls also had silk dresses silk stockings and pointed shoes, which could compete with present-day fashions. On special occasions, such as royal visits, artificial and high erection of hair-dressing was adopted, which could not be completed in one day. Tradition has it that two of the daughters had to spend a night in chairs, lest the master-work of the hair-dresser should be spoiled.

Add to this, that during the last twenty years of Linné's life, the housekeeping cares were doubled by the increase of family, both in the town and in the country, in the latter place being joined to extensive farming, with whose prosecution Linné did not concern himself, so that this demanded a thrifty house-mother's thoughts and untiring care. In this respect Sara Lisa deserved all honour, and her merits being known and recognized, the Consistory invited State Councillor Petræus, who had presented to them a valuable shell cabinet, to dinner, it was the wife of Linné who was commissioned to undertake this. Among the academic accounts there was an item for this dinner of 77 dalers in copper [38s.].

In her house order reigned everywhere. Carelessness and bad housekeeping not being tolerated, nothing which could be turned to use was wasted. At four in the morning the spinning-wheel began to hum, one of the daughters being always engaged upon it. A family tradition is preserved, that when at work on her dowry, the eldest daughter shortly before her wedding, oversleeping herself one morning, she received a box on the ear from her mother as a reminder of her unbecoming behaviour. Similar strict discipline was maintained with the servants, who, when the old woman was confined to her bed by some illness due to her years, were obliged to come into her chamber to spin under her supervision; those who, during the winter, spun the most, received the best present from their mistress. No carelessness was permitted in spinning, and so long as she was able, she examined carefully every piece. She had great pride, as the times required, in an ample linen closet, full of home-made linen.

This unresting occupation in housekeeping with the oversight of many servants, caused a certain roughness of speech and moroseness, Fabricius mentioning that she "often drove pleasure from our society." This must not be ascribed to want of good feeling, but rather to a certain unpolished straightforwardness of character. Thus, when Gustaf III. on the 12th August, 1775, visited Linné in Uppsala, and his escort remained mounted on their horses in pouring rain by the yard, she thought it ought not to be, and when the King, after a time, asked her if she had any wish which he could fulfil, answered straight out, that she wished that His Majesty would allow his people to dismount, dry their clothes, and take some refreshment. The King knit his brows at the unexpected reply, but the result was that the fellows, to their great delight, emptied the water out of their boots, and before a big kitchen fire dried their soaked garments and enjoyed a hearty meal with beer. Among her descendants the idea prevailed that "the old woman Linné" was somewhat rough and overpowering in her manner, but was good and friendly at bottom.

It is not to be supposed that dissensions and misunderstandings between husband and wife never occurred; they did sometimes happen, as was natural, for each of them was hasty and easily provoked. From this, other folk surmised that Linné "stood under the slipper," i.e., petticoat government, and that his home was not entirely happy. As has before been mentioned, Linné let his wife direct household matters as she wished, and if he was comfortable he never complained. He had so little interest in household affairs, that he did not know his own people, and once asked a man-servant "Who are you?" Once when Professor Melanderhielm came before his time to a party, he and Linné engaged in a lively discussion, quite forgetting the occasion, and drew various figures with charcoal on the clean floor, which was more than the hostess's patience could stand, outraging her sense of orderliness. When sometimes the wife let fall some hasty words, saying that it was the money she brought into the home, and therefore she had a right to be the determining party, it was only a passing cloud over their usually sunny married life. If Linné had felt himself oppressed or unhappy at home he surely would not have had that glad and free humour nor the indomitable love of work, which he retained to his old age. It is known that he sometimes wished to surprise his wife with something pleasant. Thus, when she once uttered a wish to have a little silk for a pinafore, he went out unknown to her, and came back with a whole piece. What she said is not known, but with part of it she made a cape, which is still extant.

How warmly Linné was attached to her was shown during a severe illness; on Christmas Eve of 1754 she sickened with the malignant epidemic fever, and then followed more than a month, which was the severest trial he had ever had. The whole house was in confusion, the children complaining and crying, and he himself sat by her bedside weakened by disquiet, sorrow and want of sleep. It was not till the 26th day that she showed signs of improvement, but a relapse soon followed. At last on the 31st January he was able to tell Bäck, that he was now like a prisoner liberated from prison, as his little wife was gradually mending.

With his children Linné was a too indulgent father, the elder son especially being spoiled by him for life. He said his principle was to hold a tight hand over a young man, and encourage him to fare hardly, thus making a man of him, but he did not act so with his own son. This is shown by his appointment while quite young as Demonstrator and Professor, partly because he did not oppose it nor show other wishes for a profession. He was certainly not chosen on the ground of special fitness, but for his position as his father's son. It must be specially borne in mind, that without having undergone any examination or ever disputed, by the Crown Prince Gustaf's order he was promoted honorary doctor in the medical faculty. In 1765, when he was only twenty-four years of age, this brought him into an awkward position with his equals in age, who felt a certain ill-will towards him, nicknaming him "the young Dauphin."

Still more unfortunate was the effect on the young man himself of these lightly won advancements. Although he did not lack a good disposition and possessed an uncommonly fine memory, he found it difficult to apply himself to serious work. For a short time he would do well, and seriously attack scientific problems, but his ardour soon slackened. His prepossessing appearance and polite manners made him welcome in society, and it was difficult to drag him from it. In the letter of a contemporary it is written: "The young gentleman enjoys himself every day, enquiring less after Flora than after the Nymphs; he has a proud gait, dresses and powders in the fashion, and is a constant visitor where handsome ladies are." That his father did not shut his eyes to this, and still less his mother, is well known. Probably they hoped that a change would take place if he married well. The young elegant's pretensions in this respect were meanwhile not easy to satisfy; "difficulties," he said, "arise in securing a girl who has money, and is at the same time beautiful and amiable; I have not found these qualities in one person: I was near it with Archbishop Troil's daughter, who is a beauty, and we became fast friends, but her father's sudden death showed that there was not much to divide among ten children and a stepmother." Accordingly his father dissuaded him from this marriage, and suggested another lady, S. Asp, who was rich and pretty, but had not had smallpox, so he could not be certain of her future looks. Still another lady was suggested, but she was disfigured with pock-marks, the result being that he never found his ideal, and remained unmarried.

Linné's other son, Johan, born in 1754, only lived three years, and a daughter, Sara Lina, lived but a fortnight after her birth.

The four daughters who grew up were described by Fabricius as being "all quick, but raw children of Nature, without those fine manners with which their bringing up should have provided them." How far this is true cannot be determined, but it may be taken that the children's training could only be faulty, as the mother herself was rough and uneducated, so naturally she could not direct her children wisely. Opportunities of acquiring book learning and society's polite ways were few in so small a town, consequently Linné's wish was that they should grow up hearty, strong housekeepers, and not as fashionable dolls. So when poor students, as often happened, were invited to dinner in the Linnean home, it was with the distinct proviso that they should not lend the girls any books by which they might learn French or any other useless accomplishments. This went so far, that Linné himself prevented his wife when she did her best to provide more tuition for the young girls. It is said that once when he was in Stockholm, the mother placed the youngest, Sophia, in a school, but when the father came home and missed his favourite he betook himself at once to the school, and begged the teacher to grant an hour's leave. This could not be denied to the Archiater and Knight, and he thus enjoyed the society of his girl. The same thing happening the next day, it was repeated day after day.

Towards his daughters he was very tender, almost too weak, allowing them to do whatever they liked. Great was his trouble when Sara Christina, aged three, was so ill that one day they despaired of life. Two days after, however, he wrote to Bäck " I have received my daughter up out of the grave, when both feet were in it, up to the knees. She is now out of all danger with this remittent fever." Thus he retained his beloved child, who in her liveliness and gaiety, excelled all of her own age.

But the especial favourite seems to have been the youngest, Sophia, possibly from the time of her birth. In 1757 he wrote to Bäck, " On Tuesday evening my wife was delivered of a daughter after severe labour. The girl was apparently stillborn, but we used artificial respiration and after a quarter of an hour she showed signs of life, and is now tolerably well. My wife is still weak. God help her."

Sophia became his special darling. When she grew older, he often took her to his lectures, where she remained all the time between his knees; sometimes her head, neck, and arms being bare, that she should not catch cold he tied his handkerchief round her neck. She was also protected by him from the mother's roughness. Once when going upstairs with a pile of crockery she chanced to fall down and break all the pieces. In her distress she ran to her father, who bade her not to be sorry at the accident, himself going out and buying new porcelain. It can easily be imagined that Madame Linné opened her eyes at dinner time, and suspected that Sophia had " done it again," but was assured that the old porcelain was so ugly, he had broken it and bought more in its place.

His eldest daughter, Elizabeth Christina, was married on the 24th June, 1764, to Carl Fr. Bergencrantz, a lieutenant in the Upland Regiment. The marriage was unfortunate, owing to the brutality and laxity of her husband, so the wife escaped to her parents with her little girl. As Linné's daughters were now grown up, the granddaughter, as the smallest, became the object of Linné's special regard.

Other relations visited Uppsala, as for instance Samuel Linnæus in 1741, and later, his very poor and sickly nephew S. N. Höök, who was often at Hammarby, dying there in 1773, as the result of drinking ice-cold table beer when heated by dancing. Linné also kept up a correspondence with his brother and sisters in Småland, as well as with his mother-in-law at Falun. The long visits which he made with his family at Christmas in 1743, 1752 and 1755 to Falun, and his mother-in-law's journeys to Uppsala, show the affectionate relations between them.

Mention should be made of the animals he had at various times, such as the monkey "Grinn" which was a present from the Queen, and a weasel, provided with a bell for its neck, which had its lair among the rocks at Hammarby, and hunted the rats; he even kept crickets in the bake-house, which sang him to sleep at night, to the no small disgust of his wife, who could not make out whence they came, and spared no trouble to try and banish them.

His dogs were special favourites, especially a big one named "Pompey," who was so attached to him as to follow him everywhere, even to church. When Linné was living at Hammarby, he went to Danmark church on Sundays, when he used to stop and rest a while on a certain big stone, to smoke a pipe, which on resuming his walk he hid under a bush, to take up again on the way home. Linné stayed in church about an hour, but when he thought the sermon was too long, he went out, followed by his dog This became so constant a habit of his, that the dog would go alone the same way, stop by the same stone, go into church, and then seat himself quietly on the Hammarby bench; this when Linné from some cause was prevented from going himself. The parson noticed this, and when he complained to Linné, he was answered jokingly, that he could see that his sermons were altogether too long, when even a dumb animal went out.

Linné's economic position must now be mentioned. A certain pecuniary advantage followed his removal from Stockholm to Uppsala, though on the other hand, he lost his position at the Admiralty, and his medical practice. The professorial salary was not available, as though his predecessor had resigned, he had retained the entire proceeds, which Linné would only have after Roberg's death. Until this took place, he had only a 100 ducats or 600 dalers in copper [£15] from the Mining College.

It was, however, not long before Roberg died, on the 21st May, 1742. Thereupon Linné informed the Consistory that the Mining College had intimated that as soon as he was in receipt of the professorial salary, their contribution would cease. The accounts of the University show that at all events during part of 1743, the Professor's salary was paid to Linné. This was made up of the first sum of 700 silver dalers, the value of the Prebendal farm of Törneby in Vaksala, reckoned at 53 dalers 20 öre, partly from the Royal bounty of 50, afterwards 100 barrels of grain, for which the Consistory each year fixed the money equivalent. Of this account the salary during various years differed considerably, the most being in 1773, 2,903 [£217 11s. 6d.], the lowest in 1751, 1,053 [£79]. The average for 1745-75 was 1,780 dalers [£133 10s.], small additions being received from the minor offices he held, such as Inspector of Stipends, 33 dalers [£2 9s. 6d.], and as a member of the Inspection of Monies, now the Finance Board, 75 dalers [£5 12s. 6d.]. Finally a part of income was receivable in kind, being official sales from the hop-garden, and rent-free house.

Additions to these lawfully determined amounts were receivable from examination fees in the medical faculty, also the fees as the promotor of the doctorate in medicine. Each one promoted was liable to pay down 600 dalers in copper [£15], and the surplus, after certain expenses were met, was the promoter's, later on to be divided between the professors of that faculty.

To these academic sources of income must be added the sums received for private and "most private" coaching lectures, from the office of President at disputations, which amounted to a fair sum, according to the ideas of the time. Linné's active authorship contributed also a not inconsiderable total. After his father-in-law's death, a goodly amount came to his wife, although the bulk was bequeathed to his mother-in-law, which afterwards came to the Linnean household. Consequently the Professor's income, especially during the last two decades, was decidedly comfortable, whilst the amount of the savings effected by his wife increased.

A further increment came from investments; thus in 1746 Linné paid 6,000 copper dalers [£150], (probably the bequest from his father-in-law) to Nils Kyronius for certain fields outside Svartbäckstullen, which he, or rather his wife, made use of for her farming, and several such transactions are on record.

Far more important was the purchase made in 1758, "of a little residence near Uppsala for 40,000 dalers [£3,000] with five farms and a village costing 40,000 more; 200 barrels of grain was the produce of the farms, "but four men in war time will suffice for their cultivation." Both estates, Hammarby and Säfja, were in the parish of Danmark near Uppsala. In March, 1759, he acquired by purchase from Professor John Ihre, the neighbouring estate of Edeby. In consequence of these transactions, he had to borrow in 1759, 40,000 copper dalers [£1,000] of the Consistory with his chair and the rents for security.

The purchase of these landed estates was advantageous as producing even now a larger income than before, but it was mainly due to his declining health, and his wish to provide a certain sum for his wife and children, if he should soon be called away. In letters to his friends he bewailed the small income, and added: "It was an unlucky hour when I obtained the

LINNÉ'S HAMMARBY IN 1909, AFTER BEING RESTORED TO ITS FORMER STATE.

professorate; if I had kept to the golden practice, I should have well provided for my family."

It was under these circumstances that in 1761 he gladly received a letter from Colonel Baron C. Funck, President of the State Committee on Economy and Commerce, in which he was informed that the pearl fisheries of the kingdom were under consideration. In his reply Linné stated that he had heard of people who made gold, but had never heard of any who were able to make pearls, but he knew the art, and could readily impart the simple procedure. He had busied himself with the problem since his visit to Purkijaur during his Lapland journey, though he had no opportunity of experimenting with the true pearl mussels, only using lake mussels, but even with these, he had produced beautiful pearls. In 1748 he had communicated with Haller and Harleman, but the latter had cooled his ardour by saying he must not expect help from any future government in the production of pearls, for as soon as the secret art became known, as no reliance could be placed on the officials, they would fall in value.

The Committee before mentioned not taking the same view, called Linné to a meeting in the following July, when under promise of secrecy, he described his procedure. It was to bore a small hole in the shell, and introduce a small round object of plaster of Paris or uncalcined limestone, attached to a silver wire and then fastened to the shell. Round the introduced body the pearl substance was deposited, and after five or six years, one had true pearls as large as peas. Finally he showed five mussel shells thus treated, and also nine pearls thus produced. These had been tested by a jeweller who found them to be quite beautiful, and of especial quality. The Committee reported, and recommended a national reward to Linné of 12,000 dalers in silver [£900]. A sympathetic reception was accorded to this report, less with regard to the economic gain, than for the value as a natural history fact, deserving of honour and national reward. The Committee determined to find out how the money could be raised, but wrote down the reward to one half [£450].

One person who played a not unimportant part in these negotiations, though he was not on the Committee, was Linné's old friend, Bishop Mennander, who put forward the catalogue of Linné's merits, and did his best to advance the project. But Linné's concern was not for himself but for his son's future, hoping he might obtain the chair in the University after his own death, and urging that nobody but his son could in future better take care of the botanic garden, now in such condition, that it could be valued at a "barrel of gold" [£1,400]. Mennander then told him in confidence that a reward was being discussed. Meanwhile negotiations were opened with a member of the Pearl Committee, P Bagge, a merchant at Gothenburg, with a view to his advancing the said sum, promising that he should enjoy the monopoly of artificially made pearls, except with the obligation of sharing half of the proceeds with the Crown.

The next day Mennander related to Linné how matters stood, the latter being delighted. "If I should get that, there would be nobody happier than I; free from all debt, my children well to do, what can mortal man wish for more?" The proposal passed three of the Estates, but the House of Nobles postponed it to see if the manufacture actually took place. Confirmed by the King, however, a national reward was made to Linné, which had not cost the Committee a farthing!

The right to nominate his successor in the botanic garden which Linné had gained, he used at the close of 1762. He wrote privately to the Chancellor, Höpken, "that he felt a desire for rest after his heavy academic difficulties." The Chancellor agreed that the request was reasonable, but put two considerations before him: (1) that Linné during his period of rest, should work for science and the national credit, and (2) he should during his lifetime see that his successor did not depart from the methods introduced by Linné, and upon which he built, so that science should not suffer loss for several generations. In a later letter, he added, that if Linné continued his tasks, he advised that the Archiater should not stay his important instruction, for the times were troublesome, and conspiracies and intrigues everywhere. This advice Linné hastened to follow and informed the Chancellor in an official memorial, that as he had regarded science more than his life, he had worn out his body, shortened his days, and brought on too soon the infirmity of old age. On account of this, he begged that his son, Demonstrator Carl von Linné, whom he had reared from childhood as his successor, though with the proviso that so long as his own powers lasted for academic affairs, he should continue his professorial labours, to give his son time to improve himself in his duties, to travel, and work under his own supervision, and he hoped that the science which he had worked so hard for, should also, in his time, attain widespread prominence. This letter was submitted to the King by the Chancellor, the Royal intimation expressing a hope that the Archiater would continue his duties for a time, but as he could not escape the application, the appointment of the younger Linné as professor was ordained.

This appointment, whereby a young man, who had only just completed his twenty-second year, and who had not undergone any academic examination, was made professor, could not fail to rouse in high measure both remark and disapprobation, especially among Linné's pupils. Complaints were therefore made, and with some reason. But it must not be forgotten that Linné for his method of producing pearls had never enjoyed both pecuniary reward and the right to name his successor, but declared himself that the plan of his son succeeding him might be surrendered. The complaint should therefore rather be directed against the Secret Committee, and the Estates. Further, it may be noted that a similar right, without observing the legal forms for the creation of professor at that time had been granted to others, as to Rosén von Rosenstein, and Strömer, whose merits were less than those of Linné. Consequently, it was excusable that Linné, in his paternal love, thought that his precious botanic garden would be prosperous, if its maintenance were once given to his son, who had better opportunity than anyone else of developing into an eminent botanist. "If I live three years longer," he wrote to Mennander in 1761, "I am sure that nobody in Europe will discharge his duties better than my son." When it came to a decision, he may have had some doubts, as he wrote to his pupil Solander, then living in London, inviting him to be his successor. The latter in July, 1762, answered that he would be back in Sweden in October; he did not come, however, writing in November that the authorities in London had promised him a salary in the British Museum, so Linné reverted to his original plan for his son, for whom he promised he would be responsible.

Thus no change took place in the title of the younger Linné in his professorship. His salary remained that of a Demonstrator, for the father was assured that whether he resigned or not, his salary, with its accustomed privileges, would be continued. Of the permanent release from duty which was open to him, he never availed himself, except when ill, or on some other piece of work. For more than ten years he continued with increased, rather than diminished, fervour to devote himself to all of a professor's labours. Few of his pupils would have been willing to accept the position, with the title, but without salary, and without prospect of improvement. When the younger Linné was thirty-six, he first received from the King formal permission to discharge the office to which he had been appointed fourteen years before, and it was two years before he received the appropriate salary.

Of the various estates he possessed, Hammarby pleased Linné the most, and it was there that he preferred to live. The former possessor inhabited a little wooden house of one story, and here Linné lived with his family until, in 1762, he built the larger central dwelling, simply but comfortably furnished. The best rooms being on the first floor, he used them, for there was the drawing-room, with paintings of plants from the tropics on the walls, his bedroom being adorned in the same fashion. Above the door of his bedroom was inscribed "Innocue vivito, Numen adest!" On a big boulder behind the house, he had engraved in runic letters, "Riddar Karl Linné köpte Hammarby-Säfja 1758" [The Knight Carl Linné bought Hammarby and Säfja in 1758].

Thus Linné obtained a convenient and quiet home in the country to which he could go during the holidays, away from the unhealthy part of Uppsala where his official house was situated, and where epidemics, such as the so-called Uppsala fever, often raged. No further building was intended, before an extensive fire in 1766 in Uppsala threatened to destroy his house. In haste his collections and books were taken out and stored in a barn outside the town. Writing to Bäck, he said, "Our Lord was gracious and preserved me this time. Actually one-third of the town was burnt down. I removed all my possessions to a barn outside the town and then to the country, where they now lie in the utmost confusion." Fear for a similar fate caused him in 1768 to build a museum on an eminence on his property, having the most splendid view. Here he had his herbarium, zoophytes, shells, insects and minerals, and thither all curious people came to see them. This "little backroom," "pleasure house on my hill," "my castle which I built in the air," as he called it in varying phrases, had truly, according to our present-day ideas, almost laughably small dimensions; but was nevertheless his delight and pride. Among the visitors was Lord Baltimore, who came in a great carriage he had brought from England and for which, in order to get to Hammarby in it, all the gate-posts in the byways had to be removed. Another tale is that Linné, asking why he had not stopped in Stockholm to see the King, he was answered, that he did not care to see him, as he had never even seen his own King [George III.] The size of the museum is given as a square, 4·78 m. [15 ft. 8 in.] and 2·75 m. high [9 ft. 1 in.] having three windows, but no fireplace for fear of fire. This produced its own dangers, as will be shown hereafter. Here he worked in summer and held his private lectures, unless the beauty of the weather induced him to move the lecture-chair outside. His pupils sat on benches (yet preserved), unless they preferred to throw themselves down upon the grass, or on the rocks around.

During Linné's possession, the estate naturally underwent a complete change. There was an old garden, but this, through neglect, became overgrown and tangled, so in this wild forest grove he planted rare plants without special order. This grove, or as the grandchildren called it, mother's father's bower, was a square place surrounded by trees, not far from Hammarby, towards the near-lying Hubby driving road to the left, as one comes from Hammarby. Here in summer time was placed the dinner table, when the weather permitted; here Linné sat and smoked his pipe, and listened, when the wind stirred, to the music of the glass bells hung upon the branches of the trees. How dear this place was to him appears from the following direction to his wife on a paper left behind him: "Keep my grove which I planted during your time, and if a tree dies, plant another in its place."

MUSEUM AT HAMMARBY, BUILT IN 1768.

In another contrivance, meant exclusively for scientific purposes, one may yet see traces on the slope consisting of rocky ground and open towards the south-west below the museum, he naming it " Siberia." Here he planted preferably plants from Russia, seeds being sent to him by the Czarina. In May, even yet, it is glorious from the yellow flowered *Corydalis nobilis*. Linné became more and more attached to the place, and the quiet countryside, and he hastened " to my rusticity " as he called it, as soon as the holidays began, using it even in winter, though the temperature of the unheated museum did not allow him to stay long among his collections and books. Not less did his daughters enjoy the fresh life in free nature, in the cherry and plum orchards, where each of them had her own trees. Here particular friends were often invited, and there too the silver wedding of the parents was celebrated on the 9th July, 1764.

After Linné's death, Hammarby belonged to his widow who lived there till her own death in 1806. It passed then to the youngest daughter, Sophia, and her husband, Proctor Chr. Duse, then to their daughter, Johanna Elizabeth Sophia, married to Chief Director F M. Ridderbjelke, and next to their son, Master of the Chase, Carl Ridderbjelke. In the Riksdag of 1844, a motion by Baron J. G. von Paykull that the Linnean estate should be acquired by the State as a memorial to the great naturalist, was brought forward, but eventually thrown out. The question was again raised by F Asker in 1878, which led to an appeal to the King; later, both Houses decided without a vote that the whole property should be bought for 30,000 kronor [about £1,656]. The rest of the property, not then bought, was acquired by the Uppsala University.

After the purchase was concluded, certain work had to be carried out, partly to preserve the buildings, partly for their restoration to their former condition. This was specially needed as regards the wings; the western one, mainly built before Linné's time, had a stairway at each end, and was then in a most lamentable state; in place of the eastern wing, low and turf-thatched, which was falling, had been erected a new two-storied house, which had become the principal building on the estate. All this was altered in consonance with the old drawings and trustworthy traditions. With regard to planting, an inexorable rule was made that not a single plant should be allowed which was not cultivated at Hammarby or in the Uppsala garden in Linné's time. It was seen that many of the tall grown elms, ash trees, and maples round the buildings, must be thinned, as they had in later times grown up where formerly fruit trees had flourished.

This description gives a very good representation of how the place looked in Linné's time. In front of the main building were two horse chestnuts (the last one being blown down in 1907) under which he used to smoke his evening pipe, and there was also a Siberian crab, crooked with age. On the gable of the western wing was the "porridge" bell, which used in his time to summon the labourers to their meals, and round about is a luxuriant carpet of *Aquilegia*, *Myrrhis*, *Mercurialis perennis*, *Tulipa silvestris*, *Lilium Martagon*, *Epimedium alpinum*, *Crepis sibirica*, *Asarum europæum*, *Corydalis nobilis*, *Campanula latifolia*, *Galanthus nivalis*, *Leucoium vernum*, and other plants, persisting from his time. On entering the main building there are to be seen articles of furniture, portraits, and other objects, which once belonged to him. In the dining-room there is still to be seen an unwieldy dinner table, a simple yellow-painted cupboard, and the same clock that Linné used still records the passage of time. In a room on the first floor, is a bed with bedclothes, on which he died. There is kept his big peculiar inkpot, his everyday and Sunday sticks, his leather-covered favourite chair, his bed with hangings of Chinese stuff, pieces of Chinese porcelain with trails of *Linnæa*, his doctor's hat and red velvet skull-cap, and much besides. Turning into the narrow path which threads among the boulders to the museum, rosettes of *Sempervivum globiferum* on the mossy stones are found, with scions of a Finland beam-tree, *Sorbus fennica*, dead fifty years before. On the little museum building may be seen in the corners of the walls, inserted porcelain plates on which the Linnean coat-of-arms are burnt, many of which, some decades ago, being loosened from the walls and stolen. Inside the museum are now only a few empty cupboards, a low bookcase with volumes from Linné's library at the Academy of Science, 260 of which were returned to Sweden in 1894, from the Linnean Society of London, as not pertaining to natural history, and from the roof is suspended a dried fish, *Regalecus glesne*, probably sent by Bishop N. C. Friis from Trondhjem. The whole imparts a feeling of veneration. One feels treading on classic ground, and can only with astonishment reflect that from this little insignificant spot, light was once spread over the whole field of natural research; that it was hither from all parts of the world that students came to hear from the master's own lips words of wisdom, to be cherished by them all their lives.

CHAPTER XVI

LINNÉ'S LAST YEARS AND DEATH—HIS SCIENTIFIC REMAINS—HIS SCIENTIFIC IMPORTANCE

The colossal activity which Linné exerted in various directions, seems, after the event, to have been due to an iron constitution, which endowed him with bodily strength and mental buoyancy. It is unquestionable that though on some days he could not work because of slight ailments from errors in diet, colds or mental shocks, often causing megrim (which he called "my old comrade"), yet up to his sixtieth year he could rejoice at having, as a whole, enjoyed good health. Even during that period, he had to suffer several severe attacks of illness, as shown by his letters to Bäck, whose skill as a physician often stood him in good stead. Thus in 1746 and 1750 he was laid aside by angina, which nearly suffocated him, in the latter year being followed by gout. For this ailment he found that wild strawberries were curative, and every year afterwards he ate as many as his stomach would bear, to his entire relief from that excruciating disorder. In June, 1751, when in bed, and refusing relief by the then prevalent bleeding, he was restored to health by the return of Kalm from America, bringing with him ample collections. He rose from his bed, and forgot his troubles. On another occasion, the prescribed remedies not availing, he cured himself with *Cinchona* bark. In 1752, he had a chest complaint, and in 1753 in consequence of constant writing—his "Species plantarum" appearing that year—he suffered from pain in the right side, but his cure came again from strawberries. After a few uneventful years as regards illness, he had symptoms of scurvy in 1756, and gave up coffee for a whole month. During the following years, in 1764 he had pleurisy; in 1767, he was very weak for six months, an attack of Uppsala fever following. In 1770, and each year after, he was not without some sickness. At the beginning of 1772, he informed Bäck that his end was approaching, apprehending apoplexy, as his head swam when he bent forward. Bäck prescribed special diet, rest from lecturing, lessened work and bleeding. Linné promised to follow this advice in part. After some rest, he declared that giving up lecturing was impossible, as it made him forget things, rendering him dumb, and had he continued so another term, he might have forgotten his own name.

In the early part of 1773, Linné suffered from an angina, and later from sciatica, "from the hip to the knee." May 1774, when he was lecturing in private, he had what he called his first "messenger of death," a stroke of apoplexy, so that he could not raise himself from his chair, move himself, nor hold up his head; he gradually improved, though slowly. Therefore in 1775, he asked for release from lecturing, "because I, old and tottering, can hardly bear the autumn cold, and am toothless, so it is hard to talk."

The early months of 1776 showed gradual but continuous deterioration. "Linné limps, can hardly walk, talks confusedly, can scarcely write," is an entry from one of his autobiographies. Three pupils came from Denmark and Hamburg. "But Linné is so sick, that he can scarcely speak to them, for he also has had tertian in addition to his lameness and weakness." And with this his notes close.

The increasing weakness of old age and sickness necessarily prevented him from fulfilling the duties of his chair; therefore in the spring of 1776, there was laid before the King a memorial, begging that his successor should enter upon his duties, but he wished to continue inspection of the botanic garden, and so far as his strength permitted, to take part in the Consistory and the Medical Faculty. To this appeal was added another request. Through the death of Rosén von Rosenstein, the position of an emeritus professor had become vacant. Linné asked for this position and salary in place of the ordinary professor's, and the other emoluments he had hitherto enjoyed. He ventured to hope for a gracious reception of this request, which was not unusual in other countries, and had been granted to a few Swedes, among whom might be named N. Rosén von Rosenstein, " who, besides his salary as Archiater, was granted that of an emeritus professor."

The answer was given in a letter from the King to the Chancellor, that he was unwilling that so distinguished a professor should lay aside any part of his functions; and he wished that Linné should receive the vacant position of emeritus professor, and that the Consistory should consider how far his income could be increased, without prejudice to the University.

Linné thus had obtained only part of his request, but that in so flattering a manner, that he felt himself obliged to follow His Majesty's wishes with thankfulness. In a letter to the Chancellor, he therefore recalled his resignation, and a copy was sent to the King.

The Consistory at once took steps to augment Linné's privileges, and recalling that the University's farm " Hubby " in the parish of Danmark, was close to the Archiater's country seat of Hammarby, thought it would be a great convenience for him to possess it. For this reason the Consistory resolved that the Archiater should have the right of possession, with the stipulation that the same rent as before should be paid. Linné accepted the proposal with thanks

CARL VON LINNE

(Portrait by A. Roslin, 1775, from the oil-painting at the Royal Academy of Science, Stockholm).

for the kindness of the Consistory, and it was confirmed by the King.

The economic gain which thus accrued to Linné and his family was of small importance. Only one and a half year's doubled salary was received by him before he died, and as regards the rent from Hubby, the return cannot be considered much, as it was the same which was paid by the former tenant. It is now impossible to determine how much Linné hoped for when accepting the Consistory's offer.

His health soon became lamentable. Even in May his appearance was much altered by his wasting away, and his powers of thought so weakened that his last letter to Bäck, 24th May, 1776, was practically unintelligible. His memory was so muddled, that at times he could not remember letters, but when writing, mingled Latin and Greek characters together.

A short period of improvement followed, and then he had the pleasure of a visit from the President of the Academy of Science, J. L. Odhelius, who was commissioned to acquaint himself as to his condition, and show him the tenderness which so great a man deserved. Odhelius found him actually greatly weakened in health, but not so ill as reported, and especially in good spirits and livelier than could be expected; but soon afterwards he had another apoplectic seizure, causing paralysis of the right side, and such diminished power of speech, that he could only utter words of one syllable. "Although," related J. G. Acrel, "he is now in that condition as more dead than alive, and can hardly talk, one noticed a special gladness in his face when he saw any of his pupils, or when the talk turned on natural history." This is confirmed by a letter from A. Sparrman to Wargentin: "He ventured to go a few steps from his chair without help, but with extreme difficulty. If anyone takes him into the garden, he is pleased to look at the plants, but cannot recognize any. He laughs at almost everything, but sometimes weeps, can speak only three or four words, but listens to all." And A. Afzelius adds: "All his limbs and organs, the tongue especially, the lower extremities, and his bladder, were paralyzed. His speech was unconnected, and sometimes unintelligible. Without the help of others, he cannot stir from the place where he sits or lies, cannot dress himself, eat, or carry out the least thing he wishes. Of his organic life, only his respiration, digestion, and circulation are yet in tolerably good state. Everything else is more or less destroyed. He had forgotten his own name, and mostly seems to be unconscious of both absence and presence. For a few short periods here and there, his power of thought returned, as when he found lying near him some books of botanical or zoological contents, even his own, of which he would turn the leaves with evident pleasure, and let it be understood that he would think himself happy if he could have been the author of such useful works."

The summer of 1777, which he spent at Hammarby, seems to have brought some improvement. He was carried out every day when the weather permitted, either in the garden or to his museum, where he would enjoy himself for hours together with the sight of the treasures there, to his great satisfaction, and was carried back again. He came back in the autumn to Uppsala, with better health, so that he could walk several steps supported by another person, and smoke his pipe with enjoyment. By the physician's orders, during fine weather, he drove out to obtain fresh air, but the coachman was strictly forbidden to drive outside the town. Once in December when sleighing, he ordered the servant to drive him the three miles to Säfja, and the man thought he was bound to obey his master's repeated orders. When the accustomed time of his return home passed, the family became very uneasy, and sent out messengers in every direction to seek for him. He was at last found at Säfja, where he had had the little sleigh taken into the kitchen, and there he lay in front of a blazing fire, quite happy with his little pipe in his mouth. Here he was ready to pass the night, and much trouble was experienced in driving him home safely, as it was already dark, and a thaw with steady rain had set in. This was the last time he passed out of the town's gateways, and very few times afterwards did he go outside his own house.

Shortly afterwards, his strength visibly waned, and his pains increased. The only thing which gave him any relief was beer, which he drank with such pleasure, that he did not take his mouth from the tankard, so long as a drop was left. On the 30th December he had a terrible attack of convulsions, so that each breath seemed as if it would be his last; but his wasted body still had so much resistance, that death only came on the 10th January, 1778, at eight in the morning, and freed him from his suffering; the actual cause of death being ulceration of the bladder. At his death-bed were only the University Proctor his son-in-law elect, Samuel Duse, betrothed to his youngest daughter, and his English pupil, John Rotheram.

During the last year of Linné's life, dark shadows had rested on his home, becoming darker owing to the circumstances then prevalent. An unhappy contributory reason, according to many unanimous reports, was his wife's frugality, which in later days degenerated into avarice. She was particularly blamed, for, without regard to her husband's grievous condition, she did not prevent him from giving coaching lectures, which, however, no one could understand. In the autumn term of 1776 he attempted to act as dean and examiner in the medical faculty, etc., and this only because of the insignificant pecuniary advantages which could thereby be gained. Confirmation of this accusation is strengthened by an event happening at the end of 1776, which arouses in a high degree both astonishment and compassion. When at the meeting of the Consistory on the 13th December of that year, the members of the Inspection of Finance were to be chosen, the Secretary stated that Linné, whose turn it was to serve, declared that he was not willing to decline the trouble. As his broken health was well known, the Rector and Treasurer were requested to call upon him to obtain closer knowledge of his condition.

At the next meeting, 23rd December, the Rector reported that they had found Linné so feeble, that he certainly could not fill so responsible an office. Discussion ensued as to what should be done, and after much doubt, the Consistory decided to remit the matter to the Chancellor for his decision.

Before the Chancellor had this put before him, he had received a memorial, written by the younger Linné, but signed by the father, declaring himself ready for the Finance duties, but if prevented by ill-health from discharging them, he requested that Professor Berch might be entrusted with them. No official letter came from the Chancellor, but a private note was sent to the Rector, expressing his wish that the matter should be settled amicably; so Professor Berch was named with the Archiater to fulfil the duties. This arrangement also affected the disposal of the fees, which did not amount to more than 12 riksdalers 24 skillings [18s. 9d.]. It was no doubt the son who had raised this question, for his father was now so broken down, that he could not even intelligibly write a letter to his best friend, whose name had escaped him. Perhaps it was also due to the fact, that the Hubby rent came to Linné and his younger children, excluding his son and his wife, who bitterly complained of this arrangement.

But this was not the only complaint against her; it was reported that she did not devote due care and attention to her invalid husband, who when he tried to rise from his easy-chair, fell down, and remained lying on the floor. "She neglected to assist him, entirely forgetting that it was he who had given lustre to the family name. But unquestionably it was also deplorable, that between mother and son should rage an enmity of such unnatural intensity. This was so notorious, that the younger naturalists talked about it in very contemptuous expressions, compassionately lamenting that " she should persecute and hate him so much." But the son did not escape blame, as he made use of his father's feebleness to advance his own interests, by requesting the King to appoint him ordinary professor. This being granted, on the 27th October, 1777, the younger Linné was inducted with the customary ceremonies. But as this request was not, as customary, first laid before the Chancellor, it is not surprising that it aroused his displeasure. Also in another quarter the son's intrigues attracted unpleasant comment among the community, and especially among the younger naturalists, who regarded the younger Linné as a " lazy loon in a superlative degree," and by no means worthy to take his father's professorial chair. He himself realizing his incapacity, wrote that he " wished to be separated from the whole concern." " Wretched boy," and " He seems to see that he is out of his depth," etc., were current expressions among his detractors. It was hoped that J. A. Murray would leave Göttingen and become professor at Uppsala, but his brother had a letter from him saying that he was not desirous of the change.

Several years before his death, Linné had in writing arranged about his funeral and the observances. Shortly after he passed away, the envelope was broken, and it was found that he had laid down the following for his wife's guidance.

" 1. Put me in the coffin unshaved, unwashed, unclad, enveloped with a sheet; and close the coffin immediately, so that no one may see my wretchedness. 2. Let the great bell [of the Cathedral] be tolled, but not in any of the other churches, or the Peasant Church or Hospital, but do so in Danmark's Church. 3. Let a thanksgiving be held both in the Cathedral and Danmark's Church to God, who granted me so many years and blessings. 4. Let my countrymen carry me to the grave, and give them each a little medal, one of those with my portrait. 5. Entertain nobody at my funeral." In the summer of 1775, he had a big elm at Hammarby felled, with the remark that his coffin should be made of it. These instructions were carried out except number 5. His widow, against his wishes, provided on the day after the funeral, January 23rd, an ample dinner.

An account of the funeral itself on the 22nd January, in the Cathedral, is given by two who were present. A. Afzelius relates, "It was a dark and still evening, the darkness only dispersed by the torches and lanterns carried by the mourners, the slow progression of the procession whose silence was only disturbed by the murmur of the multitude of people assembled in the streets and the great bell's majestic heavy tolling, were the only sounds to be heard. This great bell, annoying to Linné in his lifetime, now for the first time was heard only as an unusual and impressive accompaniment in the procession to the grave of a distinguished man, forming an example for future important funerals. The entire University and a great company were in the procession, many doctors of medicine (all former pupils of Linné) bearing the dust of the great man." This may be completed by the following extract from a letter by J. Hallenberg to C. G. Gjörwell: "Archiater Linné's burial taking place about six o'clock in the evening, was lit up on the way to the church by lanterns, accompanying twenty-one carriages, and in the church by candles, with which all the candelabra were filled; besides these and the organ there was no other preparation in the church. I never saw so many people in the building as on this occasion. Dean Hydrén buried the late professor, and that in so striking a manner, that everyone wondered at his ability in so old a man; he then being eighty-three years of age. The Chancellor's Secretary,

Wallenstråle, carried the Order [of the Polar Star]. In the absence of other relatives besides the son, his countryman, Professor Floderus, Adjunct Hageman and I were the mourners, Professors Melander and Sidrén being staff-bearers. About twenty people of the procession remained to supper, and the day following, the whole Småland Nation was invited to dinner." The coffin was put down in the walled-in grave, at the north side under the organ loft, between the first and second pillars behind the women's benches, which grave Linné had so far back as 1745 bought of Councillor Nils Kyronius for 100 copper dalers [£2 10s.]; the grave was at once closed down. After the son's death in 1783, the widow had an inscription made, stating that Carl von Linné, father and son and herself here had their resting place. Linné had ordered that close to his grave, a bronze medallion from his museum should be placed, with "Princeps Botanicorum," and dates of birth and death. It seems to have been taken away at the time when the memorial of Älfdal porphyry, with a bronze medallion by Sergel, was erected in 1798, by "amici et discipuli" in the Banér Chapel adjoining.

In connection with the funeral, was the sending out of notifications, of which Linné had himself prepared two, the longer being chosen by the widow.

Linné's death was not unexpected, but when it took place, there was much grief in wide circles for the great loss which science and his country had sustained, found expression in many ways. The Academy of Science hastened to decide that a Memorial Oration should be given, and Archiater Bäck was invited to undertake it. This he did on the 5th December in the round saloon in the old castle, before the King and a numerous auditory. The Uppsala students decided to have a bronze medal struck, but postponed it until Sergel should return to Sweden; perhaps it resulted in the plaque now on his memorial.

The King of Sweden, Gustaf III., made himself

the mouthpiece of the nation, when from the throne in the Estates in October, 1778, he specially referred to Linné with the words: "I have lost a man who did honour to his fatherland as a worthy citizen, being celebrated all over the world. Long will Uppsala recall the reputation which Linné's name conferred upon it." Nine years later, on the 17th August, 1787, he laid in the new botanic garden, given by the University, the foundation stone of the building of the institution, which he declared should be erected to honour Linné's memory. The King marked out the place where afterwards a marble statue by J. N. Byström was erected in 1822, by the students. But what did the Consistory do to express their feelings of gratitude and respect to their departed colleague? There was no talk about any "Parentation" which had often been held at the deaths of other eminent professors, till March, 1779, when the question was raised in the medical faculty. Meanwhile a Memorial Oration had taken place in the Academy of Science, and it was stated that the public would think it strange if the University gave one after that. For this reason the faculty decided that the best memorial would be a marble bust, but ultimately nothing was done.

After the funeral came the usual business of making an inventory, bestowing the bequests and the like. It may briefly be said that though no great wealth was willed by Linné some family bickering ensued. Thus Captain Bergencrantz, who married his eldest daughter, threatened legal process, and demanded that his wife, from whom he had lived apart for many years, should, against her will, be compelled to resume cohabitation with him. Upon this the widow of Linné petitioned the King, that as her late husband had left but little wealth, she and her unprovided-for children might receive an emeritus professor's salary and an annual pension; and as this petition was supported by the Chancellor and Consistory, the King decreed that she and her children should receive an annual pension of 200 riksdalers [£45 16s. 6d.].

The question of Linné's scientific effects, library, and collections, must be considered at some length. It must be first noted that in a document which he had tendered to the Chancellor in 1759, recommending his son's appointment as Demonstrater, he urged as a reason that as the holder of the aforesaid position, he could consult daily the valuable botanic library and collection of plants, which he himself possessed, and of which his son would eventually become the possessor.

This view of the future fate of the library and collections he seems to have abandoned soon after, probably because the son would thus have been unduly favoured at the expense of the daughters. In a will prepared on the 17th July, 1769, and witnessed by his wife, he provided that his son should receive only the library, with a simple share in the house and effects, the reason being given that as his son was to succeed him in the chair, he would thereby be benefited more than his sisters altogether; he therefore ordered that his herbarium, the largest the world had ever seen, should be sold on account of the daughters, and hoped that the University would acquire it, as it might never again have the chance of possessing a similar collection. The other parts, such as the shells, insects and minerals his son was to possess. This will was renewed on the 20th August, 1776, with a few trifling alterations; it being in 1778 produced in the Svea Court, and, in October of that year, recognized by the son.

The provisions of Linné as to his scientific belongings are summarized in a document he left behind him thus:

"*Voice from the grave to her who was my dear wife.*

"1. The two herbaria in the museum: let no rats or moth injure them. Let no naturalist steal a single plant. Be firm and careful as to whom they are shown.

Invaluable as they are, they will increase in value as time goes on. They are the greatest [collection] the world has ever seen. Do not sell them for less than 1,000 ducats [£458 6s. 8d.]. My son should not have them, as he never helped me in botany, and has no love for it, but keep it for some son-in-law who may be a botanist.

"2. The shell cabinet is valued at least at 12,000 dalers [£300].

"3. The insect cabinet cannot be kept long, because of moth.

"4. Mineral cabinet contains some valuable things.

"5. Library in my museum with all my books. The price is at least 3,000 copper dalers [£75]. Do not sell it, but give it to the Uppsala library. But my library in Uppsala, my son should have at a valuation.

"CARL LINNÉ.

"UPPSALA,
"*2nd March*, 1776."

There is extant an earlier document in which he reckoned the values thus: Herbarium 50,000 dalers [£1,250], Insects 10,000 dalers [£250], Amphibia 10,000 dalers [£250], Minerals 10,000 dalers [£250].

According to the will of 1776, the younger Linné came into possession of the great library, the natural history collections (the herbarium being excepted), and his share in the rest of the estate. To his credit it must be stated that he did all he could to obviate their sale, but at first with poor prospect of success. He wrote to Bäck, "If my mother and sisters were more reasonable and just, I could hope to prevent the sale, but they suspect everything, so soon as I want something." His mother locked up everything she could, to prevent his access to such things as manuscripts, etc. Happily the Court of Justice appointed Professors J Floderus and E. Ekman as trustees for the unmarried daughters, and they saw that the sale of the herbarium would discredit the family. The remaining brother-in-law, Duse, who had the confidence of the women, held the same view, namely that the son should get the plants. In the end, an agreement was reached, that the younger Linné should have the herbarium, library, manuscripts, and the cabinets in which they were kept, but resign to his sisters his share in the Hubby property, by paying to them altogether 6,000 dalers in copper, or 333 riksdalers 16 skillings [£150]. Fortunately, this transaction was concluded before news came that Sir Joseph Banks, in London, was willing to buy the herbarium for 14 or 16,000 dalers in silver [£1,050 or £1,200], which "cruel offer" made the son as soon as possible remove the herbarium for safety into the town. It was also necessary, for the rats had caused terrible damage to the plants, also moth and mould had destroyed some, and to avoid further loss the younger Linné "laboured from morning till night so that by the evening he was as tired as a day labourer." The remaining objects were brought from Hammarby in 1780, from the damp stone house or museum, and during Christmas in the same year, he busied himself upon those "which the wood-mice had already begun to damage."

Thus Linné's collections remained in Sweden, and were not at this time sold to England, but fate had decided that they should be in the future. After the death of the younger Linné from an apoplectic stroke on the 1st November, 1783, they again came into the possession of his mother and sisters. Their wish was that the collections and library should realize the most that could be obtained for them. To this they were inclined, partly from necessity, or at least from the desirability of the sale being effected before the deceased's house was taken over by some other person, partly because after the elder Linné's death it was found that the collections without expert handling had suffered much damage. They therefore applied to the family's intimate friend, J. G. Acrel, who took upon himself the responsibility of the sale.

His first object was to apply to Sir Joseph Banks, who five years before had made so liberal an offer for the collections. The application was made by Acrel, who also requested Dr. L. Montin in Halmstad to write to his sister's son, J. Dryander, to help them in this matter, which accordingly was done. Before any answer came from London, Acrel turned to Dr. J. H. Engelhardt, then in London, with the request that he should put the matter direct to Banks. This was effected by a letter received by Banks on the morning of the 23rd of December, 1783, while he was at breakfast with friends in his hospitable house. Among the guests was the twenty-four-year-old ardent naturalist, James Edward Smith, son of a rich manufacturer in Norwich, and to him Banks handed Engelhardt's letter, saying that it was something suitable for him, and earnestly recommending him to make use of the opportunity to gain advantage and credit. Fired by enthusiasm, the young man hastened to Engelhardt, whose acquaintance he had made during their common studies in Edinburgh, the result of their talk being that both, the same day, communicated with Acrel. James Edward Smith wrote declaring that if, after receiving a full catalogue, he found that it corresponded with his expectations, he was willing to pay the sum of 1,000 guineas. The offer thus hurriedly made was somewhat rash, for to complete the purchase money he had offered, his father would have to provide the necessary funds. To him he wrote the next day, but the reply was indefinite and advised caution. Without being cast down, young Smith, supported by Banks, endeavoured again to persuade his father, and by the middle of January, had the pleasure of receiving a letter giving him freedom of action, but at the same time containing the advice: "But await calmly the answer to your letter to Dr. Acrel, till you see and examine the catalogue with care, and then determine as circumstances require, and I hope it will please Heaven to direct you for the best in a matter of so very great importance. I would caution you against the enthusiasm of a lover; or the heat of an ambitious man."

By the middle of February, Acrel's answer, accompanied by the desired catalogue, arrived, and a still later letter came a fortnight afterwards. In this it was related that the younger Linné had ordered that the collection of plants he possessed before his father's death, the so-called small herbarium, should be given up to Baron C. Alströmer in satisfaction of 200 riksdalers [£45 16s. 6d.] which the Baron had advanced for his travels abroad, and that therefore for this, a reduction of 100 guineas should be made. At the same time, Acrel felt himself bound to say, that the heirs, to avoid dividing the collections, considered that they ought to offer the whole to Alströmer, also that a rich Russian nobleman, acting on a commission of Catherine II., offered an unlimited sum, but would wait until the decision of Smith's reply. The latter, which was at once sent off, contained some dissatisfaction at the changes made, but definitely accepted the offer just received. In May, Acrel proposed his terms, namely, half of the purchase money, i.e., 450 guineas, to be paid at once, and the remainder to be defrayed three months later. After Smith had agreed to this, and stated that the money was already in the hands of a trading firm in Amsterdam, who would transmit the same without delay, the matter was completely settled. Acrel now began the packing up, which was pushed on with such ardour, that altogether twenty-six large cases were sent off from Uppsala to Stockholm, where they lay for safety in a warehouse. Here they remained for six weeks, for it was not till the 17th September that they left Stockholm on the English brig, *Appearance*, commanded by the Swedish captain, Axel Daniel Svederus, arriving in London at the end of October. The English Government waived the customs duty, except a trifling sum for the books. For freight, Smith paid £50, with five per cent. to the captain as his fee, the costs coming in all to £1,088 5s.

During the transactions just recounted, there appeared a new speculator, namely the English botanist, Dr. John Sibthorp, who first offered to buy only the herbarium, but afterwards, all the collections, if time were allowed for him to reach Uppsala. Although Acrel stated that he was bound by his preliminary dealings with Smith (who, at Acrel's request, immediately wrote to him that the purchase was definitely concluded), Sibthorp maintained his right to become the possessor of the collections, and in his first letter, written before the sale was effected, offered to purchase the whole. Afterwards, Sibthorp acknowledged the proper transaction, and congratulated Smith on his good fortune in securing the treasures.

But—what happened in Sweden at the time to try and prevent the sale, which even then was regarded as a stain on Sweden's honour? To this it must be replied, that no one was indifferent, and that though many wished to retain the collections, it could not be effected. Especially among the pupils of Linné, who devoted themselves to the study of Nature, was there the greatest dissatisfaction at the sale, considering it a national scandal.

The first who stirred in the matter was the then Demonstrator of Botany, C. P Thunberg. As soon as he knew that Acrel had moved Montin to write to Dryander, offering the collections to Banks, he hastened to incite Montin to write another letter, in which he should either dissuade Banks from buying or persuade him to leave them the English collections, acquired by the younger Linné from Banks and others. On the 17th November, Montin reported that he had carried out these instructions. Very probably it was this action of Thunberg that caused Banks to declare that he would not buy the Linnean collection, if a purchaser were found in Sweden, because he felt that they ought not to go out of the country, but if it must be to a foreigner, he was as willing as any other.

It is certain that Acrel would not have been unwilling for the collections to remain in the country, if only the heirs had shown some strong desire on that point. He applied, probably in January, 1784, to the Secretary of State, E. Schröderheim, asking him to invite the King, then in Italy, to buy them. However, it is probable that the King never received any such application.

Acrel applied yet to another person in Sweden, namely Baron C. Alströmer. In a letter dated 26th January, 1784, he represented that the natural historian's most enlightened Mæcenas in Sweden should either permit the bereaved family to retain the small herbarium, which Linné the younger had promised, or, as the family considered far more desirable, for Alströmer to buy the collection in their entirety, in order to prevent their falling into foreigners' hands and thus arousing constant jealousy. This appeal was refused by Alströmer, because, on account of his failing health, he could not make as much use of it as he otherwise might do. This was further emphasized by Alströmer's secretary, Student A. Dahl, that as the heirs would not accept 2,220 riksdalers [£462 10s.], that is, less than half the sum which Smith was ready to give, Alströmer would not on any account abate his claim to the small herbarium.

Alströmer was not content with this negative decision regarding the entire collections, but he sent Acrel's letter to the Chancellor, Count G. P Creutz, reiterating at the same time his view that the Linnean collections should be kept by the University. In consequence of this, the Chancellor's Secretary, J. E. Noreen, requested Thunberg to state his opinion as to the value of the collections, and explain how the University should purchase them. Acrel, in May of the same year, seems to have applied to the Chancellor direct, but when he mentioned the sum that Smith had declared his willingness to pay, the Chancellor found that he could not compete with this offer, so the University must regretfully dispense with them.

About this time (the date not being ascertainable), the previously mentioned Anders Dahl came forward with a proposal to buy the collections, the Gothenburg merchant, J. Mauhle, providing the means. He wanted to obtain them at his own estimate, and was especially solicitous that the Alströmers should know that before the transaction was completed. This much is certain, that Dahl declared he was empowered to pay as much as the foreigner offered, and that Acrel was simply his commissionaire, since the heirs, though only verbally, had assured him (Acrel) precedence in this transaction. On the other hand, Acrel distinctly denied this connection between the two, saying that nobody had made this statement till three years afterwards, and that only in a Gradual disputation. But however that may be, it must be conceded that Dahl more than any other Swede, was most active in opposing the sale of the collections overseas, whether his offer came too late or not.

His ardour in this subject continued to the last minute, so to speak, even when the collections had been sent off from Uppsala. In an undated document, probably at the end of September, he considered that the King should be graciously moved to reclaim these collections, not only while they were in Stockholm, but even while they were on board ship ready to start. As a reason for this, Dahl maintained that he had the assertion from the heirs before "Herr Smidt" had made his offer, that foreigners would always taunt the Swedes with their inability to retain such precious collections; that the possessor would become a Dictator in Science, and lovers of it would be obliged to impart their discoveries to him, in order to compare them with the Linnean cabinet, and that no one but himself could put in order the late Archiater's remarkable manuscripts. With no lack of self-approbation, as the above shows, he asserted that he had had the good fortune for many years, until the summons of death, to live in the house of the late Archiater, the great von Linné, assisting him in his scientific work.

What the King was able to do after this hint, appears from a certificate issued by the Stockholm Export Sea Customs Chamber. This paper, dated the 8th October, explains that the vessel with the collections passed Dalarö, the last customs post, on the 29th September, sailing seawards, so it was quite too late for anyone to do anything in the matter.

Now the people of Sweden at last awakened to the full meaning of the shame and damage which had befallen their country through the loss of the "State Jewel," and an investigation began endeavouring, by some means or other, to fasten the blame on someone. People turned against Linné's friends, Bäck and Mennander, because they did not, or at least, did not sufficiently realize the importance of the loss to the country while the negotiations for the sale were going on; against the Academic Consistory and the medical faculty in Uppsala, to whom the question was never referred, but who were supposed to have seen with pleasure an end put to the dominating influence on studious youths, which natural history had exercised during Linné's time. These charges, however cleverly constructed, have no evident grounds for support, but are only guesses on the part of the accusers.

With yet greater force have the charges of unpatriotic dealing been directed against two persons, Acrel and Thunberg. As regards the former, people did not hesitate to accuse him of having been bribed with the sum of £100, a ridiculous accusation to bring against a man who was defined by Smith as a true gentleman, and defended also by Linné's widow, who strongly and indignantly protested against the accusation; at the present day no one could entertain such a belief. It has even been thought by some that by causing the Linnean collections to be sent out of

Sweden, he calculated he would gain a better chance of obtaining the vacant professorship after the young Linné. It will, however, be seen, that if he counted on this Thunberg would be a formidable competitor for the position. He had brought home from his long and extensive travels, rich and valuable collections, which, remembering the University's total want of botanic collections, and their small zoological ones, would certainly have weighed heavily in the nomination to the position. Professor T M. Fries, after long and careful reflection, was convinced that Acrel in his negotiations was entirely desirous that the fullest advantage should follow from his trusteeship, which as the friend of the family for many years he had taken upon himself; this view agreeing with his trustworthy character, which was generally known and accepted. How he also made attempts to conserve the collection in the country, either in the University or in the possession of private persons, has been narrated previously, the only point not being clear, is, why he did not let Dahl buy them, but in this we have his word against the latter's, and one may be as valuable as the other. It was said by some that Dahl's offer being expressed in such vague terms, Acrel, and perhaps the heirs, found themselves obliged to prefer the certainty of the English offer, especially as they felt themselves bound by the contract practically completed. With regard to the unjust judgment upon Acrel when he wrote his final statement to the heirs that he demanded an unreasonably high commission of six per cent., with, in addition, credit for half of the disbursements, it is necessary to remember that no complaint was made by the heirs for his selfishness in pecuniary matters, so that we are entitled to believe that the scale of the commission was decided beforehand, and that the other disbursements were also settled and approved by the family. This is strengthened still further by the fact that he continued to be one of the family friends.

The other who was the object of bitter censure, was Thunberg, though it was acknowledged even by his accusers, that he was free from direct participation in the sale. But the burden was laid upon him, the accusations being chiefly based on his answer regarding the value of the collections and the necessity for the Chancellor's action. Here it may be noted, that no one knows whether anyone else was asked to buy them, whose word would have been better than that of the Demonstrator Thunberg, and probably the Chancellor did not abstain from buying only because of Thunberg's dissuasive advice, but was really frightened at the amount of the sum asked. Thunberg's reply, notwithstanding all search, could not be found, but that did not prevent people from imagining its entire contents. It may have been, that in his self-satisfaction at the extent and elegance of his own collections, he may have despised the Linnean herbarium, and thereby unwittingly have helped the sale. At this time, Acrel and Thunberg stood in a somewhat strained relation to each other, but they were agreed in wishing the Linnean collections to go abroad. The remark of Thunberg as to the far better paper in his herbarium was made in his old age, thirty or forty years later, and therefore cannot serve as testimony for his thoughts immediately after the death of the younger Linné. That the threatened destruction by damp of his plants, after removal to the new building, was looked upon as a kind of Nemesis, deserves no other comment than that they still remain in good condition. Still, both before and after the sale, he bitterly attacked Acrel, who resented his attitude for two years. The ardour for collecting which he ever showed, and his warm interest in the University Museum, took shape in the splendid gift of his own valuable collections, which show that he would not have opposed the conservation of the Linnean collections in the Museum, if it could have been arranged.

If thus both Acrel and Thunberg must be acquitted from the accusations levelled against them, what were the causes which brought about this deplorable transaction? There are two which seem to be the most important.

The first must be sought in the then prevalent defective appreciation of the value of such collections, which increases with the passing of years—an ignorance which then prevailed, not only with the general public, but even with eminent naturalists. It was much later that experience showed the great value of original specimens such as the type-specimens of Linné or of other distinguished men, especially when more complete and better preserved examples were obtained. The sale of these collections was regarded as something praiseworthy, and it was even reckoned as being meritorious in Acrel, that before they left the country he put the collections in order and catalogued them. When the Consistory voted upon the question of a successor to the younger Linné, Acrel obtained eleven votes and Thunberg twelve; so the latter was subsequently appointed by the King.

The other and probably more important reason for the unhappy scientific loss was the King's absence in Italy and France, whence he returned to Stockholm on the 2nd August, 1784. It has never been known whether any intimation of the negotiations ever reached him during his absence. He remained in Stockholm till the 28th September, when he started on a journey to the southern provinces of the kingdom, returning on the 10th October to his palace of Gripsholm. How much the King during this journey learned of what had taken place is uncertain, but it is probable that it was from C. Alströmer at Gothenburg —or after his return, from Dahl's petition, that he became informed as to these transactions. It is certain that the Chancellor, then at Gripsholm, on the 11th October (the day following the arrival of the King), sent a letter to Acrel with a request for an explanation "how it could happen that these collections were sold to a foreigner when a Swede offered to pay the like amount." The answer, which was requested by return, was to be laid by the Chancellor before His Majesty for his gracious consideration. Acrel's answer, dated the 13th October, related all that had happened, and stated that the heirs had already received one half of the purchase money, adding, that since the collections were sold, a student named Dahl desired to negotiate for their purchase, but as the sale had already been concluded with the foreigner, this offer could not be considered. Probably the King considering the explanation afforded to be satisfactory, concluded that nothing then could be done, as the vessel had already sailed.

What Gustaf III. did in this matter, shrank to pure insignificance. But had he been informed in proper time it is certain that he would have strongly exerted himself and rescued these precious collections for the fatherland, especially when one considers his care for Sweden's honour, and the great admiration he entertained for Linné. Such was the conviction among the people that they for a long time believed the rumour that directly he had information of the brig's departure, he despatched a warship to follow it and bring it back, but it did not succeed. Whether this rumour arose in Sweden or England, matters not. That Dr. Smith, three years later, after receipt of the collections, had no knowledge of it, is certain, although afterwards he spoke of it and believed it. Another statement also obtained currency, that Smith, in his delight in the story, had a medal struck, which showed on one side the little English vessel pursued by the Swedish frigate, and on the other side an inscription "The pursuit of the ship containing the Linnean collection by order of the King of Sweden." No such medal has been discovered, in spite of a hundred years' search. The origin of this story is probably due to an engraving of a portrait of Smith in

Thornton's "New Illustration of the Sexual System of Linnæus," the plate being dated 1800; underneath Smith's portrait is a representation of the two vessels within hailing distance, with the legend just quoted.

This was copied into Schrader's "Journal für die Botanik," iii. (1800), and the German version of Smith's "Compendium Floræ Britannicæ," Erlangen, 1801.

Linné's collections, his delight and pride, thus came to England, where their preservation is regarded as almost a national honour. The young naturalist, Smith, previously an unknown medical student, became at once famous and esteemed. In so much honour was he held that in the following May he was unanimously elected a Fellow of the Royal Society.

However pleased he may have been at this distinction, he was still more so with the Linnean treasures, which, when putting them in order, he found more valuable than he expected. He was especially surprised when he found among them the whole of Linné's extensive and valuable correspondence, with all the manuscripts he left, said, rightly or wrongly, to be put in to fill up empty spaces in the ample cases. Among Smith's own valuable collections, the Linnean acquisitions took the place of honour. After his death on the 17th March, 1828, they were bought by the Linnean Society of London, where they are still preserved at Burlington House, Piccadilly, but not in their entirety. The specimens in spirit are entirely wanting, also all stuffed mammals, birds and so on. What zoological collections remain, are placed in three cabinets, whose drawers contain the shells and insects, and considering the age of the specimens, are in wonderfully good conservation. Nothing remains of the mineralogical collections, because on the ground of their weight and bulk, Smith decided to dispose of them, before he removed in 1796 from London to his birthplace, Norwich; they were sold by public auction on the 1st and 2nd of March, and thus were completely dispersed. Traces of them have been met with, as late as 1830.

What, on the other hand, has been preserved with pious care, is the herbarium. The three unassuming, green painted cabinets, of Swedish make, contained till recently the priceless contents of Linnean types, carefully secured against London smoke and dust by specially devised envelopes. But in 1915, when the menace of air-raids from the enemy caused much anxiety to the Linnean Society, the outside cabinets were lined with steel and asbestos, the packets of plants being put into steel boxes, so that in case of fire from enemy-bombs, they could be rescued easily. Similar precautions were taken as regards the Linnean correspondence, and his annotated copies of his own works, which were lodged in steel boxes for quick removal in case of danger.

Though the Swedes may, with sorrow and shame, reflect upon the fate of the Linnean collections, it can yet be admitted that their transference to London has contributed in no small degree to the spread of knowledge in natural history, so that the Swedes themselves share in the diffusion of Linné's beloved science. It was this event which led to the foundation of the Linnean Society of London on the 18th March, 1788, which scientific society has since then flourished and borne rich fruit.

The great reputation enjoyed by Linné both at home and abroad, not among naturalists only, has been described in the foregoing pages. Many further proofs could be adduced, but only a few more may be added.

Linné, during his residence abroad, was offered inducements to stay in foreign countries; Holland, England and France being already mentioned in this respect. These temptations to forsake his fatherland were not the only offers made. One of his correspondents was Baron Otto von Münchhausen, Chancellor of the University of Göttingen, where Haller was a professor. When the latter, because of his restless disposition, quitted his post, Münchhausen, offered it to Linné, although Haller threatened to return, an event which the Chancellor considered undesirable, as Haller had shown himself intolerant of his colleagues. Certainly, the Swedish King would have done his utmost to retain Linné, but the offer was received. Linné at first seemed somewhat irresolute; he had already heard the complaint in Uppsala that natural science was put in too high a position, and therefore feared a future decadence, but on the other hand, he hoped if he undertook this professorship, that he would be able to draw half Germany's youths to Göttingen for the sake of natural history, where, during the vacancy, the deputy teacher of botany had had no more than a dozen hearers, and these so ignorant that one can hardly credit it. These feelings soon vanished, and Uppsala retained her Linné.

Another temptation came shortly afterwards from Spain. Linné kept this offer so secret that in Sweden it only became known through a German scientific journal. Bäck, who saw this announcement, was greatly astonished, but upon his questioning Linné he confirmed it, and added, that though he could not write about it, he would impart the news verbally on a visit to Stockholm. It was explained that he was invited to become president of the medical college and museum, with the botanic garden in Madrid, with a salary not below that of a Swedish Councillor of State. Linné decided to withstand this temptation and refuse the offer which seems to have been made without any doubt as to its acceptance.

The contemporaneous idea of Linné's life-work has already been narrated; there now remains only to answer the question: how he and his work were appraised in his time, and confirmed by later generations, or, if the splendour which surrounded his name, has now faded? As answer it may be well to recall the many scientific societies, even in distant parts of the world, which bear his name: the memorial festivals celebrated in his honour in different quarters of the world, and in some places even annually; his busts and statues, especially in botanic gardens: the streets in many great towns which bear his name; and the numerous medals which at home and abroad have been struck in his memory. But more important are the panegyrics delivered by naturalists of the most distinguished eminence. These have, with very few exceptions, joined in the expressions of admiration and gratitude to him who brought order out of chaos in which science was at that time nearly drowned, and who framed settled laws, which are still valid, and probably will be so for all future time.

It must not be concealed, however, that in the years after his death, voices were raised protesting against the hymns of praise sung in his honour. But one is entitled to ask, has anyone been found on our earth, being a truly great man, who has not sooner or later been the object for attack and blame? And have not these detractors usually been insignificant persons themselves, who have come forward as iconoclasts? Themselves unable to achieve any great thing, and powerless to form an idea of their victim's actual merits, they have made themselves notorious by their daring statements and want of consideration. In most cases their names are forgotten, or only mentioned sometimes by reason of the evil reputation which they gained by their conduct.

So with regard to Linné; but few were they who in their petty power sought to diminish the regard he so generally enjoyed, and does enjoy. One may except from these people an inconsiderable minority of real scientific workers, who, though they do not dispute Linné's merits, yet, according to their own ideas, honestly criticized his methods and system. The rest of the objectors are negligible, and not seldom only repeaters of what others had said. Mention should here be made of J. Sachs, who was without doubt a very eminent researcher, but whose history of botany suffers from a one-sidedness of conception and contempt of other branches of botany, besides the few upon which he wrote. His book "Geschichte der Botanik" has been the quarry whence certain persons have obtained their information; while neglecting the actual productions of Linné, they have ventured to pronounce judgment on him and his authorship. The contempt shown by Sachs for botanists of other nationalities is lamentable and reprehensible.

If one takes the trouble to investigate a number of these anti-Linneans, the first place must be assigned to those lacking knowledge of Linné's works. Soon after his death, Vicq d'Azyr criticized his medical writings, and accused him of touching on matters he did not understand, as for instance, that he wrote on the use of the muscles. Unhappily for his own credit, he was dealing with the thesis "De usu muscorum"—the use of Mosses, not as he misread it, "De usu musculorum"; and he has not lacked successors, who negligently turned over the leaves of books which demanded attentive study. Even Professor T. M. Fries admits, that after repeated perusals of Linné's writings, he often came upon matters which had easily escaped notice.

Again, a want of reflection is the cause of some complaints levied against Linné, as without careful study, it is easy to condemn him for a statement, which later he publicly abandoned; for he never stood still, but was always learning. Thus he is reproached for his belief in the permanence of species, vital conditions, etc., and in his "Philosophia Botanica" he says, "Species tot numeramus quot diversæ formæ in primitio sunt creatæ" (We reckon so many species as were created of diverse form in the beginning), but in later writings, he speaks of species, as being "temporis filia," "filia præcedentis" (the daughter of the time, the daughter of the last mentioned), clearly showing that he believed in the appearance of evolved species. A recent author, A. Hansen (Giessen, 1902), has issued a small volume packed with blunders, in which he has the assurance to state, that Linné "from our standpoint can no more be considered a botanist."

Censure is often directed against Linné, because he did not busy himself in such important departments as the anatomy and physiology of plants. It may be noted that to be a great man in this science, it is necessary to do something in these branches, it is impossible to find anybody who can be recognized as having done so. Linné's activities fell principally in that department which was most important in his time, and in that he produced great work. That he thereby was obliged to leave on one side other branches, which then attracted but little notice, is easily explained, for with full reason one may ask: has anyone the right to demand more of a single man? Is it not marvellous that he achieved so much as he did?

He has also been accused of slight valuation of researches concerning the inner construction of plants, and it has been recalled that Linné in his "Philosophia botanica," does not reckon plant-anatomists and physiologists as botanists but as botanophilists. These are to be found in his "Bibliotheca," evidently from the order in which such works can be suitably arranged. But he put all who investigated plants as phytologists, divided into two principal groups, (1) botanists, who occupied themselves in describing plants, (2) botanophilists, who devoted themselves to plant-anatomy, etc. These complaints are based simply on this, that the word "botanist" is now the same as his "phytologist," and that Linné concerned himself chiefly with systematizing and description as a botanist is natural, for the number of the others was then inconsiderable.

The chief cause of a perverted appreciation of

Linné's scientific work lies not in a retrospective comparison between him and his forerunners, but between him and later investigators, when new, unsuspected fields of work were opened, previously neglected owing to want of means. How wrong it would be to blame Berzelius, Lavoisier and other chemists because they did not even know the composition of the atmosphere, and made use of chemical formulæ now abandoned. How insignificant are Newton, Franklin, Galvani and other physicists, who had not the slightest knowledge of spectrum analysis, telegraphy and the like, which are now part of an elementary school education? And what a bad soldier—to borrow the comparison of a Norwegian author, N. Wille—was Cæsar, who did not employ artillery? One thinks of the old story of the dwarf who sat on the giant's shoulder, and boasted that he could see farther than the latter.

Finally it may be mentioned, that false judgment of nationality has probably sometimes been the cause of detraction. For instance, persons—not naturalists—created a smoke-screen against Linné by speaking of him as "our celebrated compatriot" (French), or "the renowned German naturalist" (German). These were attempts to claim Linné as of local celebrity, "a genius the like of whom the great civilized countries could show a hundred." It recalls the tale of the birds disputing who should be king. The eagle, with powerful flight, mounted high, and leaving the others of the winged troop beneath him, cried out, "Now I am king," but unwillingly heard the protest of a little feeble kingfisher, who had crouched all the time on the eagle's back, and now fluttered some yards higher.

Probably if any Linné-censor should read this, he may compassionately or contemptuously smile at this account, but besides a Swedish author in a field where Linné was a pioneer, two writers of the highest eminence may be quoted. Franz Unger, an Austrian, in 1852 wrote: "One of the most eminent men of the previous century was the great reformer of natural history, Carl von Linné. On his shoulders the genius of that science now before us rests." And M. J. Schleiden (the universally recognized reformer, and strenuous, sometimes ruthless judge of predecessors) closes his faithful and sympathetic record of Linné thus:

"Truly if we compare the work of Johannes Müller, Agassiz, Milne Edwards, Owen and others with that of Linné, from the six folios in which he first published his system of Nature, we may see the difference as between the brilliant and luxurious New York liner of the present day and the small Spanish caravel, which in 1492, first landed at San Salvador. One must not forget that it was this caravel, guided by Columbus, which discovered the New World, and laid the way by which the captain can now travel with safety and ease, but which without Columbus he would have found difficult. No development in the knowledge of geography can obliterate the name of Columbus from the memory of mankind, so never can a step in the development of natural science be reached, when it will be possible to forget, that without Linné's 'Fundamenta,' it could never have taken place."

There now remains a report on Linné's scientific importance. To set out a complete and trustworthy account, would demand more space than is at our disposal; all that can be done, is to present in a brief form, the rôle Linné played in the history of botany, zoology, mineralogy and medicine.

Botany was his first love, and he remained true to it till death. The chief part of his unresting industry he devoted to his "scientia amabilis" [lovable science], and it is with this, therefore, that his name is indissolubly connected. Hence the illuminating epigram which admiring contemporaries used, "Deus creavit, Linnæus disposuit" [God created, Linné set in order],

The best known of all his works to the general public, is his "Sexual System," which undoubtedly of all those before and after the so-called artificial system, by its simplicity and applicability possessed an unquestionable advantage. It was put forward just at the moment when such a plan was most required; when earnest searchers for two centuries had amassed so many plants of various forms, that they resembled a planless, heaped-up mass of materials in the temple of the Flower Goddess, for the disentangling of which no thoughtful and practicable plan had until then been laid down. Now when all this material threatened to overwhelm the builder, the sexual system was produced, by which plants could easily be examined and determined, thus forming an Ariadne thread in the labyrinth. Long after Linné's death, those who proudly termed themselves "true Linneans," regarded that as his chief accomplishment, and trampled underfoot as heresy, each attempt to bring about another system.

These people were more Linnean than Linné himself. Soon he saw the weakness of his system and set himself to work upon another, in which plants would not be arranged according to a single or to a few organs; when in the same class were included forms widely different, but in which the nearer or more distant relationships of dissimilar forms should be the only determining principle. During the whole of his life he laboured to discover this, and recommended others to take part in the work. The relatively small number of discovered forms made this for him an impossibility. Linné was too honest to issue his conclusions as complete, as he himself found them wanting, and therefore he pleased himself with merely creating natural families, leaving it to others to finish these and others into a systematic whole. From that time till our own days, botanists have been framing a natural system, without attaining their aim, or even finding a ground plan for the same. Concerning this, all are agreed that the contributions to it made by Linné are of uncommon value, and bear witness to his sharpsightedness, sometimes showing the greatest power of divination possible to a mortal. They are also at one, that he was the first who (in opposition to the artificial system) clearly set forth the natural families, staked out the path of progress and made certain of its dominance. It was remarkable that Linné at once brought the sexual system to its greatest height, and also laid the firm foundations for the natural system, and strongly showed the unquestionable necessity for this, as he himself said, "The A[lpha] and O[mega] among desirable botanic objects."

With clear-sightedness, without fear, but without arrogance, he undertook to elucidate all questions which previous botanists had put forward, and thus he effected a revolution. Botanic language he sorted out from its barbaric confusion, giving the requisite precision to each botanic concept; for descriptive purposes he decided and settled simple and still valid laws, established by accurate investigation of the structure of flowers and fruits in many thousands of plants, thus laying down the only right method for circumscription of relationship. In definite opposition to his predecessors, he drew a sharp distinction between what he regarded as independent species and mere accidental former varieties. To about 8,000 then known plants, he gave not only new names, but new descriptions, in which he separated the essential from the non-essential, and to which he added critical differentiating remarks on their names by the old authors, and made reports on their native countries, occurrence, properties, application, etc. During all these labours, he constantly set himself to attain his end by the most natural, and for each, the most easy way, by which many of his most striking changes in descriptive botany and zoology, readily suggest the egg of Columbus. Briefly, in small things and great, he showed himself an unsurpassed master in bringing order, light and system where ignorance and indifference had produced obscurity and confusion. In connection with his work in descriptive botany, we must remember his activity in obtaining a wide knowledge of the plant-world in different foreign countries, which, at that time, were entirely unknown, or only insufficiently explored. What he obtained by dispatching his pupils abroad on research work has been narrated in earlier pages. We must also remember his own work concerning the flora and fauna of Sweden, for which he obtained and worked up the material placed at his disposal by others.

In various quarters there has grown up an article of faith, that Linné devoted his time and strength to such work as laid him open to the charge of regarding giving of names, describing and classifying as the only, or at least the highest, attainment. But he declared in distinct words that he held quite a different view. The works specified were only drawn up as necessary stipulations for the study of botany in its still important parts. For the acquisition of a foreign language, he says in his noteworthy speech on "Deliciæ naturæ," are needed knowledge of its letters, and grammar; only when those are learned, can one enjoy all that it offers. Such is the case with the speech in which the plant-world's history is written, being comparable to botanic letters, plant-naming to words, and system to grammar.

This was not an empty comparison idly made by Linné. On the contrary, it would be hard to find a single botanic cultivator who studied the world of plants from so many sides, and who displayed so many new points of view from which plants ought to be observed. It is quite erroneous to deny or conceal this, because if Linné, in many cases, after full investigation of details, was not able to settle certain questions, he enjoyed bringing them forward and in quick, striking words, gave a first sketch to be filled by his successors, who not seldom obtained credit for the whole explanation.

It may be pointed out that it was Linné who first laid down the lines of geographic distribution of plants, though Humboldt and Wahlenberg have usually gained the credit. Also the first to introduce the doctrine of metamorphosis, though many believe the poet Goethe to be its originator. How many important contributions on biology did he bring to light which previously had been overlooked! He set forth in his books, after renewed observations, the phenomena of fertilization, and its attendant manifestations, hybridization, seed-dispersal, development of vegetation at different times of year, the day and night position of leaves (sleep of plants), the opening and closing of flowers at different hours, anomalous growths, protectives against enemies and unfavourable weather, the formation of buds, different kinds of plant communities, the relations of plants and animals, protective covering, etc. Questions earnestly debated later, were not strange to him, such as Darwin's "Origin of Species," and "Struggle for Existence," although he, in some cases, gave a different interpretation to those now current. In some of these disputations the last word has not been said, and it may be that Linné's ideas may yet prove correct.

If we count the many blunders in the older authors which he set right, the false legends which he cleared up, the defective statements which he completed, it is easy to see how he received the epithet "princeps botanicorum."

For the wider development of zoology, Linné's activity was of fundamental value, even if he here almost entirely restricted himself to systematics, applying the principles which he made use of in his botanic writings. "The Linnean system's greatest merit," observes Wirén (Zoological principles) consists in the introduction of a definite terminology, and very practical naming of animals, partly in his excellent descriptions both of species and genera as well as the higher groups, by which it became possible to retain order in the increasing number of new forms; partly and not least, in the consistent carrying out of divisions in the upper and lower categories, and referring species to their natural genera, by which the path was opened out for the modern conception of system. Here too he subjected the statements of older authors to careful criticism, and numerous are the mistakes which he rectified, the extravagances which he pruned. The boundaries of knowledge were extended by his descriptions of a most important multitude of previously unknown forms, from the most widely separated tracts. An instance may be made of the removal of the whales from the fishes to a mammalian class, the employment as a guide of the different structures of the teeth in classifying mammals, the forming of new bases for arranging reptiles, fishes, snails and mussels and their description, etc., while for knowledge of animals, particularly insects and their habits of life, he made contributions of no small worth.

Turning to minerals, it will be found that Linné took a most important place, especially by the system he introduced. That it has been superseded by a better, is a natural sequence of the marvellous developments since his time; it made, however, a stepping-stone, without which present enquirers might not have reached their eminence. It must be emphasized that Linné's views on the origin of crystals, and their application for the classification of minerals, has been regarded of such importance, that he has been styled the founder of crystallography. Geology also owes no small gratitude to him, especially by the astonishing accuracy of his reports on that science from various Swedish provinces. Thus, drawing a correct profile of Kinnekulle, he compared its strata with the corresponding strata elsewhere, and by accepting the existence over the entire globe of a certain succession of strata, which were formed in or from the ocean, he laid the foundations of the geological system which was afterwards put forward by the celebrated A. G. Werner. The true nature of petrifactions was not unknown to him. In place of regarding them as instances of "lusus naturæ," nature's playwork, or products of unnatural origin, he studied them and concluded that they were really the remains of animals and plants, which sank to the bottom of the sea or lakes, and there were covered by mud, which afterwards hardened to stone. On these grounds he showed that Gotland is chiefly built of coral, to which he devoted special attention. That the petrification-bearing beds occurred above the present level of the sea, he attributed to an ever-proceeding diminution of water.

Still more eminent was Linné's activity in the domain of medicine. His attempt to arrange various diseases in systematic order, like everything else from his hand, bore the stamp of genius, and gives him a position higher than many of his predecessors; his "Materia medica" being always reckoned as a classic in pharmacological literature. In many respects he was ahead of his contemporaries in medicine, as proved subsequently. Thus he wrote on the subject of certain skin affections caused by parasitic "small animals" or bacteria, on the proper nursing of young children, on public health, on tuberculosis infection, and conveyance of the infective particles in the clothes of patients, on the hurtfulness of unnecessary bleeding (then so universally practised), the value of electricity in certain complaints, on polypus, on the treatment of ague by quinine, etc., etc. Bacteria in his writings, appear as the cause of many diseases, especially small-pox, measles and other eruptive fevers, also of fermentation and putrefaction. Probably he himself never saw these microscopically small organisms, but he had no doubt that the above-mentioned diseases were due to "nothing else than living particles."

Great spirits impress their stamp on their times, and it is not difficult to discover many a Linnean influence in the eighteenth century, especially in its latter half. Such, for instance, was the prevailing lively desire for increased knowledge of nature's productions in different quarters. Linné had imparted his glowing naturalist ardour to those seeking education, both men and women, high and low, and beyond the boundaries of Sweden. False impressions prevailed even among the learned of his own country, for instance, that nature could only be understood from hair-splitting interpretation of Biblical Hebrew and Greek texts, or from the classical writers. Then appeared Linné, and his activity can be likened to a fresh wind, driving away mists and showing a free prospect over a sunny landscape. Natural science, formerly a neglected child, who seldom came into view, soon became a cherished possession of high and low, old and young. The consideration now bestowed on natural history forms a sharp contrast to the neglect it often encountered in former days, and proceeds in no small degree from the glory of its days of rejoicing during the Linnean period. The memory of Sweden's celebrated son will be treasured by his countrymen,

> So long as a flower its scent shall exhale,
> On mountain, in woodland, in calm-sheltered vale.
>
> —*After Frondin.*

ADDITIONAL NOTE ON RÅSHULT

The foregoing pages had been set in type, when an interesting article by Dr. Emil Lindell appeared in the "Svensk Linné-Sällskapets Årsskrift," vi. (1923) pp. 136 *seqq.* entitled "Råshult Sodregård" —the southern house at Råshult—which gives additional information concerning the history of the birthplace of Linnæus. From this paper we learn that when Nils Linnæus in 1705 became assistant to the rector of Stenbrohult, he took up his abode in the rectory, but during the winter of 1705-6, he had the modest dwelling at Råshult erected, which became famous for all future time as the Linnean birthplace. To this house Nils Linnæus moved soon after his marriage, and in the following May, his elder son was born here. Less than two years later, in 1709, N. Linnæus having succeeded his father-in-law as rector of Stenbrohult, he and his family moved into the rectory, being succeeded in the small house four years later by his brother-in-law, P Zelander. On his death in 1725 his widow in 1726 married the new comminister, T. Nicander, who, however, died in 1748, a few months after the death of Nils Linnæus. In 1731, the birthplace of Linnæus was rebuilt after a fire which destroyed the original house, but for economical reasons being rebuilt of old wood, it was condemned in 1751. It is strange that Carl Linnæus, who came home on Christmas Eve, 1731, and stayed there till the spring of 1732, never alludes to the changes at Råshult, nor did he in 1741, even when mentioning the burning of the rectory.

For illustrations of the Råshult buildings refer to the above named journal, "Ett besök vid Råshult"—a visit to Råshult—in vol. iii. (1920) pp. 103-116, by Professor C. A. M. Lindman, and in vol. iv. (1921) pp. 34-64, "I Linné's fotspår"—In Linné's footsteps—by Professor R. Senander.

B. D. J.

June, 1923.

APPENDIX I

LINNE'S AUTOBIOGRAPHIES

ALMOST the only sources from which Linné's biographers can draw, are some autobiographic notes, which he made in leisure hours, chiefly to please his memory with the distant events, sometimes brought down to later years. Unquestionably these must be regarded as especially weighty and valuable, but they cannot be reckoned as perfectly true materials. The various biographies concerning the same occurrences vary much, probably because they were written long after the events.

Nevertheless, these autobiographies are of such great importance as to deserve an account of them, and so much the more, as that published by Afzelius is not only incomplete, but somewhat confused and partially erroneous. The autobiographies known to the author are the following:

1. Vita Caroli Linnæi, with the heading: Ens entium miserere mei (Being of Beings, have mercy upon me). This is the oldest and therefore for a knowledge of his younger years, the most important; it extends to 1734. The original manuscript is in the library of the Linnean Society of London, where it was discovered in 1881, by the unwearied and enthusiastic enquirer, Dr. E. Ährling, who afterwards published the same in "Carl von Linné's Juvenile Writings," in 1888. That it remained so long unregarded is due to the fact that it is bound in at the end of an interleaved copy of J. Scheuchzer's "Operis Agrostographici Idea," Tiguri, 1719, with many annotations of Linné.

2. Historiola vitæ meæ, contributed by Linné in a letter to Haller, dated 12th Sept., 1739, which was, without the writer's permission, printed in Haller's correspondence, vol. i., Bern, 1773, a want of tact which greatly hurt Linné. It is of small extent, embracing the years 1730-39, but notable by its lively style and certain small details which do not appear elsewhere.

3. Vita Caroli Linnæi. This, apparently begun before

1745, and added to in 1751, is entirely written in Linné's hand, as a note on the back of the title-page states, and was meant to serve as a guide for the " Parentation," or address, which should, according to the custom of the time, be delivered after his death, partly in Uppsala (by Professor Beronius) partly in the Academy of Science, Stockholm. It belonged to Linné's son-in-law, S. C. Duse, the Proctor of the University; and remained with his family till it was sold by his daughter's son, Engineer M. Ridderbielke, to the British Museum in London. Two copies are in Sweden; one by Afzelius in the library of Uppsala University (MS. X. 274 b.), the other by E. Ährling, in the Academy of Science.

In Afzelius's work: "Egenhändiga anteckningar," it is ranked as III., and a " complete " extract is there given on pages 101-123, but the differences between this and the next (No. 4) are many and greater than appears from Afzelius's statement, through alterations, omissions, and additions, he allowing himself great freedom in quoting Linné's words, though he has marked these with quotation marks.

4. Vita Caroli Linnæi, was published by Afzelius and ranked by him as VI. It was found in the house after the death of the younger Linné, but when the other manuscripts were sent to England, it was kept back by J. G. Acrel, whether by permission or not is uncertain, together with Linné's " Nemesis divina." The loose sheets of paper of which it was composed, were arranged by him, added to here and there, a written title-page provided, and all bound in one volume. After Acrel's death, P von Afzelius became the owner, and he presented it to the library of Uppsala University, where it is now kept (MS. X. 274 a.). It may be added that a somewhat fragmentary title-page, in Linné's own handwriting (MS. G. 152 a.), certainly belongs to it.

This autobiography is the fullest, coming down to 1776. The beginning, about 1751, is written by another person, probably Linné's pupil, P. Löfling, to whom, when recovering from a severe illness in that year, he was accustomed to dictate what he wanted written down. The rest is in his own hand, clearly betraying the weakness of old age. A note on the back of the title-page expresses his wish that this autobiography should serve as a basis for the Memorial Oration, which he hoped " Archiater Bäck, my truest friend in life," should give after his death, in the Academy of Science. This wish was duly carried out.

A copy of this down to 1771, was taken by Linné's pupil, A. Murray; this is numbered as V by Afzelius, and is now in the Uppsala University library (MS. X. 274 d.). In this is a genealogy palpably prepared by Linné, which is wanting in the original.

Yet another copy, noted by Afzelius as IV and coming down to some time in 1769, was made by Linné's pupil, J. Lindvall, but has here and there between the lines, small corrections and additions in Linné's own hand. This was delivered to Bishop Mennander in January, 1770, with the request that he would make a Latin translation to be sent to the French Academy, of which Linné had been elected a member. In 1799, it was sent by Mennander's son to a Mr. Robert Gordon, a merchant or banker in Cadiz, to be published. Gordon died soon afterwards, and Dr. W G. Maton bought the manuscript and included an English translation of it in his 1805 edition of Pulteney's "General View of the Writings of Linnæus." This manuscript now belongs to the Linnean Society of London; another copy down to 1764, with part of an English version, being in the library of the Academy of Science.

5. Fragment of a life-description, clear copied by some other person, with an addition in Linné's own hand. The contents are practically the same as the foregoing, though with a few errors here and there, unhappily stopping at 1728. It was in the possession of the late Professor T. Tullberg.

These Linnean manuscripts can be regarded as life-sketches, but there are, besides, small notices by himself, in some degree of biographic style. Usually, they are merely notes of important events in his life, authorship, teaching, etc. The following may be named:

(*a*) "Memorial concerning my small services," sent in January, 1762, to Bishop Mennander to be delivered by him to the Secret Committee, with reference to a national payment to Linné; a copy is in the possession of the Linnean Society, presented by Baron Oscar Dickson.

(*b*) "Merit list, concerning those who were in 1767 on the Academic list"; the original seems lost, but a transcript is possessed by the Academy of Science.

(*c*) A similar list for the Academy of Science, the original being in the library of Gothenburg.

(*d*) Caroli Linnæi Vita, translated by Professor C. Aurivillius, and intended for the French Academy. A copy, sent in 1776, was used by Condorcet for his "Eloge" in 1778.

(*e*) Two short notes about his appearance, mode of life, properties, etc., written by Linné.

(*f*) A leaf in folio; no title, but with biographic notes down to 1753; ends with the characteristic words "God has graciously shown me more of His handiwork than any other person; I cannot say that I am free from faults, but I was never a parricide."

It need hardly be said that all these Linnean manuscripts have been diligently made use of in the present work.

APPENDIX II

GENEALOGIES

THE tables here printed have been compiled from others previously published, but are now corrected; they are supported by statements by Nils Linnæus from the archives of the parish of Stenbrohult, by Samuel Linnæus, from the same; at a later period, from the Växjö library, church entries, and official documents. Among the various members named more fully than in the tables, it may suffice to extract this notice concerning Carl Linnæus's only brother Samuel. He was born in 1718, became student at Lund in 1738, visited his brother in Uppsala in 1743, ordained priest 1744, the following year was made Philosophiæ Magister at Lund, and succeeded his father as Rector of Stenbrohult in 1749, where he died in 1790. He was celebrated for his skilful and successful management of bees, of which he published a complete account, printed at Växjö in 1768, wherefore he was generally called the "bee-king" or "bee-priest."

(A) CARL VON LINNÉ'S PATERNAL ANCESTRY

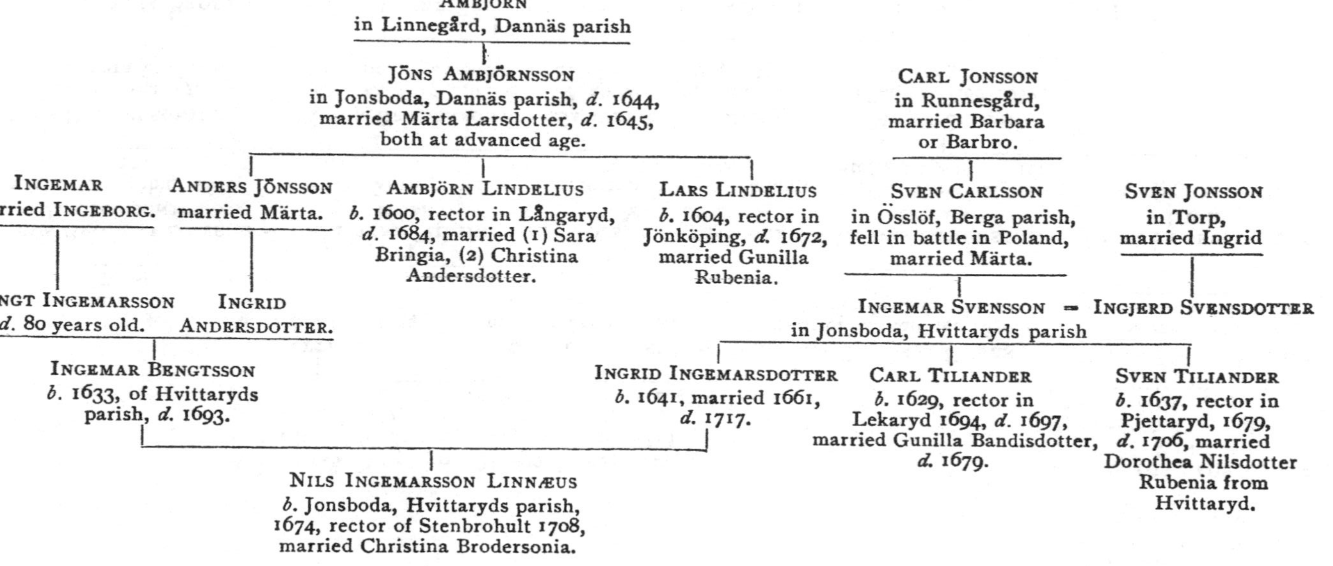

(*B*) CARL VON LINNÉ'S MATERNAL ANCESTRY.

LAURENTIUS BIGH
rector in Wisseltofta,
Skåne, *c.* 1540.

NICOLAUS LAURIDSSON BIGH
succeeded his father,
rector 1565.

MATTHIAS PETRI = MÄRIT LARSDOTTER
rector in Wislanda 15 years, *d.* 1617. — afterwards married Petri's successor, H. S. Almosius, *d.* 1647.

PEDER NILSSON BIGH
succeeded his father,
rector 1623, *d.* 1640.

MATTHIAS
priest, *d.* before 1634.

INGRID or INGJERD = NICOLAUS TORGERI KLEEN
married about 1630, *d.* 89 years old. — Komminister in Blädinge 1630, rector in Stenbrohult 1641, *d.* 1660.

BROR
married Botela.

PETRUS BRODERI UNBERG or BRODERSONIUS = CHRISTINA NILSDOTTER KLEEN
rector of Stenbrohult 1662, *d.* 1687. — *d.* 60 years old.

MATTHIAS NICOLAI MOSELIUS
Komminister in Blädinge 1665, *d.* 1667.

JORGEN SIMONSSON SCHEE = ANNA PEDERSDOTTER BIGH
b. 1612, rector of Wisseltofta 1640–86, *d.* 1692. — *d.* 1703, aged 81.

NILS BRODERSONIUS
b. about 1659, *d.* 1725,
married Sara Lindelia.

BROR BRODERSONIUS
b. 1666, *d.* 1738, marr.
Ingeborg Andersdotter.

SAMUEL BRODERSONIUS = MARIA SCHEE
b. about 1658, rector of Stenbrohult 1688, *d.* 1707. — *b.* 1664, married 1687, *d.* 1703.

CHRISTINA BRODERSONIA
b. 1688, *d.* 1733, married in
1706 to Nils Ingemarsson
Linnæus.

Carl von Linné's mother.

PETER BRODERZONIUS
b. about 1691, Komminister
in Nydala 1724, *d.* 1767,
married — Wettermark.

ANNA MARIA BRODERSONIA
b. 1692, *d.* 1772, married to (1)
Petrus Zelander, Komminister
in Stenbrohult, *b.* 1670, *d.* 1725,
(2) his successor, Erland Nicander, *b.* 1691, *d.* 1742.

JOHAN BRODERZONIUS
b. about 1694, Komminister
in Solberga 1726, *d.* 1754.

(*C*) LINNEAN GENEALOGY.

NILS INGEMARSSON LINNÆUS
b. 1674, rector of Stenbrohult,
d. 1748, married Christina Brodersonia
in 1706, *b.* 1688, *d.* 1733.

CARL LINNÆUS
ennobled VON LINNÉ,
b. 1707, *d.* 1778, married
Sara Elizabeth Moræa,
b. 1716, *d.* 1806.

ANNA MARIA
bapt. 1710, *d.* 1769,
married in 1730 to
Gabriel Höök, *d.* 1769,
rector of Wirestad.
10 children.

SOPHIA JULIANA
bapt. 1714, *d.* 1771,
married to Johan
Svensson Collin, rector
of Ryssby, *d.* 1766.
10 children.

SAMUEL
b. 1718, rector of
Stenbrohult 1749,
dean 1777, *d.* 1797,
married in 1750 to Anna
Helena Osander, *b.* 1731,
d. 1790. 12 children.

EMERENTIA
b. 1723, *d.* 1753,
married to Carl
Amman Branting in
1749. 4 children.

Children of Carl Linnæus:

CARL VON LINNÉ
the younger,
b. 1741, *d.* 1783,
Unmarried.

ELIZABETH CHRISTINA
b. 1743, *d.* 1782, married
to Major Carl Fredrik
Bergencrantz (grandson of
O. Rudbeck, the younger),
b. 1725, *d.* 1792. 2 children.

SARA MAGDALENA
b. and *d.* 1744.

LOVISA
b. 1749, *d.* 1839.
Unmarried.

SARA CHRISTINA
b. 1751, *d.* 1835,
married to Capt.
Hans Henrik Duse,
b. 1740, *d.* 1811.
No children.

JOHANNES
b. 1754, *d.* 1757.

SOPHIA
b. 1757, *d.* 1830,
married to University Proctor
Samuel Christoffer Duse,
b. 1748, *d.* 1826.
One daughter.

APPENDIX III

LIST OF LINNÉ'S PUPILS

The first dates are when instructions began; an asterisk denotes private lectures.

Norwegians

Anker, Peder, 1764 (1749-1824).
Ascanius, Peder, 1769 (1723-1803). Professor at Copenhagen.
*Borchgrevink, Janus Finne, 1766 (1736-1819).
Hagerup, Eiler, 1762 (1736-95).
*Tislef, Johannes, 1771.
*Tonning, Henrik, 1766-68 (1732-96).
*Vahl, Martin, 1769-74 (1749-1804). Professor at Copenhagen.
Wellemsen, Gert von der Lith, 1754 (b. 1738).
Wellemsen, Wellem Friedrich Kölner, 1754 (1740-94).

Danes

Berger, Johann Wilhelm von, 1776 (1754-1779).
*Eding, Peder Wilhelm, 1771 (b. 1746).
*Fabricius, Johan Christian, 1762-64 (1745-1808). Professor at Copenhagen, afterwards at Kiel; famous entomologist. See p. 219.
Hagen, Mathias, 1760-62 (d. 1802 aged 63).
Holm, Jörgen Tyge (1726-59). One of Linné's best pupils.
Horrebow, Peder, 1776 (1756-78).
König, Johan Gerard (1728-85). In Iceland and India.
Mangor, Christian Elovius, 1760-62 (1739-1801).
Moltke, Friderik Ludvig, 1764 (1745-1824).
Rottböll, Christian Friis (1727-97). Professor at Copenhagen.
*Zoega, Johan, 1762-64 (1742-88). See p. 220.

Germans

*Beckman, Johann, 1765 (1739-1811). Professor in Göttingen.
*Giseke, Paul Dietrich. See p. 224.
Grüno, Isaac, 1776-77 (1756-83).
Leppertin, — (Stöver, i. p. 347).
*Meyer, Johan Karl Friedrich, 1764 or 1766 (1739-1811).
Nathorst, Theophil Erdmann, 1755-56 (1734-1804).
*Schreber, Johann Christian Daniel (1739-1810). See p. 223.

Swiss

Ehrhart, Friedrich (1742-1795). See p. 225.
Valltravers, Johann Rodolph von, 1760-61 (b. 1723).

English

Rotheram, John. See p. 224.

Dutch

*Burman, Nicolaus Laurens, 1760 (1734-93). Professor at Amsterdam.

French

Missa, Henri. See p. 222.

Italian

Manie, — 1758.

Russians

*Aphonin, Mathæus, 1761-69 (b. 1740). Professor at Moscow. See p. 223.
Demidoras, — (Stöver, i. 31).
Demidoff, — 1760-61. Three brothers, Paul, Gregorey and ——. See p. 219.
Hoppius, Christian Emanuel, 1757-62 (b. 1736).
Hornborg, Bogislaus, 1757.
*Hornborg, Henrik, 1768 (b. 1745).
*Hornborg, Johan, 1768-74 (b. 1750).
Karamyschew, Alexander, 1761-67 (b. 1744). See p. 223.
Lepechin, Ivan (1737-1802.) Afterwards director of the botanic garden at what is now known as Petrograd.

Viborgian

Lada, Christian, 1760.

African (Algiers)

*Logie, Fredrik, 1756-58 (1739-68).

American

*Kuhn, Adam, 1761-65 (1741-1817). Professor at Philadelphia.

Prominent Swedes and Finns

(In addition to those enumerated on pp. 218-244.)

Afzelius, Adam (1750-1837). Professor at Uppsala.
Afzelius, Johan (1753-1837). Professor at Uppsala.
Bergman, Torbern (1735-84). Professor at Uppsala; famous chemist.
Bjerkander, Clas (1735-95).
Bjerchén, Pehr, ennobled as Bjerkén (1731-74). Eminent physician.
Blom, Carl Magnus (1737-1815).
Casström, Samuel Niklas (1763-1827).
Dalberg, Nils (1736-1820).
Dubb, Pehr (1750-1834).
Fagræus, Jonas Theodor (1729-97).
Ferber, Johan Jacob (1743-90).
Fornander, Anders Niclas (1715-94).
Gyllenhaal, Leonard (1752-1840). Famous entomologist.
Hallman, Johan Gustaf (1726-97).
Haartman, Johan Johansson (1752-87). Professor at Åbo.
Hagström, Anders Johan, ennobled as Hagströmer (1753-1830). Professor and eminent physician.
Hall, Birger Martin (1741-1814), the only mycological student of Linné.
Hallenberg, Jonas (1748-1834).
Hedin, Sven Anders (1750-1821). See p. 397.
Hellenius, Carl Niklas (ennobled as von Hellens) (1745-1820). Professor at Åbo.
Hoffberg, Carl Fredrik (1729-90).
Hoffman, Anton (1739-82).
Holmberger, Pehr (1745-1807).
Jörlin, Engelbert (1733-1810).
Ljungh, Sven Ingemar (1757-1828).
Martin, Anton Rolandsson (1726-88). See p. 236.
Odhelius, Johan Lorens (1737-1816).

Ödman, Samuel (1750-1829).
Samzelius, Abraham (1723-73).
Scheidenburg, Daniel (b. circa 1720).
Schulz, David, ennobled as von Schulzenheim (1732-1823).
Swederus, Nils Samuel (1751-1833). Entomologist.
Tidström, Andreas Philip (1723-79). See p. 303.
Westring, Johan Peter (1753-1833).
Winbom, Johan (1746-1826).
Wollin, Christian (1730-98).
Wåhlin, Anders Magnus (1731-97).

APPENDIX IV

NOTES ON "NEMESIS DIVINA" [DIVINE PUNISHMENT]

THESE were drawn up for his son's guidance; after the father's death they were lost sight of, till in 1844 they were found at Kalmar in the house of the deceased Olof Acrel, the son of J. G. Acrel. The manuscript now is in the library of Uppsala University.

When found, they consisted of slips in a case, but are now bound in a volume. The work was first published in 1848 by E. M. Fries, in the "Botaniska Utflygter"; again by T M. Fries in 1878. The manuscript was written in a mixture of Swedish and Latin. A few extracts are given; the style is terse and abrupt.

Laws

1. Be persuaded by nature and experience in God, who made, maintains and governs all; who sees, hears, knows all, thou art in his sight.
2. Never take God to witness in an unjust thing, nor swear falsely.
3. Look upon God's purpose in creation; believe that God guides and keeps thee daily, and all good and evil comes from His hand.
4. Be not ungrateful, that thou mayest live long.
5. Beware of manslaughter, sin is not suffered, unless restitution is done, and that cannot be, so not forgiven.

6. Dishonour no woman, and steal no man's heart.
7. Procure no unjust advantage.
8. Be honest and a man of ancient virtue and faith, then you will be loved of all.
9. Plot not to overturn others, that thou fall not into a pit.
10. Have nothing to do with intrigue.

Aphorisms

Revenge persecutes him; all things go against the guilty. No calamity by itself.

Everything went badly with me, when I harboured revenge, but [I] changed, and left everything in God's hands: since then all happily. 1734. See p. 124.

Crimes punished by Nemesis

Beware of great sins.

Sin is not forgiven, unless restitution is made.

1. Beware of manslaughter or murder.
2. ,, ,, blood guiltiness.
3. ,, ,, contempt of God.
4. ,, ,, ingratitude to parents.
5. ,, ,, ruining one's welfare.
6. ,, ,, injuring the defenceless.

Many of my colleagues took holidays from lectures, came up for half-time, enjoyed themselves in society every day (Frondin), many had double salary.

I gave myself no repose night or day, had no rest, lectured, wrote, examined. What had I more? Name is wind, annihilated by others. Obs., what I did, others copied as their own. Titles are wind; Noble, Knight, Archiater.

Münnich in Russia, Premier Minister, had Birong taken to Siberia and lodged permanently in prison. The house in which B. was, M. had boarded up so that no sunlight should get in to lighten his solitude. After some years, Birong came out, and Münnich was put in the same prison. Münnich caused the Russian captain Keller to murder Cincler [Sinclair] on his journey back from Turkey.

Examples of " Nemesis "

Måns of Sannaböke in Småland, a hard man; evil against his father. Mån's son dragged his father by the hair to throw him out of his own house. When the father came to the door, he cried out, " Drag me no farther; I did not drag my father any farther." The son answered " God's death, thou dragged thy father to the door, I will drag thee out of the door." This happened in my birth-place, in my childhood.

Two unmarried women, Friesendorf, lived at Hammarby before me, died 1725. Were always so perverse that they could not bear each other; they divided the estate in two. When one died, the other rejoiced and said she would mourn in scarlet, but in four days she too lay dead; they were buried on the same day in one grave. Then they first agreed.

Captain Cincler [Sinclair] when imprisoned, stabbed to death an under officer, Lod, and by legal process escaped [judgment for] the murder. Captain Cincler had so mortal a hatred for the Russians, that he said he did not wish to go to heaven if any Russians went there. (Similarly Artedi had mortal hatred of the Dutch, when he was drowned at Amsterdam.) He was sent with despatches to Turkey, to stir up the Turks against the Russians. Münnich, Premier Minister in Petersburg, had his portrait made, set four officers in ambush at Ingerstedt in Germany, one of whom, Keller, murdered him.

A man freed a thief from the gallows. The same man was taken by enemies and was to be hanged, but a rope was wanting; the thief came and gave a rope.

Divinations

Carl XII. had his fortune told by General Carl Cronstedt by " puncture " that he would be killed before the end of November, though amongst his trusty friends, the officers. One of Cronstedt's friends said to him on the last day of November: " It is now the last day of November, and the King is still alive." Cronstedt answered: " That is so, but the time is not past." At night he was killed at Fredrikshald. But some think that it was the same Cronstedt who shot the King at Fredrikshald, but really the French colonel Stickart.

A woman was carried round to all houses as sick and poor, but could tell fortunes. She said that the house [at Stenbrohult] stood in danger of destruction by fire. My mother was alarmed; she [the soothsayer] said pray God to postpone it in your time. The house was burned after her death. My brother Samuel, brisk, was at Wexiö school; I was newly come to Lund. Everybody called my brother Professor, and said he would become a Professor. She, who had seen neither of us, asked to see some of our clothes, and said of brother Samuel that he would be priest; of me, he would be professor, travel far, be more famous than anyone in the kingdom, and swore thereto. My mother to deceive her, showed another coat, saying it was my brother's. " No, that is his which will be professor and live far away."

My father saw one night as it were a human form in a sheet sitting by the fireplace; talked about it to everybody. Two days after came the dancing master, Sobrant, who sickened the next day and died.

A week before my wife was confined of our daughter Helena [Sara Magdalena], the neighbours saw at night, lights in all our windows, as if illuminated; they talked about it to everybody My wife got to know about it, and feared that it portended she should die in childbed; but she came through. The girl died soon after her birth.

1765 at midnight between 22nd and 23rd July, my wife heard [somebody] outside our bedroom; it went into the upper chamber, my museum. Something went heavily to and fro. Wakened me, and I heard it also. I knew that nobody was there, the doors were locked, and the keys with me. After a few days I learned that my special friend and trusty commissary, Carl Clerck, died the 22nd July at nine at night, and really the walk was so like his, that if I had heard it at Stockholm, I should have known him by his walk, but I was then at Hammarby, six miles from Stockholm [really about thirty-six English miles].

When Löfling before starting for Spain, came to take leave of me, he stumbled [on the threshold] came not back. Forskål likewise.

[A long account of Alexander Blackwell, a native of Aberdeen, will be found translated in the " Journal of Botany," xlviii. (1910) pp. 193-195.]

APPENDIX V

SWEDISH TITLES, MONEY AND MEASURES

Adjunct, Assistant Professor; sometimes a Curate.
Akademiska forsámling, Convocation of the body of Professors.
Archiater, a Chief Physician.
Assessor, primarily an assistant to a Board or Council, but used in various senses; in Linné's time usually an honorary title.
Auscultant, a special assistant in the Medical College.
Candidate, see Kandidate.
Consistorium, Academic Consistory, greater or lesser, according to the number of Professors on it.
Decanus, or Dean, the presiding officer in a faculty.
Dissertation, a formal discourse supported by a Respondent before the President (Præses).
Docent, lecturer; Docentstipendier, lecturer fellowships.
Doctor, a recipient of the highest academic honour.
Härad, hundred or district; Häradsrätt, district court.
Höstermin, autumn term, 1st September—16th December.
Hovrätt, Court of Appeal.
Informator, private tutor; Information, private tuition.
Justitieråd, Chief Justice.
Kakelugn, earthenware stove.
Kandidate, Bachelor; a graduate in any faculty.
Kansler, Chancellor, the chief officer of a university; an honorary office, the duties usually devolving on the Vice-Chancellor.
Kansli, Secretary's Office.
Kollegier, coaching lectures, for extra payment by pupils.
Komminister, Perpetual Curate.
Kondition, private tutorship.
Kronofogde, Crown Bailiff.
Kyrkoherde, Rector of a parish.
Kyrkomotet, the Convocation of Clergy.
Laborator, Demonstrator.
Läkare, Physician, Medical Practitioner.
Län, county or district, twenty-five in number.

Landmarschalk, Speaker of the House of Nobles, in the Riksdag.
Landshofding, Governor of a Province.
Landskap, see Nation.
Lärare, Teacher, Professor, or Schoolmaster.
Lärjungar (pl.), pupils.
Lasåret, the academic year.
Lektor, native teacher of a foreign language in a school or university.
Licentiate, a graduate, having passed the final honour schools.
Lifmedicus, a Body-Physician.
Nation (Landskap, Nationsförening), a provincial club at a university.
Öfverstathållare, the High Governor at Stockholm.
Ombudsman, Commissioner or Proctor.
Opponent, the critic of a thesis.
Parentation, a memorial oration on the death of a person of eminence.
Præses, President at a disputation; in the time of Linné he usually dictated the substance of the thesis to be defended by the Respondens.
Prokansler, Vice-Chancellor.
Prorector, Deputy for the University Rector.
Prosector, Demonstrator in Anatomy.
Provins, Province, the old divisions of Sweden.
Räntmastare, Bursar (or Treasurer).
Rektor (Rector Magnificus), Head of the University; held six months during Linne's time, now for three years.
Respondens, one who maintains or answers for the thesis or disputation put forward.
Ridderhus, House of Knights or Nobles.
Riksdag, Parliament, organized in four estates down to 1809, the Nobility, Clergy, Burgesses and Peasants.
Riksråd, Councillor of the Realm.
Secret Committee [Sekret utskottet], consisted of 100 delegates of the Estates or four houses of the Riksdag.
Stadsphysicus, old term for Stadsläkare, a borough officer of health.
Statsråd, Cabinet of Ministers.
Ständer, the Estates in the Riksdag.
Stipendium (pl. Stipendier) a fellowship bursary or scholarship; not an exhibition.
Student, undergraduate; Student examination, matriculation examination.
Thesis, the proposition to be maintained by the respondent.

Tullhus, Custom-house; Tullport, the town gate where customs were levied.
Vice-Rektor, the Dean or Prorektor.
Vårtermin, spring term, 15th January—1st June.

The value of Swedish money varied during the "Era of Liberty," but the following values have been assigned as the average:

1 daler, copper coinage, equalled 6 pence.
1 daler, silver coinage, equalled 18 pence.
1 plåt (pl. plåtar), 2 silver dalers, equalled 3 shillings.
3 silver dalers equalled 1 riksdaler; in 1761, 4 shillings and 7½ pence (later about 4 shillings and 5 pence).
1 ducat equalled 9 shillings and 2 pence.

The length of the Swedish mile was 6·6423, or nearly 6⅔ of an English statute mile; as the Swedish mile was so long, it was often reckoned by the quarter mile, rather less than one mile and three-quarters English.

Other measures were:

1 inch (tum)	= 0·97	English	inch
1 foot (fot)	= 11·69	,,	inches
2 feet = 1 (aln) ell	= 1·95	,,	feet

1 quarter (of wine), slightly more than half-a-pint.

APPENDIX VI

SKETCH OF SWEDISH HISTORY DURING LINNÉ'S TIME

THE following short sketch of Swedish history, including the life-time of Linné, may help to recall the circumstances under which his work was done, and some of the influences which moulded public life at that period.

In 1682 the Diet had entrusted Carl XI. with sole executive power, and one of his latest decrees was intended to reform the Calendar (see p. 2). Dying in 1697, he was succeeded by his son Carl XII, whose personal absolutism was fraught with disastrous consequences. He was clever, well educated, and energetic, but reckless and self-willed; succeeding his father when only fifteen, his autocratic power became exercised from early youth. In 1699, an alliance between Denmark, Russia and Poland was concluded against Sweden, leading to the great northern war, but helped by Britain and the Netherlands, Carl successfully compelled the Danes to make peace at Travendal in 1700. He defeated the Russians at Narva, took Curland from the Poles, and obliged the Elector Augustus of Saxony to sign the peace of Altranstädt. Meanwhile Tsar Peter (Peter the Great) had, on the other hand, gained Kexholm, Ingermanland, and Esthonia. Instead of a direct attempt to regain these provinces, Carl, tempted by a promise of help from Mazeppa, a Cossack chief, marched to the Ukraine, but sustained a signal defeat at Poltava in 1709, largely due to the fact that the Swedish army had expended all its ammunition, and had only cold steel for its defence. Though wounded, Carl escaped into Turkey and resided at Bender, while Denmark and Saxony again declared war against Sweden, and the Tsar occupied Finland. Swedish resources being now exhausted, the nobility began to plot against their king. Carl escaped from Turkey, and returning to Sweden in 1715, found that Britain, Hanover and Prussia had also declared war against him. Having with great difficulty raised some money, Carl invaded Norway and besieged Fredrikshald, midway between Christiania and Gothenburg, but was shot in the trenches, dying at the early age of thirty-six. With his death absolutism came to an end, and the "Era of Liberty" succeeded, lasting from 1719 to 1772.

Carl XII. was succeeded by his sister, Ulrika Eleanora, who, shortly afterwards, abdicated in favour of her husband, the Crown Prince of Cassel, who ascended the throne as Fredrik I., a new constitution being framed in 1720. In the course of two years, peace being concluded with the surrounding nations, a period of repose followed; a new code of laws was drawn up in 1734, and efforts were made to revive commerce. Meanwhile the people became divided into two parties, namely the "Hats," who, under Counts Gyllenborg and Tessin, advocated an alliance with France and war with Russia, and the "Caps," who preferred to form an alliance with Britain and keep France at a distance. In 1741, the "Hats" were supreme in power, and after the death in 1751 of Fredrik I., who had no issue, they plunged in 1757 into the Seven Years' War, with ruinous results to themselves, and by 1760, impeachment was imminent. Adolf Fredrik had succeeded Fredrik I. as King, his queen being the masterful Lovisa Ulrika, sister of Frederick the Great. With these personages Linné was in constant communication, as both King and Queen were amassing rich collections of natural history, and often invoked the aid of the great naturalist in arranging and classifying them.

The "Caps" came into power in 1765, but their administration (though taking an opposite course to that of the "Hats" whom they displaced) was so unfortunate, that, three years later, they had to yield once more to the supremacy of the "Hats."

Adolf Fredrik died in 1771, and his son Gustaf III. succeeded. In a few months, by a bloodless military revolution in 1772, he ended the "Era of Liberty," and acquired the sole executive power. He, however, used his victory with moderation, abolished torture, brought in liberty of the press, and promoted commerce, science and art. During his early years on the throne, his relations with Linné were cordial and generous until the death of the great naturalist, who had been in his grave for nearly twenty years, when the King was shot at a masked ball, by Count Ankarström, dying a few days afterward. The assassin was scourged during three successive days, and then executed.

A paragraph may be devoted to reminding the reader of current European history outside Sweden during the period of 1707-78.

The political union of Scotland with England took place in 1707; the Treaty of Utrecht in 1713, closing the war of the Spanish succession; the death of Louis XIV occurred in 1715, and in the same year the Earl of Mar conducted the rising in Scotland in favour of the son of James II., the "Old Pretender." The excitement of the South Sea Bubble reached its height in 1720; the Quadruple Alliance was formed in 1721, and war with Spain began in 1739; battles were fought at Dettingen in 1743, and Fontenoy in 1745; the latter year also witnessed the attempt of the "Young Pretender," who suffered a total defeat at Culloden in 1746. The Treaty of Aix-la-Chapelle closed the war of the Austrian succession, the Seven Years' War lasting from 1756 to 1763, the battle of Rossbach being fought in 1756; the conquest of India began under Clive in 1757 culminating in his victory at Plassey; the victory and death of Wolfe took place at Quebec in 1759, with the conquest of Canada the year after, and the war for the independence of North America began in 1773. The period thus briefly traced extended from the middle of the reign of Queen Anne, the whole of those of George I. and George II., and the first eighteen years of George III.

APPENDIX VII

SELECT BIBLIOGRAPHY

Part I—Linnean Books

Fuller particulars and titles of minor works are given in Dr. J. M. Hulth's "Bibliographia Linnæana," 1907 (see Part II). All 8vo unless otherwise noted.

LINNÆUS (after 1761 VON LINNÉ), Carl.

D. botanica de planta Sceptrum Carolinum dicta Præs. L. Robergio . . Auctor J. O. Rudbeck, Upsalis, 1731. 4to.

De febrium intermittentium causa. Harderovici, 1735. 4to. Reimpr. in Am. Acad. i. 1749. 1-19. ib. ed. III. x. (1790), 1-22.

Systema Naturæ. Lugd. Bat. 1735, fol. Repr. Paris, 1830. 8vo; ib. 1881. Facsimile, Stockholm, 1907, fol. Ed. II. Stockh. 1740. 4to; ed. VI. ib. 1748; ed. X. Holmiæ, 1758-59, 2 vols.—Reimpr. Regnum animale. Ed. X. cura societatis zoologicæ Germanicæ iterum edita. Lipsiæ, 1894. Ed. XII. Holmiæ, 1766-67. 3 vols.

(The intermediate editions were not revised by Linné.)

Bibliotheca botanica. Amstel. 1736—ed. II. ib. 1751.

Fundamenta botanica. Amstel. 1736; ed. II. Stockh. 1740. 4to; ed. III. Amstel. 1741. 4to.

Musa Cliffortiana florens Hartecampi, 1736, prope Haarlemum. Lugd. Bat. 1737.

Critica botanica. Accedit J. Browallii, De necessitate historiæ naturalis discursus. Lugd. Bat. 1737.

Flora lapponica. Amstel. 1737; ed. II. cura J. E. Smith. Lond. 1792.

Genera plantarum. Lugd. Bat. 1737; ed. II. ib. 1742; ed. V Holmiæ, 1754; ed. VI. ib. 1764—Corollarium Generum plantarum, ib. 1737.

Methodus sexualis sistens genera plantarum. Lugd. Bat. 1737.

Hortus Cliffortianus. Amstel. 1737, fol. (Paged 1-231, 301-507).
Viridarium Cliffortianum, ib. 1737.
Classes plantarum. Lugd. Bat. 1738.
P. Artedi, Ichthyologia . . . postuma, ed. a Linnæo. ib. 1738.
J. E. Ferber. Hortus agerumensis. Holmiæ, 1739.
Tal, om märkwärdigheter uti insecterne. Stockh. 1739. (Reimpr. multoties.)
Orbis eruditi judicium de C. L. scriptis [Stockh. 1741]. Reimp. in facsimile a W. Junk. Berol. 1901 et 1907.
Oratio de necessitate peregrinationum intra patriam. Accedunt J. Browallii Examen epicriseos Siegesbeckianæ . et J. Gesneri D. partium vegetationis. Lugd. Bat. 1743.
Flora suecica. Stockh. 1745; ed. II. ib. 1755.
Öländska och Gothländska resa. Stockh. och Uppsala, 1745; ed. II. Stockh. 1907. (In German, Halle, 1764; in Dutch, Dordregt, 1770.)
Gothländska resa . 1741 Ny Upplaga, Visby, 1890.
Fauna suecica. Stockh. 1746; ed. II. ib. 1761.
Flora zeylanica . fuere a P. Hermanno. Holmiæ, 1747.
Wästgöta resa. Stockh. 1747.
Lars Robergs Tal. ib. 1747.
Hortus Upsaliensis. Vol. i. (all issued), ib. 1748.
Materia medica. Holmiæ, 1749; Materies medica, lib. 2 [et 3] ib. 1763.
Amœnitates Academicæ. Holmiæ, tom. i.-vii. 1749-1769; ed. Schreber, i.-x. Erlangæ, 1785-89.
Miscellaneous Tracts . transl. by B. Stillingfleet, Lond. 1759; ed. IV ib. 1791.
Select dissertations . . transl. by the Rev. F. J. Brand, ib. 1781.
Druce, G. C. Linnæus's " Flora anglica." In Scot. Bot. Rev. (1912) 154-161.
Philosophia botanica. Stockh. 1751. Reimp. Viennæ Austriæ, 1755, 1763, 1770, 1783; cura Gleditsch Berol, 1780; ed. III.; cura Willdenow, ib. 1790.
Skånska resa. Stockh. 1751. Reimpr. Lund, 1874, ib. 1907. Ed. II. av J. Sahlgren. Stockholm, 1920. In German, Leipzig, 1756.
Museum Tessinianum. Holmiæ, 1753, fol.
Species plantarum. Holmiæ, 1753, 2 vols.; facsimile, Berlin, 1907. Ed. II. Holmiæ, 1762-63; ed. " III."

Vindobonæ, 1764, 2 vols. Index perfectus ad C. L. Species plantarum . 1753 collatore F. de Mueller, Melbourne, 1880. Index abecedarius; an alphabetical index to the first edition of the "Species plantarum" . compiled by W. P Hiern (Journ. Bot. 1906). Indices nominum trivialium ad Linnæi species plantarum, ed. I. Berlin (Junk), 1907. Hulth, J. M. Linné's första utkast till Species plantarum (Sv. Bot. Tidsk. vi. 1912).

Museum S. R. M. Adolphi Friderici regis. Holmiæ, 1754, fol.; Tomi secundi prodromus, ib. 1764.

Hasselquists Iter palæstinum. Stockh. 1759. In German, Rostock, 1762; in English, Lond. 1766; in French, Paris, 1769; in Dutch, Amstel. 1771.

P. Löfling. Iter hispanicum. Stockh. 1758. In German, Berlin, 1766 and 1776.

Disquisitio de quæstione proposita . sexum plantarum, etc. Petropoli, 1760. 4to. In English, Lond. 1786.

Museum S. R. M. Ludovicæ Ulricæ Reginæ Holmiæ, 1764.

Clavis medicinæ duplex. Holmiæ. 1766. Reimpr. Longosalissa, 1767; Uppsala, 1907

Mantissa plantarum. Holmiæ, 1767

Mantissa altera. Ib. 1771.

Deliciæ Naturæ. Stockh. 1773.

Systema vegetabilium a J. A. Murray, ed. XIII. Gottingæ et Gothæ, 1774; ed. XIV Gottingæ, 1784.

Supplementum plantarum ed. a C. à Linné [filio]. Brunsvigæ, 1781.

[D. Vandelli]. Floræ Lusitaniæ specimen. Et epistolæ a Carolo a Linné. Conimbricæ, 1788. 4to; ed. Rœmer.

Haller, A. von. Epistolarum ab eruditis viris ad A. H. scriptarum [partes i.-vi.]. Bernæ, 1773-75. (The letters from Linné are contained in vols. i.-iii.)

Fabricius, J. C., and P D. Giseke. C. a Linné, Prælectiones in ordines naturales plantarum, ed. P D. Giseke. Hamburgi, 1792.

Collectio epistolarum . . ed. D. H. Stoever. Hamburgi, 1792.

Smith, Sir J. E. A selection of the correspondence of Linnæus . 2 vols., Lond. 1821.

Epistolæ ineditæ annis 1736-93 ed. H. C. van Hall. Groningæ, 1830.

Epistolæ ad N. J. Jacquin præf. S. Endlicher, Vindobonæ, 1841.

C. a L. ad Bernardum de Jussieu ineditæ, ab mutuæ Bernardi ad Linnæum epistolæ, curante Adriano de Jussieu. In: Mem. Amer. Acad. N. S. v. (1853) pp. 179-234.

Lettres inédites de Linné à Boissier de la Croix de Sauvages, rec. par M. le Baron D'Hombres-Firmas, ed. par M.C.C. Alais, 1860.

Lachesis lapponica, transl. by C. Troilius ed. by Sir J. E. Smith, 2 vols. Lond. 1811.

Exercitatio botanico-physica de Nuptiis et sexu plantarum latine vertit M. J. A. Afzelius. Upsaliæ, 1828.

Systema, Genera, species plantarum uno volumine sive Codex botanicus Linnæanus edidit H. E. Richter, Lipsiæ, 1835. In Codicem botanicum Linnæanum Index alphabeticus composuit G. L. Petermann, ib. 1840.

Anteckningar öfver Nemesis divina. Uppsala, 1848, fol.; Reimpr. in Bot. Utflygter, ii. (1852) 299-344; Ny uppl. Uppsala, 1878.

Flora dalekarlica ed. E. Ährling. Orebroæ, 1873.

Hortus uplandicus af T. M. Fries. Uppsala, 1899.

Hortus uplandicus 1730: Manuscriptum autoris quod in bibliotheca Degeeriana Leufstadiensi adservatur arte photo-lithographica expressum [Holmiæ, 1907]. 4to.

Adonis uplandicus manuscriptum. ib. 1907. 4to.

Catalogus plantarum rariorum Scaniæ, item Catalogus plantarum . Smolandiæ, 1728, manuscriptum expressum [ib. 1907].

(These three MSS. were facsimiled by Count Carl De Geer for the Linnean bicentenial celebration of 1907.)

Carl von Linné's Ungdomskrifter samlade af E. Ährling och efter hans död med statsunderstöd av K. Vetenskapsakademien.

Första Serien. Stockh. [1888].

En sjelfbiografi till år 1734.

Skånes sällsyntare växter, 1728.

Spolia botanica, 1729.

Hortus uplandicus, 1729.

Id. 1730 et 1731.

Adonis uplandicus, 1731.

Andra Serien, ib. [1889].

Iter lapponicum; Iter ad fodinas.
Iter dalecarlicum; Iter ad exteros.

Skriften af Carl von Linné utgifna af Kungl. Svenska Vetenskapsakademien.
I. Flora lapponica, öfversatt till svensk språket af T. M. Fries. Uppsala, 1905.
II. Valda smärre skrifter, ib. 1906.
III. Classes plantarum, ib. 1906.
IV. Valda smärre skrifter af botaniska innehåll, ib. 1908.
V Iter lapponicum; andra upplagan . af T. M. Fries, ib. 1913.

[Letter to P. Arduino.] Proc. Linn. Soc. 1906-7 (1907) pp. 83-87.

Pluto svecicus, utgifna af C. Benedicks. Uppsala, 1907.

Beskrifning öfwer stenriket, utgifna af [densamme], ib. 1907.

Methodus avium sveticarum, utgifna af E. Lönnberg, ib. 1907.

Vorlesungen über die Cultur des Pflanzen, utgifna af M. B. Swederus, ib. 1907.

Linné och vaxtodlinger, af [densamme], ib. 1907.

Linné's Dietetik . . ordnad och utgifna af A. O. Lindfors, ib. 1907.

Bref och skrifvelser af och till Carl von Linné. I. Afdeln. Svenskar; Stockh. i.-viii., 1907-22. II. Afdeln. Utlänskor, af J. M. Hulth, Uppsala, i. 1916. (Ed. by T. M. Fries, and after his death in 1913, by J. M. Hulth; in progress.)

Föreläsningar öfver djurriket utgifna och forsëdda med förklarende anmärkningar af E. Lönnberg, ib. [1913].

Part II

Works relating to Linné; arranged in chronological order of writers.

Siegesbeck, J. G.

Botanosophia verioris brevis sciagraphia . . Accedit Epicrisis in Cl. Linnæi nuperrime evulgatum Systema plantarum sexuale. Petropoli, 1737 4to.

Vaniloquentiæ botanicæ specimen a M. J. G. Gleditsch in Consideratione Epicriseos Siegesbeckianæ, ib. 1741. 4to.

BROVALL, J.

Examen Epicriseos . auctore J. G. Siegesbeck. Aboæ [1739]. Reimp. in Linnæi Oratio de necessitate peregrinationum in patriam. App. 1-53. Lugd. Bat. 1743.

Curriculum vitæ Caroli Linnæi. Den första biografien öfver Linné. Utg. på svenska af Joh. Bergman. Stockh. 1820.

GLEDITSCH, J. G.

Consideratio Epicriseos Siegesbeckianæ Berol, 1740.

WALLERIUS, J. G.

Decades binæ thesium medicarum. Uppsal. 1741. Reimpr. De injurioso contra Linnæum libello In: Stoever, Coll. Epist. 119-158.

BÄCK, A.

Aminnelse tal. . Stockh. 1779. In German, ib. 1779.

PULTENEY, R.

A general view of the writings of Linnæus. Lond. 1781; ed. II. by W. G. Maton, with "Diary"; 1805. 4to. In French, Paris, 1789, 2 vols.; in German, 1798. 2 vols.

[RATTE, E. H. DE].

Éloge de M. de Linné. In: Assemblée publique de la Société royale des sciences, tenue a Montpellier, le 28 Décembre, 1779 (1780) pp. 100-116. 4to.

HEDIN, S. A.

Quid Linnæo patri debeat medicina. D. . proponit S. A. H. . resp. C. Carlander, 1784.

SAINT-AMANS, J. F. B.

Éloge de Charles de Linné. Agen. 1791.

STOEVER, D. H.

Leben des Ritters Carl von Linné. Hamburg, 1792, 2 vols.; Nachtrag, ib. 1793. In English by J. Trapp. Lond. 1794. 4to.

ACREL, J. G.

Tal om läkare-vetenskaps grund-läggning och tillväxt vid rikets älsta larosäte i Upsala. Stockh. 1796.

Carolus von Linné; Nils Rosén von Rosenstein; Carl von Linné, sonen; Olof Rudbeck, sonen; Lars Roberg.

Turton, W
Some account of the life and writings of Sir Charles Linné. In Gen. Syst. Nature, vii., 1806.

Aikin, J.
Linnæus, Carl. In: General Biography vi. (1807) 267-293.

Thunberg, C. P.
Tal vid invignings acten af den nya Akademiska trädgården d. 25 Maji, 1807. Uppsala, 1807.

Hedin, S.
Minne af von Linné; fader och son. Stockh. 1808.

Smith, Sir J. E.
Linnæus. In Rees's Cyclopædia xxi. " 1819 " [recte 1812]. Lond. 4to. Chalmers, A. Linnæus. In: Biographical Dictionary. Lond. 1815, 294-312 [Abbrev. from preceding] Rose, Rev. H. J. Linnæus. In: New General Biographical Dictionary. Lond. 1857, ix. 282-286 [Entirely derived from Smith].

Afzelius, A.
Egenhändiga anteckningar af C. L. om sig sjelf med anmärkningar och tillägg. Uppsala, 1823. 4to. In German, Berlin, 1826.

Wahlenberg, G.
Linné och hans vetenskap. In " Svea," 1822, pp. 66-130.
Historisk underrättelse om Upsala Universitets botaniska trädgård. 1836 [pp. 51-74]. [Uppsala? 1836.]

Agardh, C. A.
Antiquitates Linnæanæ. Programma. Lund, 1826, fol.
Äreminne. In: Sv. Akad. Handl. x. (1826) 49-108. Stockh.

Fée, A. L. A.
Vie de Linné, redigée sur les documents autographes. Paris, 1832 (Lille, imp.). In: Mém. Soc. Royale des Sciences de Lille. 1832. i.

Smith, Pleasance, Lady, née Reeves.
Memoir and correspondence of the late Sir J. E. Smith. Lond. 2 vols. 1832. (Linnean Collections, i. 92-106, 110-118, 122-128).

Myrin, C. G.
Om Linné's naturhistoriska samlingar. In: Tids. Skandin. ii. (1833) 242-288.

Jardine, Sir W.
Memoir of Linnæus. In the Naturalists' Library: Ornithology, Edinb. 1833. i. 13-48.

Carr, D. C.
The Life of Linnæus [with] a short account of the botanical systems of Linnæus and Jussieu. Holt, 1837.
[—] Linnæus and Jussieu; or, the rise and progress of systematic botany. Lond. 1844.

Martius, C. F. P von.
Reden und Vorträge [über Linné u.s.v.]. Stuttg. 1838.

Pontin, M. af.
A visit to Hammarby, the country seat of Linnæus in the spring of 1834. (From the Transactions of the Swedish Horticultural Society for 1834) German by Dannfelt . English by J. C. Loudon. In: Loudon, Gard. Mag. xiv. (1838) 98-101.

Hartman, C.
Annotationes de plantis scandinavicis herbarii Linneani in Museo Societ. Linnæanæ Londin. asservati. In. Sv. Akad. Handl. (1849) 145-193; (1851) 211-426.

Brightwell, Miss C. L.
The life of Linnæus. Lond. 1858.

Fée, A. L. A.
On the question whether Linnæus in a spirit of ill-will, altered the spelling of the name of the genus *Buffonia?* In Journ. Linn. Soc. Bot. ii. (1858) 183-187. (" The error was due to Sauvages, not Linné ": J. J. Bennett—Note on the preceding communication, ib. 188-190.)

Hanley, S. C. T
Ipsa Linnæi conchylia. Lond. 1855 [In English].
On the Linnean manuscript of the Museum Ulricæ. London, 1859.

Fries, T. M.
Anteckningar rörande en i Paris befintlig Linneansk växtsamling. In: Stockh. Vet. Akad. Öfvers. xviii. (1861) 255-272.

FRIES, T. M.
Om ett Linneanskt herbarium in Sverige. In: Bot. Notiser, 1887. 141-143.
Eulogium on Linnæus. In: Proc. Linn. Soc. 1887-88. [1890] 45-54.
Bidrag [1-8] till en lefnadsteckning öfver C. v. L. Uppsala, 1893-98.
C. L. Hortus uplandicus. Uppsala, 1899.
Linné: lefnadsteckning. Stockh. 2 vols. [1903].
Linnéminnen i Uppsala botaniska trädgård. Antikritic. In Ark. bot. iv. (1905) No. 5. 1-54 (cf. Kjellman).
Ett och annat om *Linnæa borealis*. In: Fauna och Flora, iii. (1908).
Några akademiska befordringsfrågor i Uppsala under Linné's tid [Mathesius såsom Præses]. In Nordisk Tidskrift, Stockh. 1910. 99-117. Cf. Hemmensdorff; Kjellman.

SCHIØDT, J. M. C.
Af. Linné's Brefvexling. Aktstykker til Naturstudiets Historie i Danmark. In: Naturhist. Tidsskr. IIIe. Række, vii. (1870-71) 333-522.

ÄHRLING, J. E. E.
Studier i den Linneanska nomenclaturen och synonymiken. Örebro, 1872.
Några af de i Sverige befintliga Linneanska handskrifterna, kritiskt skärskåde. Akad. Afh. Lund, 1878.
Linnés Svenska arbeten. Stockh. vols. i. and ii. part I. 1878.
Carl von Linné's Svenska arbeten i urval och med noter utgifna [vol. i. in five parts]. Stockh. 1879.
Carl von Linné's brefvexling, ib. 1885.
Om Karl von Linné; Linné d.y., Linnean Society of London; Linnéska Institut. Linnéska samfundet och Linnéska samlingar. In: Nordisk Familjebok, ib. 1885.

AHNFELT, A.
Carl von Linné's lefnadsminnen tecknade af honom sjelf. Stockholm, 1877.

HJELT, O. E. A.
Carl von Linné som läkare, och hans betydelse för den medicinska vetenskapen i Sverige. Helsingfors, 1877. In German, Leipzig, 1882.

GISTEL, J. F. X.
Carolus Linnæus, ein Lebensbild. Frankfurt am Main. 1873.

SWEDERUS, M. B.
Botaniska trädgården i Upsala, 1655-1807. 1877.

LUNDSTRÖM, A. N.
Carl Linnæi resa till Lappland, 1732; några skizzer. [Uppsala, 1878.]

SAHLIN, Y.
Inbjudning-skrift till den festen K. Univ. i Uppsala kommer att fira minnet af C. v. Linné. 1878.

LJUNGGREN, G.
Carl v. Linné's vistande i Lund. Lund, 1878. 4to.

[LJUNGSTEDT, A. L.]
Linné i Upsala och Amsterdam; Skizz af Claude Gerard. Stockh. 1878.

AGARDH, J. G.
Om Linné's betydelse i botanikens historia. Lund, 1878.
Linnei lära om i natura bestämda och bestående arter hos växterna. 1885.
Festen till C. v. Linné's minne i Upsala den 10 Januari, 1878. Upsala [1878]. (Contains amongst other items: Tullberg, T Familje traditioner om Linné.) Reimpr. 1919; cf. p. 405.

MALMSTEN, P H.
Minnesord öfver Carl von Linné. Stockh. 1878.

OUDEMANS, C. A. J. A.
Rede ter herdenking van den sterfdag van Carolus Linnæus, eine eeuw na diens verscheiden in felix meritis op den 10 den Januari, 1878. Amst. 1878.

SÄTHERBERG, H.
Blomster konungen; bilder ur Linné's lif [Poems]. [Stockh. 1879].

JACKSON, B. D.
Linnæus. In: Encyc. Brit. ed. IX. vol. xiv. (1883) 671-674 [revised and compressed in] ed. XI. vol. xvi. (1911) 732-733.
History of the Linnean Collections. . . . In: Proc. Linn. Soc. 1887-88 (Centenary Celebration of the Foundation of the Society).
On Linnean specimens presented to Sir J. Banks [by Sir J. E. Smith] ib. 1902-3.

JACKSON, B. D.

On a manuscript list of the Linnean herbarium in the handwriting of C. v. Linné . . 1755. ib. 1906-7.

Index to the Linnean herbarium . . ib. 1911-12 (1912) [With bibliography, 1805-1912].

Catalogue of the Linnean [zoological specimens] ib. 1912-13 (1913).

Correspondence between C. v. Linné and C. R. Tulbagh, ib. 1917-18 (1918).

Notes on a catalogue of the Linnean herbarium [in manuscript]. (Bibliography revised, brought down to date and printed, ib. 1921-22 (1922).

ALBERG, A.

The floral king; a life of Linnæus. Lond. 1888.

CARRUTHERS, W.

[On the portraits of Linnæus.] Proc. Linn. Soc. 1888-90 (1890) 14-31.

On the original portraits of Linnæus, ib. 1905-6 (1906) 59-69; pl. 1-8.

LOVÉN, S. L.

On the species of Echinoidea described by Linnæus in his work "Museum Ludovicæ Ulricæ." 1887

LINNEAN SOCIETY OF LONDON.

Catalogue of the Memorials of Linnæus exhibited at the Conversazione . of the Linnean Society 25th May, 1888. Lond. 1888.

Catalogue of the Library. Lond. 1896. (Pp. 388-403 contain an enumeration of Linné's own copies of his works, many of them interleaved and with copious annotations.)

LÖNNBERG, E.

Linnean type specimens of birds, reptiles, batrachians and fishes in the zoological museum of the R. University in Uppsala. Stockh. 1896 [1897]. 4to.

Peter Artedi [and his close friendship with Linnæus] transl. by W E. Harlock. Uppsala and Stockh. 1905.

ANDERSON, L. G.

Catalogue of Linnean type specimens of Snakes in the Royal Museum in Stockholm. Stockh. 1899. 4to.

JUNK, W

Bibliographia Linnæana. Berlin, 1902. 4to.

C. v. Linné und seine Bedeutung für die Bibliographie. Festschrift. Berlin, 1907. 4to.

JUNK, W.
In memoriam bisæcularem C. a Linnæi scientia naturalis usque ad finem seculi XVIII. [Berol. 1907.]

MIDDLETON, R. M.
[Two letters from Linnæus to R. Warner, 1748, and to D. van Royen, 1769.] In: Proc. Linn. Soc. **114** (1902) 48-51.
[Letter from Linnæus to Haller, 1747] ib. **116** (1904) 41-42; now in possession of B. D. Jackson.

OLSSON-SEFFER, P.
The place of Linnæus in the history of botany. Journ. Bot. (1904).

KJELLMAN, F. R.
Linneminnen i Uppsala botaniska trädgård. In: Ark. bot. iii. (1904) N.7. pp. 1-33, 3 pl.

PARLATORE, F.
Sopra l'erbario di Linneo, manuscritto inedito pubblicato da E. Borone. In: Webbia (1903) 75-83.

WITTROCK, V B.
Linnæus, Carolus, nobilit. von Linné. In: Act. Horti Berg. iii. II. (1903) 49-60, tt. 2-3 [=8].
Några ord om Linné och hans betydelse för den botaniska vetenskapen, ib. iv. 1. (1907).

VINES, S. H.
[The career of Linnæus.] In: Proc. Linn. Soc. **116** (1904) 22-30.

HERDMAN, W. A.
[Linnæus on pearls] ib. **117** (1905) 25-30.

RÁDL, E.
Die Linnésche Systematik. In: Gesch. der biologischen Theorien. Leipzig, 1905, i. 129-149.

Linneska Institutets skrifter ånyo utgifvet [af J. M. Hulth] Uppsala, 1906.

DAHL, O.
Carl von Linné's Forbindelse med Norge (K. Norske Vidensk Selsk.). Trondhjem, 1907 4to.

ELFVING, F.
Carl von Linné. Ett tvåhundredårsminne. In: F. Tidskr. Helsingfors, lxii. (1907) 337-354.

ENANDER, S. J.
Studier öfver *Salices* i Linné's herbarium. Uppsala, 1907.

FRIES, R. E.
Carl von Linné. Zum Andenken an die 200ste Wiederkehr seiner Geburtstages. In: Bot. Jahrb. 41 (1907) 1-54.

HOLLAND, W J.
Address of the Carnegie Museum . 23rd May, 1907 [with plate of microscope presented by C. v. Linné to Bernard de Jussieu, August, 1738, now possessed by the Carnegie Museum, Washington, D.C.].

HOLM, T
Linnæus: 23rd May, 1707—10th January, 1778. In: Bot. Gaz. xliii. (1907) 336-340.

HULTH, J. M.
Bibliographia Linnæana; matériaux pour servir à une bibliographie Linnéenne. Partie I., livraison 1. Uppsala, 1907 [Tout paru].
Uppsala Universitetetsbiblioteks förwärf av Linnéanska originalmanuskript. In: Uppsala Univ. Minneskr. 1621-1921. pp. 407-424.

LEERSUM, E.
En souvenir du jour de naissance de C. Linné [2 lettres. Amst. 1907].

LEVEILLÉ, A. A. H.
Linneo en Espana; Homaje . 1707-1907 In: Bot. Soc. Arag. Cienc. Nat. vi. Zaragoza, 1907 [55 papers].

LEVERTIN, O.
Carl von Linné [a fragment]. Stockh. [1907].

London; British Museum. A Catalogue of the works of Linnæus . . preserved in the libraries of the British Museum (Bloomsbury) and . (Natural History) South Kensington [by B. B. Woodward and W R. Wilson]. Lond. 1907. 4to.

Carl von Linné's betydelse såsom naturforskare och läkare. Uppsala, 1907.
I. C. v. L. såsom läkare och medicinisk författare af O. E. A. Hjelt. Bilag: Clavis medicinæ [reimpr.].
II. C. v. L. såsom zoolog af E. Lönnberg och C. Aurivillius.
III. C. v. L. såsom botanist af C. A. M. Lindman (also in German).
IV C. v. L. såsom geolog af A. G. Nathorst.
V C. v. L. såsom mineralog af H. Sjögren (also in German).

[RENDLE, A. B.].
Memorials of Linnæus; a collection of portraits, manuscripts, specimens and books exhibited to commemorate the bicentenary of his birth (Special Guides, No. 3, British Museum—Natural History). Lond. 1907.

TULLBERG, T. F. H.
Linnéporträtt; vid Uppsala Universitets minnefest på tvåhundredårsdagen af Carl von Linné's födelse. Stockh. 1907. 4to.
Linné's Hammarby. In: Sv. Linné-sallsk. i. (1918) 1-79, tt. 1-5.
Familjetraditioner om Linné upptecknad af T. T. In: Sv. Linné-sallskapet ii (1919) ib. 1919 [Reimpr., cf. 1878].

Göteborg [Gothenburg]. Gemensamma fest till Linné's minne, den 23 Maj, 1907 [Tal af L. Stavenow och L. Wolff]. Göteborg [1907]. 4to.

RIBBING, S.
Inbjudning till 200 årsminnet af Carl v. Linné's födelse. Lund, 1907. 4to.

[STJERNSTRÖM, L. E.].
Karl von Linné's levnadsaga af K. Blint. Mariestad, 1907.

WITTMACK, L.
Linné und seiner Vorgänger; Festrede. In. Sitzb. Ges. Naturf. Berlin, 1907. 120-156. 2 Taf.

WITTROCK, V. B.
Några ord om Linné och hans betydelse för den botaniska vetenskapen. In: Act. Hort. Berg. Stockh. iv. No. I, 1907.

GREGORY, W K.
Linnæus as an intermediary between ancient and modern zoology; his views on the class Mammalia. In: New York Acad. Sc. xviii. (1908) 21-32.

LUCAS, F. A.
Linnæus and American Natural History. ib. 52-56.

RYDBERG, P. A.
Linnæus and American botany. ib. 32-40.

SURINGAR, J. V.
Linnæus. 's Gravenhaag, 1908.

GREENE, E. L.
Linnæus as an evolutionist. In: Wash. Acad. Proc. xi. (1909) 17-26.
Carolus Linnæus. With introd. by B. W. Everman. Philadelphia [1912].

Minnesfesten öfver C. v. Linné den 25 Maj, 1907 . utarbetad af J. A. Bergstadt. Uppsala och Stockh. [1910].

HOLMBOE, J.
Linné's botaniske "Prælectiones privatissimæ" paa Hammarby, 1770. Utgit efter M. Vahls referat. In: Bergens Mus. Aarbok, 1910. No. I. 1-69, 1 Tafl.

BECKMANN, Johann.
Schwedische Reise in den Jahren 1765-66. Tagebuch herausgegeben von T. M. Fries. Uppsala, 1911.

HEDBOM, U.
Den gamla svartpoppeln i Linné's botaniska trädgård fallen [9th Sept., 1911]. In: Sv. Bot. Tidskr. v. (1911) 378-379. med 1 fig.

RUDBECK, Count Johannes.
Campus Elysii eller Glysis wald af Olof Rudbeck, far och son. Några bibliografiska anteckningar [pp. 49-62, ex "Samlaren"]. Uppsala, 1911.

MIALL, L. C.
Carl Linnæus (Linné). In: The early naturalists. London, 1912, 310-336.

FORSSTRAND, C.
Några anteckningar om Linné's Stockholmstid. In. Sv. Bot. Tidskr. vi (1912) 657-672.
Linné i Stockholm. Stockholm, 1915.

[HEMMENDORFF, E.].
In Memoriam T. M. Fries (1832-1913) [Bibliography by J. M. Hulth]. In: Sv. bot. Tidskr. viii. (1914) 109-146.

FLODERUS, M.
Ett litet bidrag till frågan om Linnésamlingarnas öde ib. x. (1916) 47-52.

ALMQUIST, E.

Linné's Vererbungsforschungen. In: Engler, Bot. Jahrb. lv. (1917) 1-18.

Linné und das natürliche Pflanzensystem. ib. lviii. (1922). Beibl. p. 128. Heft 1. S. 1-16.

JUEL, H. O.

Bemerkungen über Hasselquists herbarium. In: Sv. Linné-sallskap, i. 1918.

Hortus Linnæanus. An enumeration of plants cultivated in the botanical garden of Uppsala during the Linnean period. Uppsala, 1919.

Studien in Burser's Hortus Siccus. In: Nov. Act. Reg. Soc. Sc. Upsal. Ser. IV v. (1923) No. 7 (cf. Plant. Martino-Burserianæ, in Amœn. Acad. i. 141-171).

WILLE, J. N. F

Linné som Læge. In: Naturen xlv. (1921) 97-106.

Stockholm: Svenska Linné-sallskapets årsskrift. i. 1918. (In Progress)

Svenska Vetenskapsakademiens protokoll för åren 1739, 1740 och 1741. Med anmärkningar utgifven af E. W. Dahlgren. Stockh. 1918. 2 vols. [Minutes of the Swedish Academy of Science for the years 1739-41, ed. with notes by E. W. D.]

INDEX

(Arranged in order of the English alphabet)

Printed for Messrs. H. F. & G. Witherby by the Northumberland Press, Ltd., Newcastle-on-Tyne.

Printed in the United States
By Bookmasters